Advanced Structured Materials

Volume 6

For further volumes:
http://www.springer.com/series/8611

Common engineering materials reach in many applications their limits and new developments are required to fulfill increasing demands on engineering materials. The performance of materials can be increased by combining different materials to achieve better properties than a single constituent or by shaping the material or constituents in a specific structure. The interaction between material and structure may arise on different length scales, such as micro-, meso- or macro-scale, and offers possible applications in quite diverse fields.

This book series addresses the fundamental relationship between materials and their structure on the overall properties (e.g. mechanical, thermal, chemical or magnetic etc.) and applications.

The topics of Advanced Structured Materials include but are not limited to

- classical fibre-reinforced composites (e.g. class, carbon or Aramid reinforced plastics)
- metal matrix composites (MMCs)
- micro porous composites
- micro channel materials
- multilayered materials
- cellular materials (e.g. metallic or polymer foams, sponges, hollow sphere structures)
- porous materials
- truss structures
- nanocomposite materials
- biomaterials
- nano porous metals
- concrete
- coated materials
- smart materials

Lucas Filipe Martins da Silva
Alessandro Pirondi · Andreas Öchsner
Editors

Hybrid Adhesive Joints

Springer

Prof. Dr. Lucas Filipe Martins da Silva
Departamento de Engenharia Mecânica
Faculdade de Engenharia da
Universidade do Porto
Rua Dr. Roberto Frias
4200-465 Porto
Portugal
e-mail: lucas@fe.up.pt

Prof. Dr. Andreas Öchsner
Department of Applied Mechanics
Faculty of Mechanical Engineering
Technical University of Malaysia
81310 UTM Skudai, Johor
Malaysia
e-mail: andreas.ochsner@gmail.com

Assoc. Prof. Alessandro Pirondi
Dipartimento di Ingegneria Industriale
Universitá degli Studi di Parma
Viale G. P. Usberti 181/A
43124 Parma
Italy
e-mail: alessandro.pirondi@unipr.it

ISSN 1869-8433 e-ISSN 1869-8441
ISBN 978-3-642-16622-8 e-ISBN 978-3-642-16623-5
DOI 10.1007/978-3-642-16623-5
Springer Heidelberg Dordrecht London New York

Cover design: eStudio Calamar, Berlin/Figueres

Printed on acid-free paper

Springer is part of Springer Science+Business Media (www.springer.com)

Preface

Adhesively bonded joints are increasingly being used in aerospace and automotive industries. The use of adhesive bonding rather than mechanical fasteners offers the potential for reduced weight and cost. However, there are still a number of issues that need to be solved before this technique is fully trusted by the industry. Due to the polymeric nature of the adhesive, the joint resistance to harsh environments (temperature and humidity) and its durability must be taken into account. Another point is the fact that the stress distribution is concentrated at the ends of the overlap giving rise to premature failures, especially with composites. However, when used in combination with traditional methods of joining such as bolts, rivets or spot welds, part of the problem can be solved. That is why adhesively bonded joints with different methods of joining are being considered. The idea is to gather the advantages of the different techniques leaving out their problems. Some of the advantages of hybrid joints are a higher static and fatigue strength and a higher stiffness with respect to simple joints, a two-stage cracking process before the final failure and improved durability. They also offer the technological advantage of keeping parts in place while the adhesive is curing. That is why hybrid joining is increasingly used by the automotive industry. Another possibility to increase the joint strength of simple adhesive joints is to use more than one adhesive along the overlap. All these joints are treated here: weld – adhesive, rivet – adhesive, clinch – adhesive, bolt – adhesive, and adhesive – adhesive. A section dedicated to threadlocking and interference-fit adhesive joints is also included. All sections are treated from a scientific point of view with modelling issues supported by simple coupons testing and a technological point of view where the idea is to present more applied results with practical cases.

The editors would like to thank the authors for their patience with the preparation of this book. Finally, we especially thank Dr Christoph Baumann, Springer editor, who made this book possible.

Dr. Lucas F M da Silva
Prof. Alessandro Pirondi
Prof. Andreas Öchsner

Contents

Science of Weld: Adhesive Joints

Saied M. H. Darwish

Abstract Weldbonding is a combination of resistance spot welding and adhesive bonding. It is a hybrid process that combines the advantages of the two individual processes. There are two techniques for conducting weldbonds: the flow-in and the weld-through techniques. The weldbonding procedure provides many advantages, e.g.: improved crash performance, fatigue behavior and corrosion resistance. The degree of acceptance of weld-bond applications has been increasing, as the process has been understood and its mechanical properties developed. The present chapter addresses both techniques (flow-in and weld through) from implementation to modeling, including analysis and characterization. Practical implications as well as fields of application are also highlighted.

1 Introduction

Bonding of metals is becoming increasingly important, both in absolute terms and relative to mechanical fastening. The applications of adhesive bonding are found in the assembly of many products including aircrafts [36], cars, trucks, and office furniture [9]. This is because adhesive bonding offers more uniform distribution of stresses, increased fatigue life, weight saving and the prevention or reduction of corrosion between dissimilar materials, added to the ability to join small and delicate parts .

Darwish et al. [19, 20, 23, 26] used adhesive in manufacturing a bonded gear box. In machine tool manufacturing, adhesives are used [5] for bonding gear

S. M. H. Darwish (✉)
Department of Industrial Engineering, Faculty of Engineering, King Saud University, KSA, 800, Riyadh 11421, Saudi Arabia
e-mail: darwish@ksu.edu.sa

Adv Struct Mater (2011) 6: 1–36
DOI: 10.1007/8611_2010_35

Published Online: 11 September 2010

wheels, joining elements of hydrostatic bearings and transmissions with load bearing surfaces. In the electrical and electronic industry, adhesives are used in a variety of ways. These range from holding microcircuits to bonding coils in mammoth electrical generators. In addition to mechanical fastening, adhesives are required in electrical applications to seal and protect substrates and to conduct or insulate heat and electricity [5]. Darwish and Davies [19, 20] developed adhesively bonded metal cutting tools, while [27] have tested the performance of adhesively bonded metal cutting tools. Bonded structures can be of two types based on either purely adhesive or on adhesive/mechanical connections [44]. The purely adhesive connections include shaft–hub joints, laminated metal–metal joints and three-layer honeycomb structures. The bonded mechanical types include bonded-welded, bonded-riveted and bonded-screwed connections. The combined connections (bonded-welded, bonded-riveted and bonded-screwed) ensure high fatigue strength of the structures and are extremely economical, because they do not require any fixtures during the polymerization process.

The manufacturing and sale of aerospace, transportation and civil structures invariably require joining components of different materials and compositions. Because manufacturers are constantly being pressured for lower cost and reduced weight, different materials are being combined in many products. As different materials are used within a given structure, there will most definitely be a need to somehow join them.

While methods such as adhesive bonding and mechanical fastening have been used in the past to join dissimilar materials [7] work in this area is now moving towards less labor intensive methods, geared more toward automation [6, 34, 39, 40]. Neither adhesive bonding nor mechanical fastening is particularly well suited for inline application with the substrate fabrication methods. Moreover, adhesive bonding may present questions from a long term joint durability standpoint, while mechanical fastening may present the risk of in-service galvanic corrosion, as well as the potential to introduce stress concentrations at the joint.

One of the promising methods being considered for the joining of dissimilar materials is resistance welding. The advantages of resistance welding process include typically short process times and heat focused specifically at the material interfaces. This process is economical since no weld material is needed. Resistance welding can easily be automated and in most cases only assistants or robots are needed to supply the material. The process also has the potential for a sophisticated type of control that cannot be implemented for adhesive bonding and mechanical fastening. The combination of dissimilar materials or dissimilar thicknesses may cause large problems which often can be solved by the application of solid phase welding processes. Studies of the bonding mechanisms in solid phase welding processes made by [31, 39], have revealed that large improvements can be obtained by introducing special surface treatment techniques like strategic interlayers, in order to facilitate bonding of metal combinations which would otherwise be difficult or impossible to join. This strategic surface coating can be passive or active surface coating.

Weldbonding is a combination of resistance spot welding and adhesive bonding. A paste adhesive is normally applied to one sheet and then the joint is closed [44]. A spot weld is then made through the adhesive. The electrode force displaces the adhesive to obtain electrical contact between the sheets and the weld is made in the normal way. As the heating of the weld is very localized, little damage occurs in the adhesive around the weld. The adhesive is finally cured to complete the assembly. Heat curing paste type adhesives are normally used as these are stable and have a consistent viscosity at room temperature. Typically, such adhesives are cured in an oven at up to 180°C for 30 min. Some adhesives are available in tape form and incorporate metal particle filler which allows initial electrical contact to be made for spot welding.

Weld-bonded joints were first developed and used in the Soviet Union in manufacturing planes of the type AN-24. The weld-bonding process is essentially the spot-resistance welding of parts that subsequently have their overlapping areas adhesively bonded. The Soviet Union initially perfected this technology, which is known there as "glue welding". The approach was a "flow-in" method (Fig. 1a), whereby parts were first welded together, and then the adhesive was flowed into the joint. A low-viscosity adhesive is used which penetrates the overlap joint by capillary action and is subsequently cured. The technique used in the United States is the weld-through method (Fig. 1b), whereby the adhesive is applied to the parts to be cured, spot welded and subsequently cured.

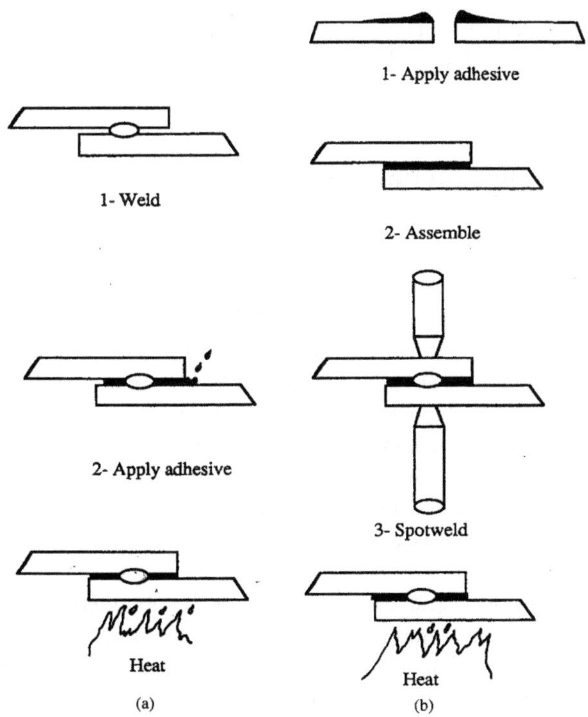

Fig. 1 Weld bonding: **a** the flow-in technique, **b** the weld-through technique

In comparison with mechanical fasteners, weld-bonding offers the following benefits: (1) reduced manufacturing costs and suitability for automation; (2) high static strength; (3) improved fatigue strength; (4) improved corrosion resistance; (5) elimination of sealing operations; and (6) reduction of the shop noise due to riveting.

The process combines the benefits of adhesive bonding for joining thin sheet assemblies, together with a fast, automated point joining technique, which eliminates the need for jigging during the curing of the adhesive. Adhesive bonds have limited strength in peel or cleavage, so the spot welds provide peel-stopping points in the event of overloading, for example in impact tests.

In the lightweight body frame mass production of automobiles, the application of hybrid bonding techniques becomes more and more important. The weldbonding procedure provides many advantages, e.g. improved crash performance, fatigue behavior and corrosion resistance [33]. Therefore, the technology is state of the art in many branches of joining metal sheets. Especially for newly developed high strength multiphase steels, also called advanced high strength steels (AHSS). This joining technique has more advantages than other joining procedures for thin steel sheets. The application of AHSS materials in conjunction with economically efficient and reliable joining processes, helps saving costs and conserving resources (weight reduction, energy minimization) and provides at the same time improved safety of the passenger cell (crash optimization). The present chapter is concentrated on experimental and theoretical aspects of spot welded and weld-bonded joints. These range from preparation of specimens, metallographic examination, tensile as well as dynamic characteristics of spot welded and weld-bonded joints. These also include finite element modeling, analysis of tensile-shear and tees adhesively bonded and weld-bonded specimens.

2 Experimental Details

Across the engineering spectrum, of joining techniques welding is an essential process in the manufacture of components, assemblies or complete machines. Spot welding is the best known and the most frequently used method of resistance welding, as it can be applied to a wide variety of articles instead of rivets, soldering and brazing [42, 43]. The advantages of the spot-welding process [1, 11, 30, 35, 38, 41] include typically short process times and heating focused specifically at the material interfaces. This process is economical since no weld metal is needed. Resistance welding can easily be automated and in most cases, only assistants or robots are needed to supply the material. Control of the quality of welded joints is essential, if the fabricated parts have to perform satisfactory in service.

Uniform spot weld quality can be obtained through the use of proper equipment, rigid adherence to established welding procedure, close control of machine settings and the use of qualified personnel. Destructive and non-destructive testing is generally essential to check the quality of spot welds. Destructive testing are those tests which require destruction of a sample part in order to evaluate the quality of the weld. Weld metal can be tested in the same way as any other form of metal. Mechanical tests are used to qualify welding procedures, welders, and welding processes and to determine if electrodes meet the requirements of the specifications. Welds in weldments are often tested for soundness, strength, and toughness by mechanical tests.

Resistance spot welding [2, 4, 8, 10, 32] is a process in which faying surfaces are joined at one or more spots due to the heat generated by resistance to the flow of electric current through the workpieces that are held together. The contacting surfaces in the region of current concentration are heated by a short-time pulse of low voltage, high amperage current to form a fused nugget of weld metal. The size and shape of the individually formed welds are limited primarily by the size and contour of the electrode forces. The weld nugget forms at the faying surfaces, and does not extend completely to the outer surfaces. In cross-section, the nugget in a properly formed spot weld is approximately oval in shape; from outside, it has the same shape as the electrode tip, and is of approximately the same size. The spots should be at a sufficient distance from the edge of the joint so that there is enough base metal to withstand the electrode force and to ensure that the pressure distribution during welding does not allow expulsion of metal from the weld. Darwish and Al-Dekhial [15, 16] built a group of statistical models for spot welded commercial aluminum sheets; these models facilitate the prediction of spot welding variables for minimum energy consumption.

In making a spot weld between dissimilar metals, a heat balance must be achieved that compensates for the differing properties of the two metals. More heat must be supplied to the more thermally and electrically conductive metal, which generates less resistive heat and has greater loss of heat by conduction. The higher melting of two dissimilar metals of approximately equal conductivity also needs more heat. In spot welding when the metals to be joined are different, or the thicknesses of the sheets to be joined are different, or both, it is best to keep the weld nugget symmetrical about the interface between the sheets. Weld nugget symmetry needs two conditions: first, that the heat generated in both plates be the same; second, that the heat transfer out of the plates be the same. Since the heat is generated by electrical resistance, the first requirement means that the electrical conductance of the two sheets must be equal. The equality of heat loss means that the thermal conductance of both plates must be the same. The electrical conductance and the thermal conductance are given by:

$$G = \left(\frac{g}{l}\right)A, K = \left(\frac{k}{l}\right)A \tag{1}$$

where l is the path length or sheet thickness, A the electrode bearing area, G and g the electrical conductance and conductivity, respectively; K and k the corresponding thermal units.

If we are working with the same metal but with sheets of different thickness, Eq. 1, gives the requirements for electrode area in terms of relative sheet thickness. Thus

$$K_1 = \left(\frac{K_1}{l_1}\right)A_1, K_2 = \left(\frac{k_2}{l_2}\right)A_2 \tag{2}$$

Since we want $K_1 = K_2$

$$\frac{A_1}{l_1} = \frac{A_2}{l_2} \text{or} \frac{l_1}{l_2} = \frac{A_1}{A_2} \tag{3}$$

In other words, the electrode areas against each plate must be proportional to the thicknesses of the sheets. A large electrode will then be placed against the thicker sheet; in effect the larger electrode provides a large heat sink so that the weld nugget will be centralized at the interface.

2.1 Weld-Bond Test Specimens

The configuration and dimensions of the weld-bond specimen used in the present work are shown in Fig. 2. The specimens were cut from 1.5 mm thick low-carbon steel sheets, the interfaces of which were prepared by grinding with silicon carbide paper up to 220 grit, degreased, dried and kept in a desiccator, ready for weld-bonding. The structural epoxy resin adhesive used was a room temperature curing,

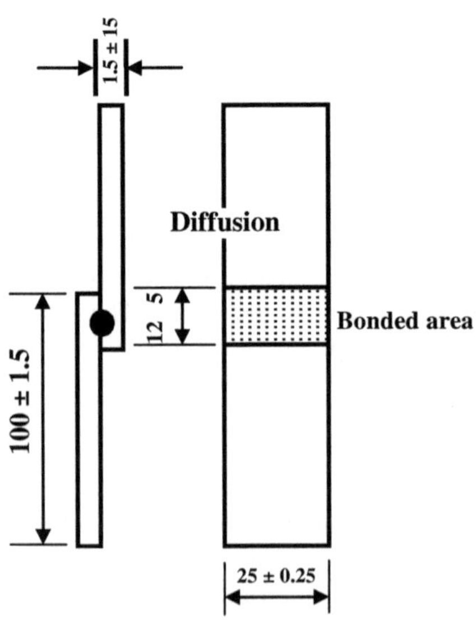

Fig. 2 Configuration and dimensions (mm) of the weld-bond lap joint

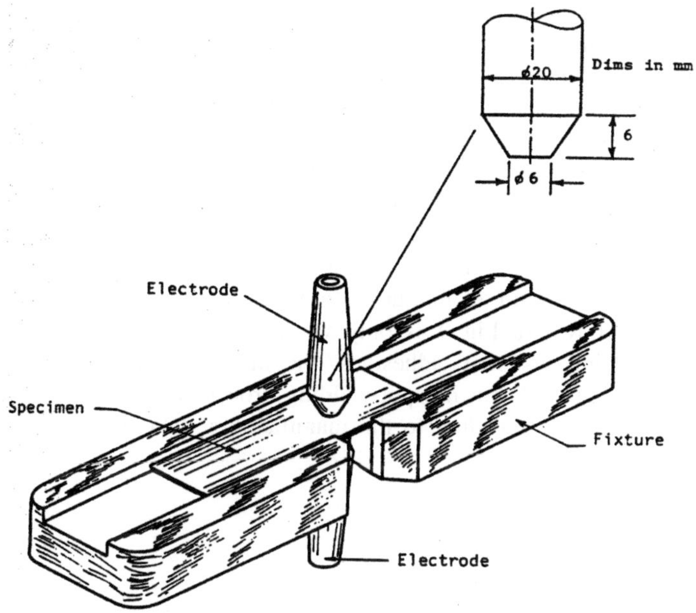

Fig. 3 Spot-welding fixture

two-part epoxy resin with a mix ratio of 100 of resin to 40 parts by weight of hardener. This adhesive had a pot life of 30 min at 23°C and a cure cycle of 24–28 h at 24°C. The procedure followed throughout the manufacturing stage of flow-in weld-bond specimens was as follows: (1) removal of unwanted adhesive components such as solvent and water; (2) spreading of the adhesive during processing to ensure the wetting of surfaces and to promote contact.

To ensure the positioning and alignment of the weld-bonded joints, a specially designed fixture was used [15, 16]. This fixture was made of hard wood in order to avoid undesirable connection between the welding electrodes. The fixture was designed to fulfill the following functions (see Fig. 3):

- To provide accurate positioning of the overlap area during the welding process.
- To establish a good alignment between the electrode tip and the center of the overlap area of the specimen.
- To ease the fixing and handling of the specimens to the machine, especially when the adhesive layer is applied.

Adhesives normally work as good insulators; however, effective improvements in the electrical conductivity were evident when fillers (powder) were mixed with the adhesive. From 1998 to 1990, [22, 25, 37] mixed a lot of fillers (metallic powders, chopped wires and metallic fine meshes) and came up with recommended formula for improving electrical as well as thermal properties of structural epoxy adhesives. The addition of 10% graphite, 15% aluminum or 20% copper powder by weight normally improves electrical conductivity) [25].

For weld-bond specimens (flow-in technique), the adhesive was spread over the overlap length, with the thickness recommended by the manufacturer (0.5 mm). Afterwards the specimens were immediately spot welded. For the weld-through weld-bond specimens, the adhesive material, plain or filled (with 10% graphite, 15% aluminum or 20% copper), was applied over the overlap length, then the specimens were cured at 24°C for 48 h, and spot-welded thereafter. The specimens (plain or filled adhesive) cannot be welded after the adhesive material has been cured, except, the specimens containing 15% Al which can be welded following cure after being heated to 100°C. It is worth mentioning that the implementation of flow-in technique is a viscosity dependant. Therefore, the flow-in module shown in Fig. 1 is only workable with low-viscosity adhesive. However, with high viscosity adhesives, the uncured adhesive is first spread along the overlap area. Then the spot-welding process is carried out after which the adhesive is cured, according to the manufacturer data sheet.

2.2 Metallurgical Examination

After welding was completed, the weld-bonded specimens were cut into two halves, and prepared for metallographic examination. The photomicrographs were conducted at 37.5× for the welding nugget, and at 300× for the microstructure of each zone.

2.3 Microhardness Measurements

Vickers microhardness measurements were conducted on a microhardness tester. The microhardness measurements were taken along the major and minor axes of the weld nugget from the center and outwards (Table 1.)

Table 1 Microhardness measurements

Condition of spot weld (spot weld)	Average microhardness (VHN)	
	Longitudinal	Transverse
$V = 220$ V		
$I = 1$ (3.3 kA)	286,211,286,220	286,211,232,201
$I = 2$ (4.5 kA)	362,257,286,211	362,201,286,201
$I = 3$ (6.6 kA)	321,210,201,192	321,201,257,192
$V = 220$ V, $I = 3$ (weld-bond)		
15% Al (weld-through)	184,201,210,221	184,184,211,175
15% Al (flow-in)	257,221,232,201	257,184,175,157
10% C (flow-in)	303,168,244,201	303,168,257,221
20% Cu (flow-in)	244,175,192,183	244,140,168,192

2.4 Tensile Strength Measurements

An Instron testing machine was used to carry out the tensile-shear strength tests. A crosshead speed of 50 mm min^{-1} and a load cell of 100 kN are recommended. The machine is provided with a chart recorder, which enables load-elongation diagrams to be recorded during the test.

2.5 Dynamic Response of Spot-Welded and Weld-Bonded Joints

The experimental setup used for dynamic response measurement consists of a 4332 B&K (Denmark-Copenhagen) accelerometer, a 2635 B&K amplifier, A/D converter and Snap-Master software for Windows® Snap-Master (HEM Corporation, USA) is PC-based data acquisition, analysis, control and display software. This performs a Fourier analysis on both the original and the filtered signals. Snap-Master combines advanced data acquisition and storage capabilities with time and frequency domain analysis and near real time plotting.

The weld-through weld-bonding technique was adopted for manufacturing weld-bonded commercial aluminum sheets. Dynamic response (natural frequency and damping), nugget size, and tensile-shear strength were tested and reported [12, 13]. The material used for the work was the B.S. 1050, commercially pure aluminum sheets, having 1 and 2 mm thickness. No surface coatings were applied. The nominal composition of this material is listed in Table 2.

The configuration and dimensions of the test specimen used throughout the dynamic response measurements are shown in Fig. 4. The specimens were cut from 1 to 2 mm thick sheets. These specimens were cut parallel to the rolling directions of the sheets, then degreased, dried and kept in desiccators, ready for weld bonding. A small circle was marked in the center of the upper face of the specimen in order to help positioning of the electrode tip on the center of the overlap area during welding, using a specially designed fixture (Fig. 3). The levels of welding parameters are shown in Table 3 [12, 13].

At the beginning it was decided to hit the specimen in the horizontal direction, since this was believed to stimulate the spot-welded region rather than the whole specimen. However no response was detected in that direction even with the maximum amplification available. This may be related to the high shear strength of the specimen when loaded under tensile forces in the horizontal direction. So, it was decided to hit the specimen in the vertical plane (since the specimen is weak in peel stresses in the vertical direction). The specimens were fixed at one end and stimulated with a hand held hammer at the other end. The output signal was

Table 2 Composition of B.S. 1050 commercial aluminum

Element	Si	Fe	Cu	Mn	Mg	Zn	Ti	Other	Al
Weight (%)	0.25	0.04	0.05	0.05	0.05	0.07	0.05	0.03	99.5 min

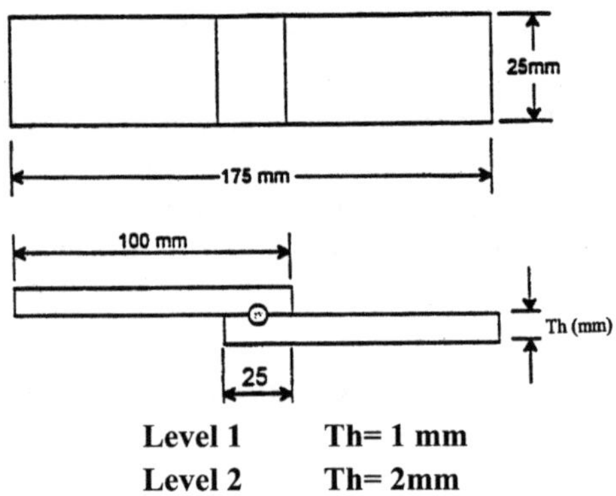

Fig. 4 Configuration and dimensions of weld-bonded and spot-welded aluminum specimens

Table 3 Levels of welding parameters for aluminum joints

Process variable	Level	
	–	+
I = welding current I (A)	15,160	17,900
F = electrode force (N)	1,000	1,428
T = welding time (μs)	145	265
S = sheet thickness (mm)	1	2

detected by the accelerometer and fed to the Snap-Master program through the A/D converter. Figure 5 shows the block diagram of the experimental setup used.

3 Experimental Results

3.1 Microstructure of Spot-Welded Steel Joints

Darwish and Ghanya [21] investigated experimentally both techniques of weld-bonded ranging from metallographic up to micro hardness and strength, for low-carbon steel sheets. Figure 6 shows a polished and etched cross-section of a spot-welded 1.5 mm low-carbon steel sheet. The figure shows different structural

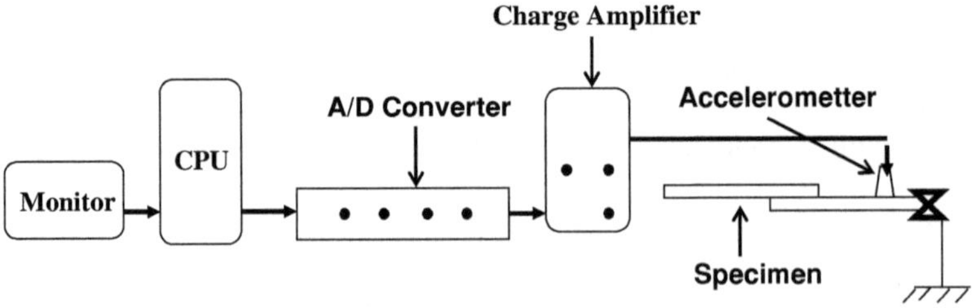

Fig. 5 Block diagram of the experimental setup

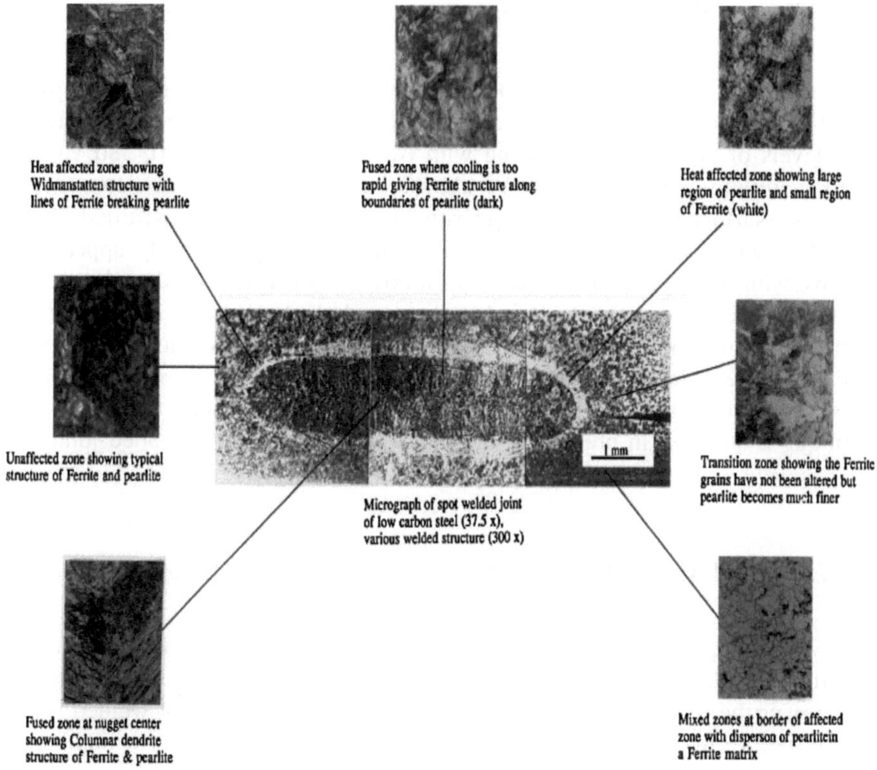

Fig. 6 Microstructure of a spot-welded low-carbon steel nugget at various positions

regions: unaffected, transition, refined, coarsened, and fused zones. The unaffected zone represents the typical grain structure of the parent low-carbon steel sheet. This was not heated enough to reach the critical zone. The transient zone was subjected to a temperature between the A1 and A3 transformation temperature [37], where incomplete recrystallization takes place. The refined zone was heated to just above A3, where the fine austenite transforms to fine ferrite and pearlite. At higher temperatures (above A3), and adjacent to melting, coarse austenite was produced and consequently transformed to a coarse structure which yields a coarsened zone. At temperatures above the solidus, at the faying surface, actual melting of the parent metal had taken place and resolidified rapidly epitaxially, which leads to perfect coalescence.

In summary, the structure of spot-welded low-carbon steel had the following features: (1) a columnar dendritic structure in the fusion zone; (2) a heat-affected zone showing gradual transition from a coarse overheated structure through a normalized region (refined) to the original structure of the unaffected base metal; and (3) a narrow ferritic zone in the interface between the overheated and unaffected zones, which is not always well defined. The above characteristics are demonstrated in Fig. 6.

3.2 Effect of Welding Parameters on Weld Quality (Steel)

Nugget formation depends on many variables, but the welding heat input is probably the most effective parameter. The available spot-welding machine had three levels of current density along with two voltage levels (220, 380 V). It is worth mentioning that with 380 V, almost all of the specimens demonstrated expulsion (surface fusion) and spot tearing, so this voltage was discarded. As for the 220 V with different current densities, the weld nugget quality appeared to improve with increasing current density (almost all specimens produced were free of surface fusion, deep electrode indentation, and electrode deposits). As a result of these preliminary tests, it was decided to work with 220 V and the maximum available current density. Table 1 shows the micro hardness measurements associated with each condition. From Table 1, it could be concluded that the flow-in technique is superior in preserving the micro hardness of weld-bonded joints.

3.3 Effect of Welding Technique on Weld Bond Quality (Steel)

The two techniques of weld-bonding, namely weld-through and flow-in, were conducted and studied. In the flow-in technique, the adhesive material was uniformly spread over the overlap length, afterwards the substrates were spot welded immediately (before the adhesive layer was cured). In the weld-through technique, the adhesive material, plain or filled (10% graphite, 15% aluminum or 20% copper), was spread on the overlap area, then the bonded specimens were left for curing at 24°C for 48 h, then subjected to spot welding after curing. It is worth noting that none of the specimens (plain or filled adhesive) could be spot welded due to the higher resistivity of the adhesive material. In order to reduce the electrical resistivity of the bonded specimens, these specimens were heated in an oven until they reached the temperature of 100°C (the maximum working temperature of the epoxy resin used is 120°C). However, only the specimens containing 15% aluminum could be spot welded. Figures 7 and 8 show micrograph of two weld-bond specimens manufactured using the different techniques of weld-bonding, namely weld-through and flow-in respectively. The figures show that with the weld-through technique, the weld nugget is of a pronounced irregular shape, especially at the ends, also the heat-affected zone was not well defined [37]. The flow-in weld-bond specimens show a uniform elliptical shape, a clear and distinguished microstructure, and a more defined heat-affected zone. Also, there is a more complete dissolution of the thin film of adhesive absorbed by the weld. As for the microhardness of the weld nugget, the weld-through weld-bond (15% aluminum) specimens demonstrated 42%, while the flow-in (15% aluminum) weld-bond demonstrated only 20% reduction in the maximum microhardness when compared with a spot-weld specimen. Therefore it is recommended to use the flow-in technique since it ensures a better quality and economy (no filler added and heat is not required to reduce resistivity).

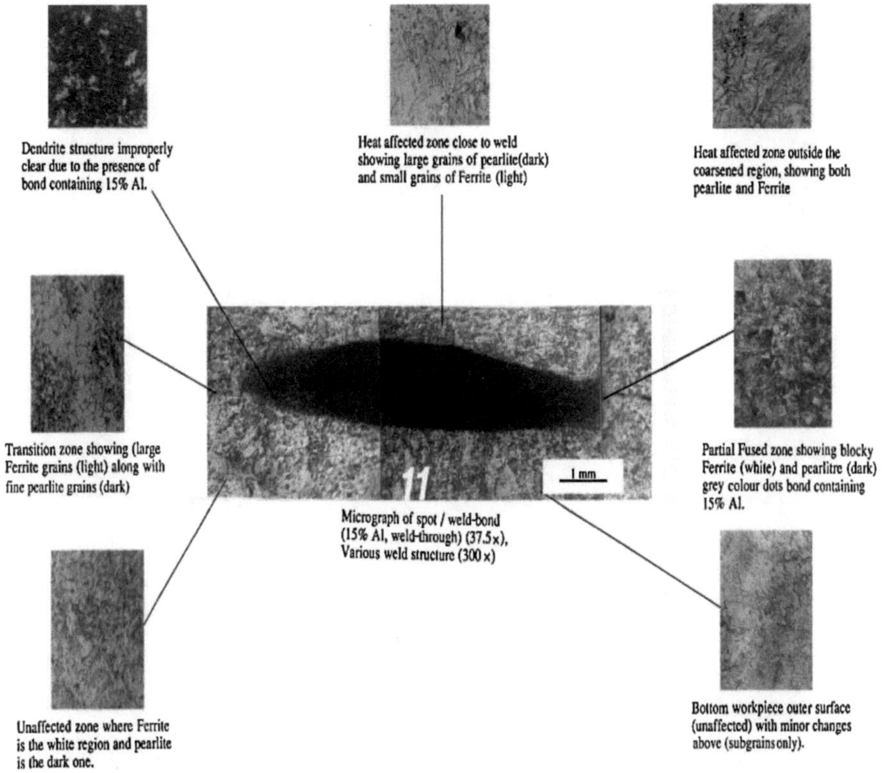

Fig. 7 Microstructure of a spot/weld-bond (15% Al, weld-through) nugget at various positions

3.4 Effect of Filler Type on Weld Bond Quality (Steel)

Figure 9 shows different structural regions of a polished and etched weld-bond (flow-in) having 10% carbon powder mixed with adhesive. Figure 10 shows two flow-in weld-bond specimens both of which were filled with metallic powder. The former contained 10 wt% carbon powder, while the latter contained 20 wt% copper. Both specimens demonstrated a well-defined nugget with a pronounced heat-affected zone. This indicates that the thin film of adhesive material in both specimens was dissolved in molten metal, and diffused into the heat-affected zone, indicating that welding was complete. The only difference that can be realized was that the heat-affected zone was wider in specimens having 10% carbon. This may be attributed to the lower conductivity of the first specimen (10% C) when compared with the second specimen (20% C). Based on the above experiments, the following conclusions could be drawn:

1. The flow-in weld-bond technique is much easier to implement when compared with the weld-through technique, and it better preserves the microstructure and hardness of the produced weldments when compared to the weld-through process.

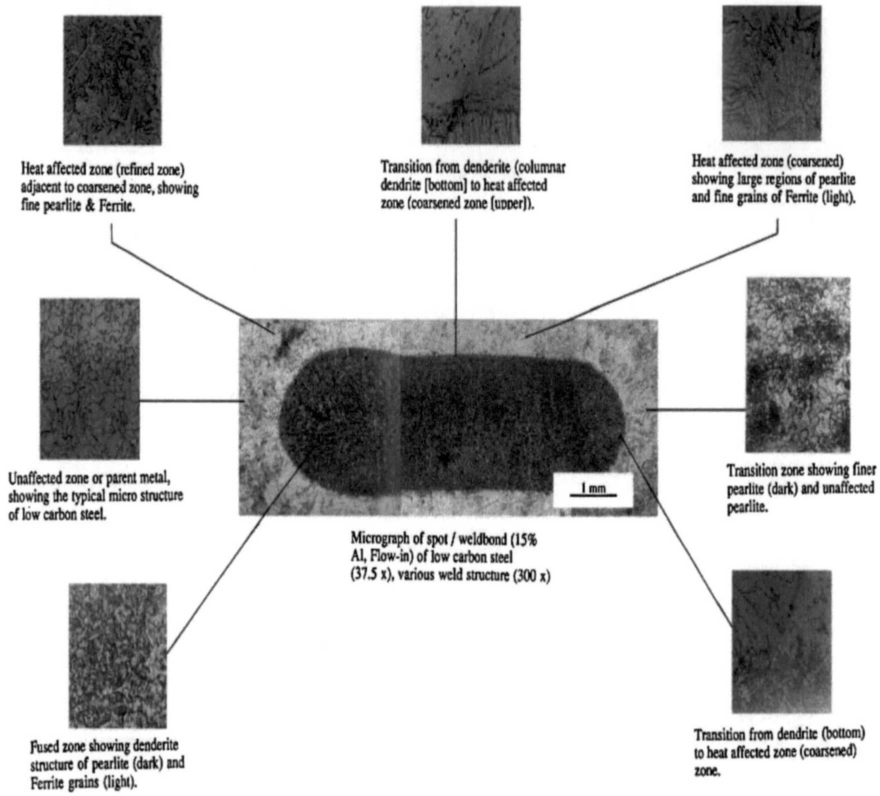

Fig. 8 Microstructure of a spot/weld-bond (15% Al, flow-in) nugget at various positions

2. From the economical point of view, flow-in weld-bond technology is much cheaper when compared with weld-through technology, because in weld-through technology the welded specimens (only those filled with metallic powder) must be heated in an oven to reduce the electrical resistivity; additionally, they should be left to cure for 48 h before welding.

3.5 Dynamic Characteristics of Weld-Bonded Aluminum Joints

Tables 4 and 5 give the dynamic response of spot-welded and weld-bonded joints (natural frequency and damping), respectively. Table 4 indicates that the deviation from the average natural frequency (24.66 Hz) for the first eight specimens having 1 mm thickness is irrelevant, which indicates that there is no direct relation between the natural frequency and the strength of the joint. The deviation from the average natural frequency (56.03 Hz) for the last eight specimens having 2 mm thickness is also irrelevant, which confirms also that there is no direct relation

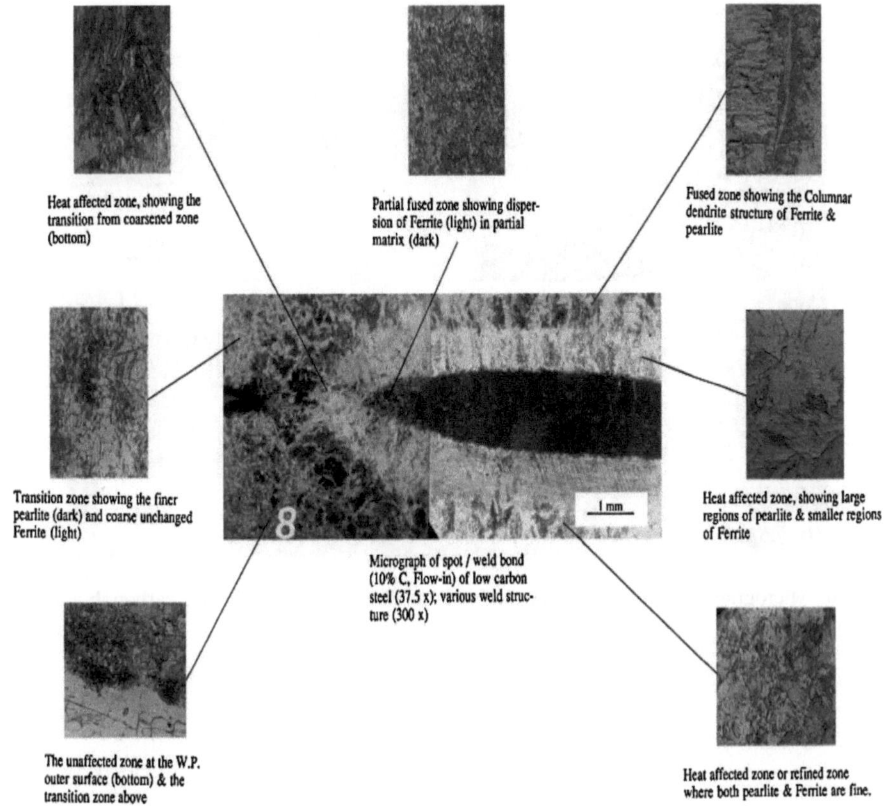

Fig. 9 Microstructure of a spot/weld-bond (10% C, flow-in) nugget, at various positions

between the natural frequency and the strength of the joint. The same conclusions can be drawn from Table 5 concerning weld-bonded joints.

Figure 11 shows the natural frequency of spot-welded and weld-bonded joints, while Fig. 12, gives the damping capacity for spot-welded and weld-bonded specimens. Figure 12 indicates that weld-bonded joints possess higher damping capacity which is reflected in the lower amplitudes associated with the natural frequency when compared with spot-welded joints. The decaying amplitude for the free vibration of a typical weld-bonded joint is shown in Fig. 13. This reflects the advantage of using weld-bonded structures in order to absorb induced vibrations.

3.6 *Tensile Strength of Spot-Welded and Weld-Bonded Aluminum Specimens*

Specimens were manufactured according to the experimental design shown in Tables 6 and 7 for spot welded and weld-bonded joints, respectively. The averages

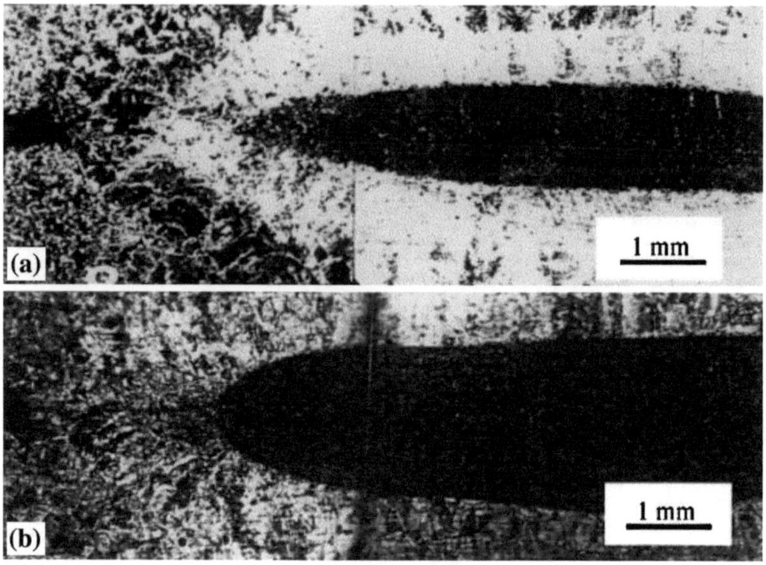

Fig. 10 Micrograph of a flow-in spot/weld-bond: **a** adhesive filled with 10% carbon, **b** adhesive filled with 20% copper

Table 4 Dynamic responses of spot-welded aluminum joints

Run no.	Factor				Frequency (Hz)	Amplitude (V)
	Force (N)	Current (A)	Time (μs)	Sheet thickness (mm)		
1	–	–	–	–	26.85	0.49
2	+	–	–	–	24.9	0.31
3	–	+	–	–	23.44	0.32
4	+	+	–	–	23.44	0.24
5	–	–	+	–	23.92	0.32
6	+	–	+	–	26.36	0.29
7	–	+	+	–	23.92	0.39
8	+	+	+	–	24.41	0.27
9	–	–	–	+	55.17	0.44
10	+	–	–	+	56.64	0.36
11	–	+	–	+	55.66	0.32
12	+	+	–	+	56.15	0.35
13	–	–	+	+	54.69	0.27
14	+	–	+	+	55.66	0.51
15	–	+	+	+	57.62	0.34
16	+	+	+	+	56.64	0.24

Table 5 Dynamic responses of weld-bonded aluminum joints

Run no.	Factor				Frequency (Hz)	Amplitude
	Force (N)	Current (A)	Time (μs)	Sheet thickness (mm)		
1	–	–	–	–	22.94	0.37
2	+	–	–	–	24.41	0.33
3	–	+	–	–	26.37	0.38
4	+	+	–	–	26.37	0.24
5	–	–	+	–	25.39	0.32
6	+	–	+	–	27.83	0.41
7	–	+	+	–	26.85	0.28
8	+	+	+	–	25.39	0.27
9	–	–	–	+	57.13	0.38
10	+	–	–	+	54.68	0.34
11	-	+	–	+	57.12	0.31
12	+	+	–	+	58.49	0.23
13	–	–	+	+	59.08	0.22
14	+	–	+	+	58.59	0.31
15	–	+	+	+	59.57	0.16
16	+	+	+	+	56.64	0.24

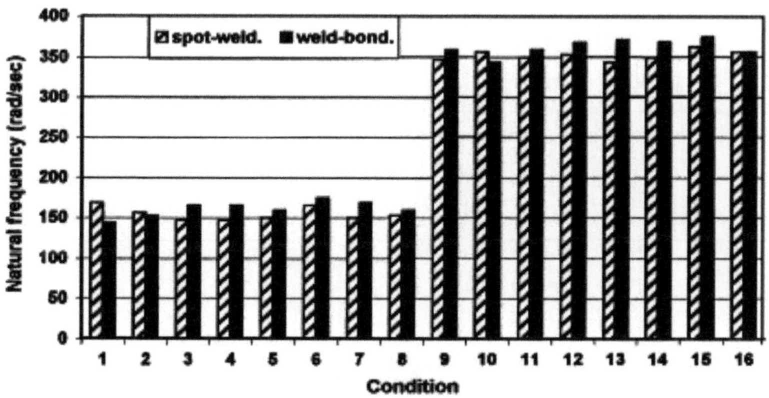

Fig. 11 Natural frequency of spot-welded and weld-bonded aluminum joints

of three specimens were taken to represent the tensile-shear strength for each welding technique (spot welding and weld bonding). Tables 6 and 7 list the average failure load associated with each spot-welding and weld-bonding condition, respectively, along with the standard deviation.

These tables show that weld-bonded joints sustained higher failure loads compared with the corresponding spot-welded joints, due to the uniform distribution of the stresses associated with the bonded layer.

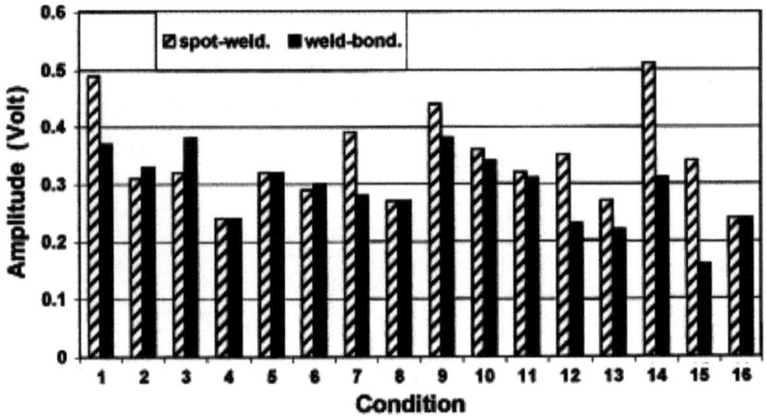

Fig. 12 Damping capacity of spot-welded and weld-bonded aluminum joints

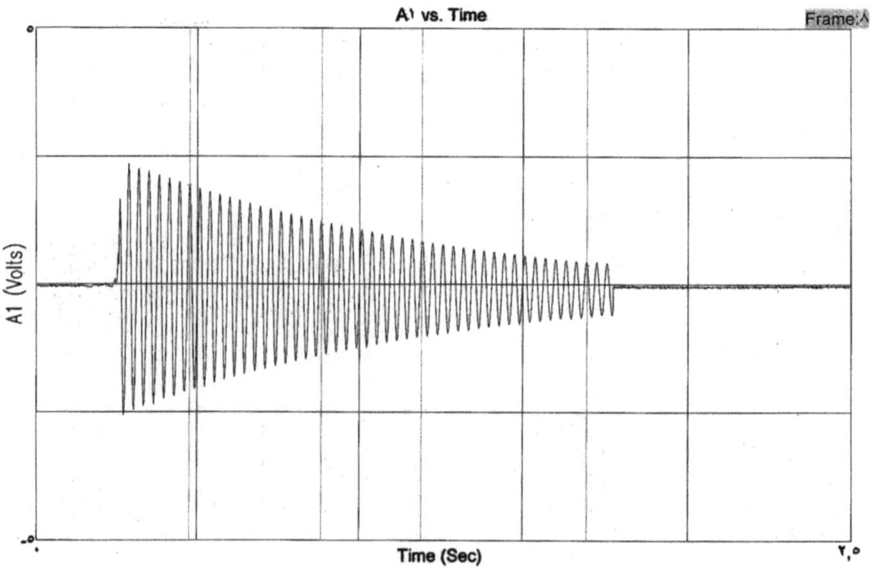

Fig. 13 Decaying amplitude of the free vibration of a weld-bonded aluminum specimen

3.7 Nugget Size of Spot-Welded and Weld-Bonded Aluminum Joints

Tables 8 and 9 show the nugget size of the spot-welded and weld-bonded joints, respectively. Figures 14 and 15 show the correlation (linear) between the failure load and nugget size, for spot-welded and weld-bonded joints, respectively. The figures indicate that as long as the nugget size increases the failure load for both spot-welded and weld-bonded joints increases.

Table 6 Average failure loads of spot-welded aluminum joints

Test condition	Factor				Average failure load (N)	Standard deviation (N)
	Force (N)	Current (A)	Time (μs)	Sheet thickness (mm)		
1	−	−	−	−	580.0	43.6
2	+	−	−	−	534.0	38.6
3	−	+	−	−	680.0	30.0
4	+	+	−	−	657.3	40.6
5	−	−	+	−	826.7	50.3
6	+	−	+	−	661.7	50.1
7	−	+	+	−	981.3	68.6
8	+	+	+	−	829.7	45.3
9	−	−	−	+	557.3	56.0
10	+	−	−	+	493.3	66.6
11	−	+	−	+	715.0	69.6
12	+	+	−	+	686.7	25.2
13	−	−	+	+	684.0	57.2
14	+	−	+	+	646.7	47.3
15	−	+	+	+	947.0	24.6
16	+	+	+	+	856.7	66.6

Table 7 Average failure loads of weld-bonded aluminum joints

Run no.	Factor				Micro hardness (HVN)	Average failure load (N)	Standard deviation
	Force (N)	Current (A)	Time (μs)	Sheet thickness (mm)			
1	−	−	−	−	29.05	737.5	10.61
2	+	−	−	−	26.69	600	70.71
3	−	+	−	−	27.08	1,160	56.57
4	+	+	−	−	30.23	1,162.5	3.54
5	−	−	+	−	27.03	1,090	42.43
6	+	−	+	−	26.68	1,080	141.42
7	−	+	+	−	27.30	1,215	120.21
8	+	+	+	−	27.03	1,180	113.14
9	−	−	−	+	26.83	655	7.07
10	+	−	−	+	25.33	435	233.35
11	−	+	−	+	27.16	799	55.15
12	+	+	−	+	29.61	705	91.92
13	−	−	+	+	31.00	854	118.79
14	+	−	+	+	28.85	625	21.21
15	−	+	+	+	29.31	1,405	162.63
16	+	+	+	+	28.78	1,180	28.28

Table 8 Nugget sizes of spot-welded aluminum joints

Run no.	Nugget size (mm^2)			Average nugget size $(mm)^2$
	Spec. 1	Spec. 2	Spec. 3	
1	4.25	4.50	3.50	4.08
2	2.90	2.75	2.63	2.76
3	5.00	4.00	5.00	4.70
4	3.00	3.75	3.50	3.42
5	4.25	4.50	4.50	4.42
6	3.00	5.00	3.00	3.70
7	6.00	4.50	5.25	5.25
8	5.75	6.00	4.50	5.42
9	3.50	2.50	3.50	3.17
10	3.50	2.50	2.25	2.75
11	4.50	4.50	5.25	4.75
12	4.50	3.00	4.50	4.00
13	4.25	3.75	4.25	4.10
14	3.50	3.25	2.25	3.00
15	5.50	6.00	5.00	5.50
16	5.50	6.00	4.50	5.30

Table 9 Nugget sizes of weld-bonded aluminum joints

Run no.	Nugget size (mm^2)			Average nugget size $(mm)^2$
	Spec. 1	Spec. 2	Spec. 3	
1	4.50	4.00	4.5	4.33
2	1.50	1.50	1.75	1.56
3	3.75	4.125	4.50	4.13
4	2.50	2.65	2.50	2.55
5	4.00	4.00	3.50	3.83
6	3.75	3.00	3.00	3.25
7	4.75	5.00	4.90	4.88
8	2.50	3.50	3.00	3.00
9	3.50	4.25	3.75	3.83
10	3.5	3.38	4.00	3.65
11	4.50	5.00	3.50	4.33
12	2.25	3.00	4.50	3.25
13	4.50	4.25	4.25	4.33
14	2.50	2.00	2.25	2.25
15	5.13	5.25	5.75	5.38
16	5.00	5.00	4.50	4.83

4 Joints Modeling

Little attention been has given in the past to the analysis of weld-bonded joints, consequently the present work aimed at predicting the strength of weld-bonded joints, including those having adhesive layers with spew fillet [28, 29].

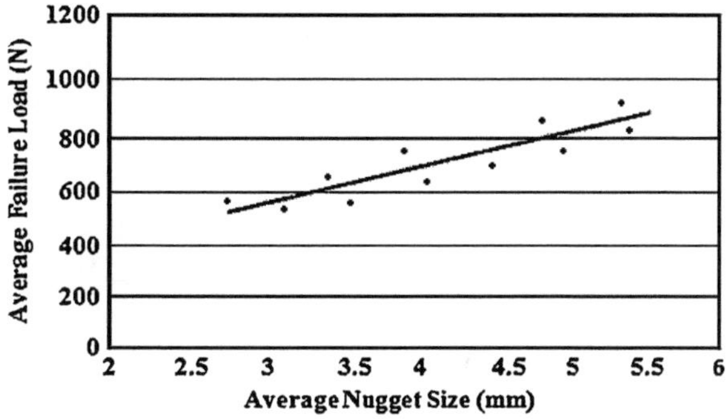

Fig. 14 Correlation between failure load and nugget size (spot-welded aluminum joints)

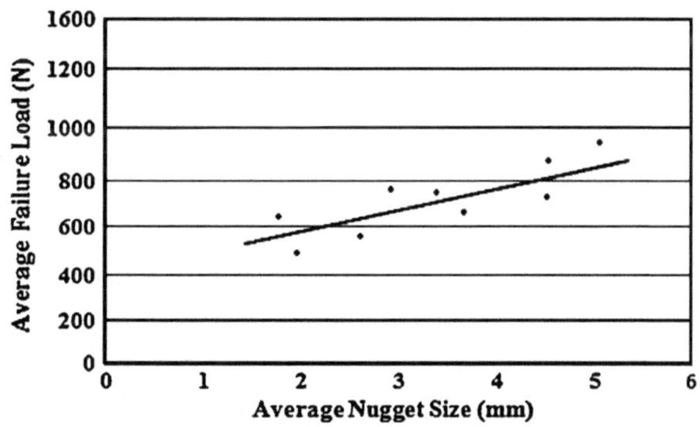

Fig. 15 Correlation between failure load and nugget size (weld-bonded aluminum joints)

The finite element technique was used in the present work for modeling spot welded and weld-bonded joints as it avoids the approximations of the closed-form solutions in neglecting the strain energy of certain stresses within the joint, so enabling more accurate results to be obtained. Darwish [12, 13] modeled spot welded and weld-bonded joints of dissimilar thickness and dissimilar materials, while [3, 17] analyzed weld-bonded joints for peel strength.

Figure 16 shows the configuration and dimensions of the considered joints. Three finite elements (FE) models were considered: a spot-welding model, an adhesive bonded model and a weld-bonded model. The configurations of these models including constrains and loading conditions are shown in Fig. 17. The material properties of the strips and adhesive materials are listed in Table 10.

The restrained points as well as the applied load (position and value) were kept constant to enable fair comparison between the data made [28, 29]. The applied load was taken to be the relatively small value of 500 N, where the displacement and accordingly the stresses were assumed to be proportional to the load. Figure 18 shows the deformed mesh overlaid on the un-deformed mesh for comparison given for weld-bonded FE model.

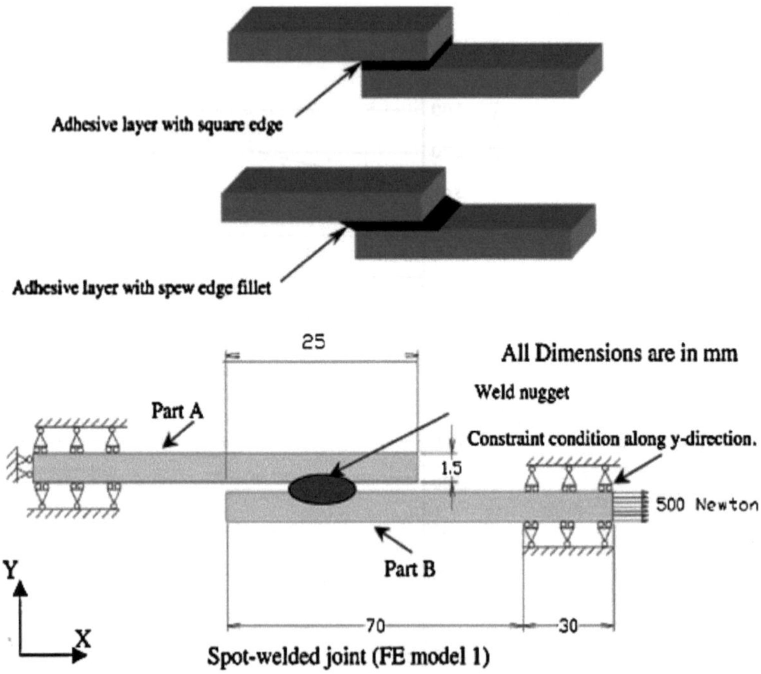

Fig. 16 Overall dimensions of the weld joint used in finite element model

4.1 Stress Distribution of Spot-Welded, Adhesively Bonded and Weld-Bonded Joints

The predicted normal stresses (σ_x) through the mid-layer of the spot-welded adhesively bonded and weld-bonded joints are shown in Fig. 19. From this figure, it can be observed that the normal stresses are concentrated at the far ends of the weld nugget in case of spot-welded and weld-bonded joints. It is worth noting that the spot-welded joint showed a 160% higher stress concentration, when compared with the weld-bonded joint. However, the normal stresses were concentrated at the free ends of the adhesive bonded joint showing a lower stress concentration, when compared with the weld-bonded joint.

The predicted normal stresses (σ_y) through the mid-layer of the spot-welded adhesively bonded and weld-bonded joints are shown in Fig. 20. From this figure, it can be observed that the normal stresses are concentrated at the far ends of the weld nugget in case of spot-welded, and weld-bonded joints. It is worth noting that the spot-welded joint showed a 280% higher stress concentration, when compared with the weld-bonded joint. However, the normal stresses were also concentrated at the free ends of the adhesively bonded joint and showed nearly the same stress concentration as the weld-bonded joint.

The predicted shear stresses (τ_{xy}) through the mid-layer of the spot-welded adhesively bonded and weld-bonded joints are shown in Fig. 21. From this figure,

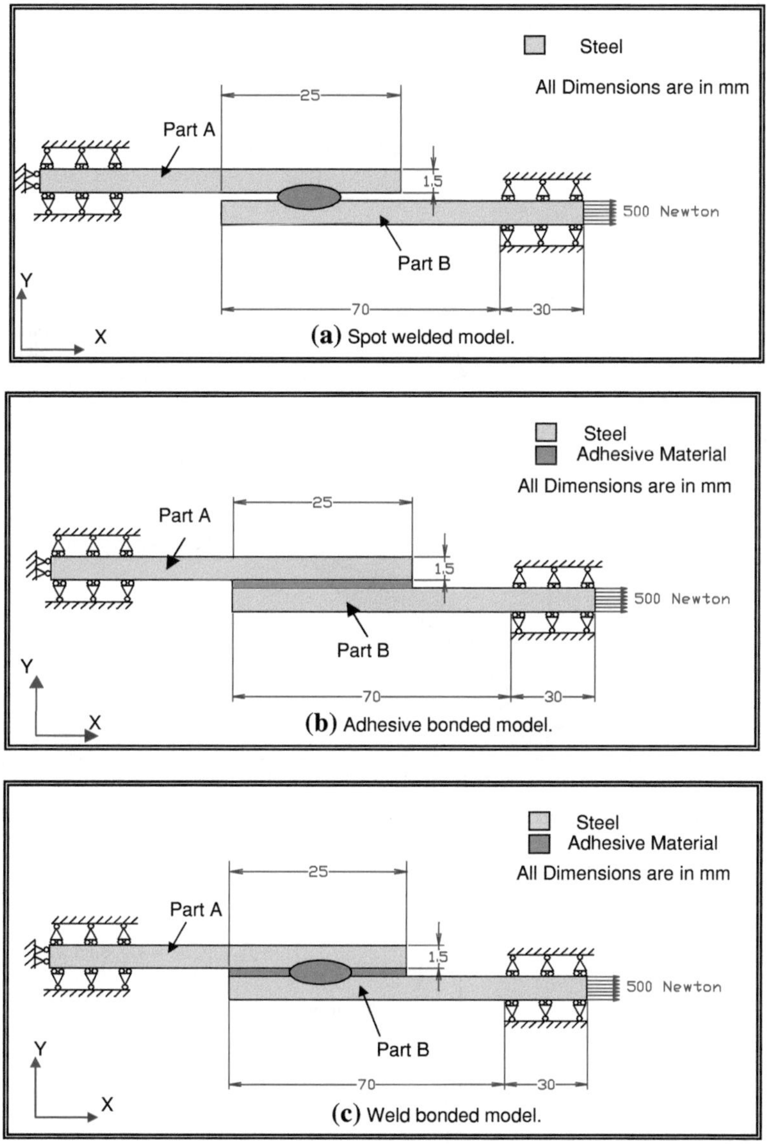

Fig. 17 Spot, adhesively bonded and weld-bonded models

Table 10 Material properties for joined strips and adhesive material studied in the finite element analysis

Material	Young's modulus (GPa)	Poisson's ratio
Steel	200	0.3
Adhesive	2.5	0.38

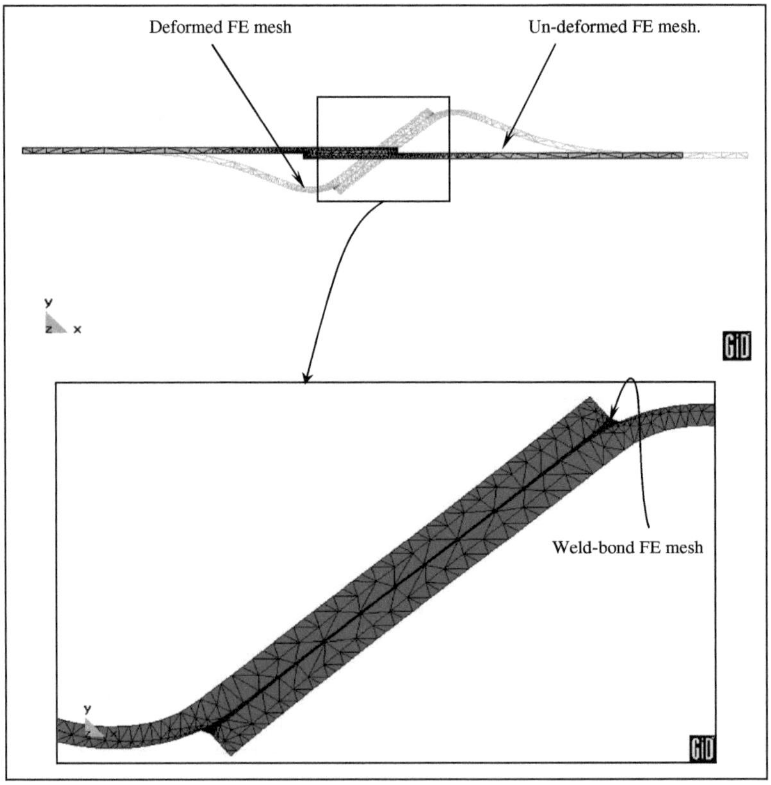

Fig. 18 Post-processing output of deformed and un-deformed FE meshes for the weld-bonded FE model

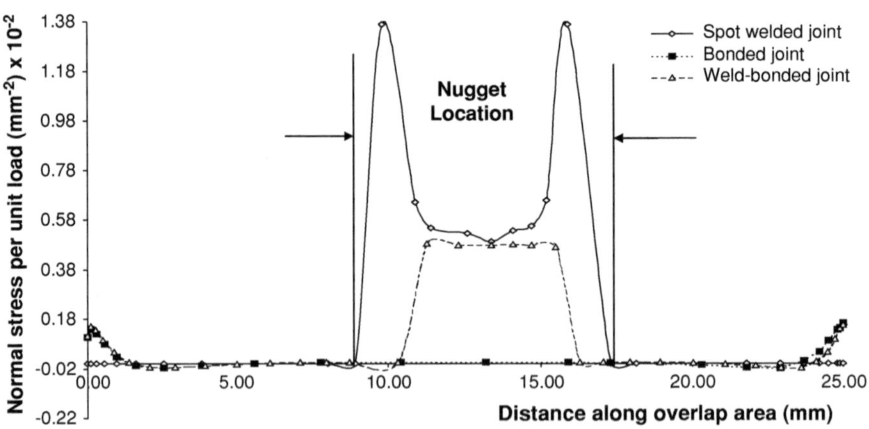

Fig. 19 Normal stress, σ_x, distribution along the mid-layer for the three models (spot welded, bonded and weld-bonded models)

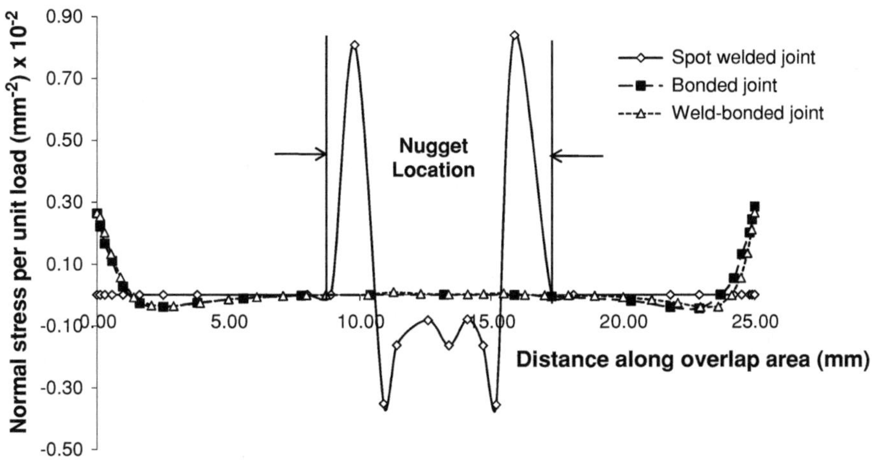

Fig. 20 Normal stress, σ_y, distribution along the mid-layer for the three models (spot welded, bonded and weld-bonded models)

it can be seen that the shear stresses are concentrated at the far ends of the weld nugget in case of the spot-welded joint. It is worth noting that the spot-welded joint showed a 625% higher stress concentration, when compared with the weld-bonded joint. The shear stresses were concentrated at the free ends of the adhesively bonded joint, however, and they showed nearly the same stress concentration as the weld-bonded joint.

The predicted major principal stress through the mid-layer of the spot-welded adhesively bonded and weld-bonded joints are shown in Fig. 22. From this figure, it can be observed that the stresses were highly concentrated in the spot-welded

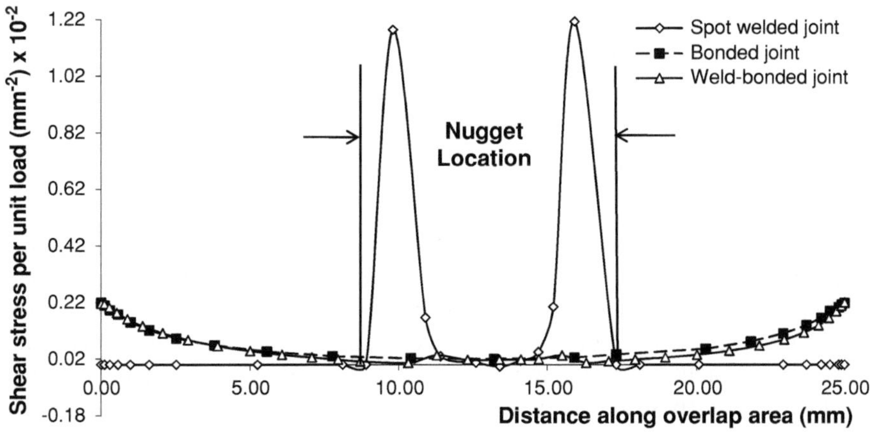

Fig. 21 Shear stress, τ_{xy}, distribution along the mid-layer for the three models (spot welded, bonded and weld-bonded models)

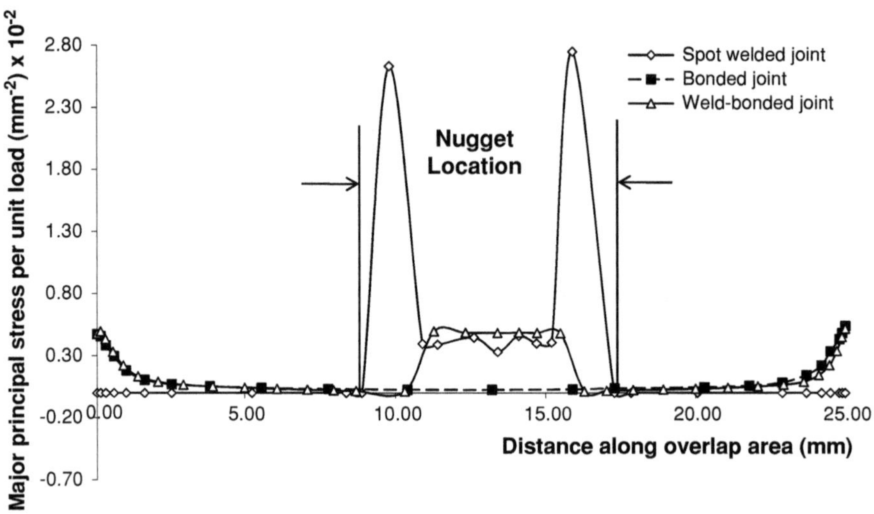

Fig. 22 Major principal stress, σ_1, distribution along the mid-layer for the three models (spot welded, bonded and weld-bonded models)

joints, when compared with the weld-bonded and the adhesively bonded joints. The major principal stress of the spot-welded joint was nearly five to six times that of the weld-bonded and adhesively bonded joints, respectively. The lower stress level associated with weld-bonded joints may be attributed to the use of adhesive, which minimizes the stress concentration located at the edges of spot welds in spot-welded joints.

In practice, there are two possible configurations of a bonded joint: a lap joint which ends in square edge and a lap joint which ends in filleted edges.

The presence of the spew fillet resulted in the strengthening of the weld-bonded joints, considerably. This is demonstrated for example in the reduction of the major principle stress by nearly 14% (Fig. 23).

4.2 Weldbonding and Spot-Welded Dissimilar Joints

Three finite element models were considered: a spot-welded model having similar sheet thicknesses (1.5–1.5 mm), a spot-welded model having dissimilar sheet thicknesses (1–2 mm) and a spot-welded model having dissimilar thickness sheets (1.5–2 mm) [12–14]. Figure 24 shows the configuration, dimensions, constraints and loading conditions of the models considered. The restrained points as well as the applied load (position and value) were kept constant to enable fair comparison between the data.

The applied load was relatively small, 500 N, where the displacement and accordingly the stresses are assumed proportional to the load.

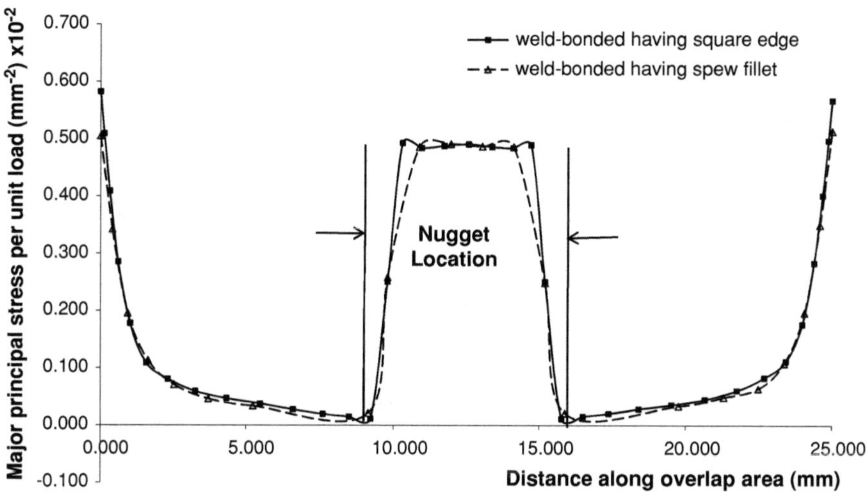

Fig. 23 Major principal stress, σ_1, distribution of weld-bonded joints having square end and spew fillet adhesive layer

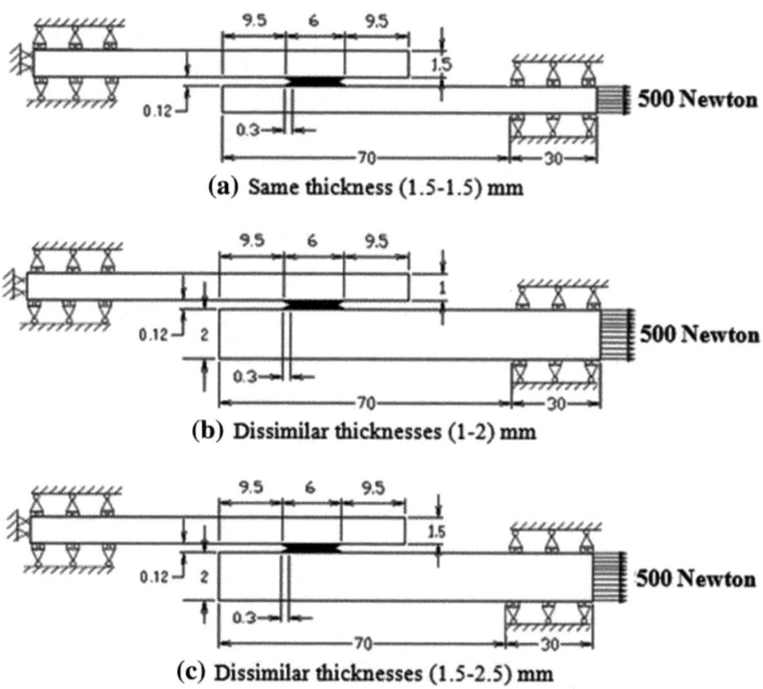

Fig. 24 Configuration, dimensions and loading conditions of the dissimilar thicknesses spot welded joints

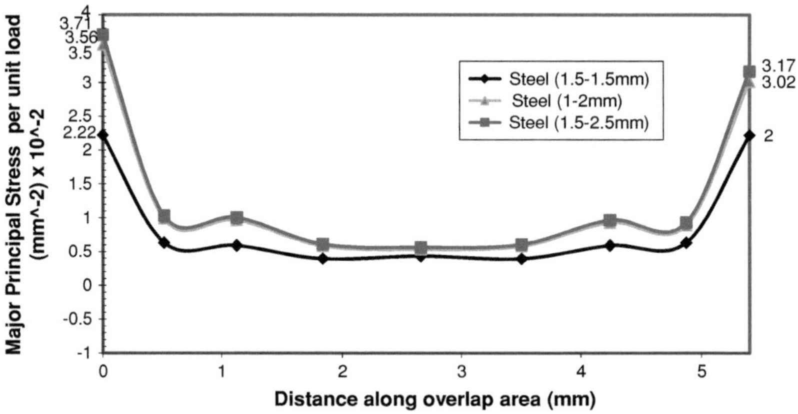

Fig. 25 Major principal stress (σ_1) distribution along overlap area for different FE models considered (equal and dissimilar joints)

The predicted maximum principal stress through the mid-layer of the overlap area of different joints, having similar and dissimilar thicknesses, are shown in Fig. 25. From this figure, it can be observed that the stresses were highly concentrated in the spot-welded dissimilar thickness joint (towards the thinner part of the joint), when compared with the spot-welded equal thickness joint (1.5–1.5 mm). It is worth noting that several dissimilar thickness joints were manufactured and tested experimentally in shear under tensile loading (Table 11). Table 12 gives the welding parameters used for spot-welding brass specimens, while Tables 13 and 14 give the average failure loads for the similar and dissimilar brass joints (1–2 mm). In almost all cases, joints having similar adherend thicknesses sustained higher failure loads, when compared with those having dissimilar

Table 11 Material properties of adherend and adhesive used for FE analysis

Material	Young's modulus (GPa)	Poisson's ratio
Steel	200.0	0.3
Brass	101.0	0.35
Aluminum	61.50	0.33
Adhesive	2.5000	0.38

Table 12 Levels of welding parameters for brass sheets used for FE analysis

Process variable	Level	
	–	+
Welding current, I (A)	17,863	19,289
Electrode force, F (N)	900	1,100
Welding time, T (μs)	265	400

Table 13 Average failure loads for brass specimens having similar sheet thickness (1.5–1.5 mm)

No.	Welding current (A)	Electrode force (N)	Welding time (μs)	Average failure load (kN)
1	+	+	+	5.75
2	+	+	–	6.0
3	+	–	+	6.68
4	+	–	–	7.2
5	–	+	+	5.2
6	–	+	–	4.68
7	–	–	+	6.18
8	–	–	–	5.35

Table 14 Average failure loads for brass specimens having dissimilar sheet thickness (1–2 mm)

No.	Welding current (A)	Electrode force (N)	Welding time (μs)	Average failure load (kN)
1	+	+	+	4.28
2	+	+	–	4.43
3	+	–	+	5.03
4	+	–	–	5.23
5	–	+	+	4.11
6	–	+	–	4.12
7	–	–	+	4.64
8	–	–	–	4.63

thicknesses. Rupture happened in almost all cases within the thinner part of dissimilar thicknesses joints, around the welding nugget.

Figure 26 shows the major principle stress developed within the welding nugget along the overlap area, for the steel, aluminum and brass models. This figure shows that the stresses were more concentrated towards the thinner part of the joint, irrespective of the adherend material involved. This figure also shows that the effect of the material type on the stresses developed within the weld nugget is irrelevant.

In order to strengthen the spot-welded dissimilar joint, it was decided to introduce an adhesive layer, in conjunction to the welding nugget. Figure 27 shows the stress distribution of principal stresses in spot-welded, adhesively bonded and weld-bonded dissimilar joints, having a 1 mm thick sheet joined to a 2 mm thick sheet. The figure reveals that bonded joints distribute the stresses uniformly along the joint with a lower concentration towards the free ends of the joint. The figure also shows the stress distribution within the spot-welded dissimilar joint, which demonstrates a 28% higher stresses towards the thinner member of the joint, when compared with the thick member of the joint. The effect of introducing an adhesive layer in conjunction with spot-welding resulted in a

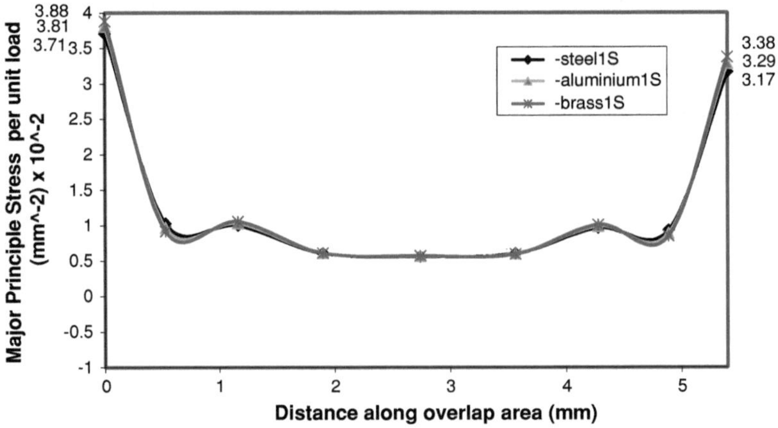

Fig. 26 Major principal stress (σ_1) distribution along overlap area for different materials considered (steel, aluminum and brass adherends)

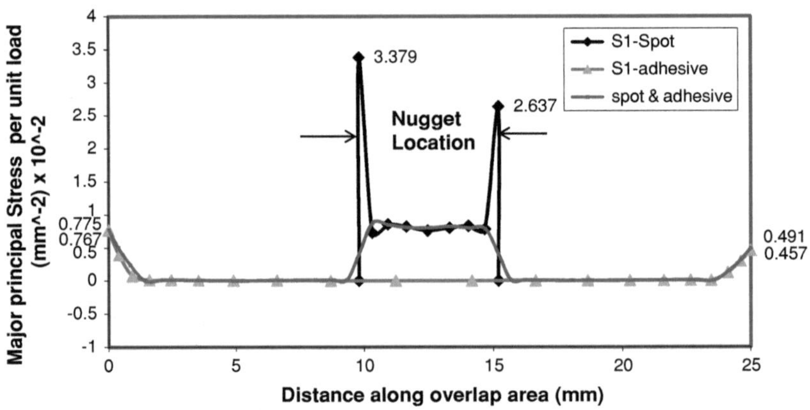

Fig. 27 Major principal stress (σ_1) distribution along overlap area for spot-welded, weld-bonded and adhesive bonded joints

uniform distribution of stresses along the joint. Both the thin and thick members of the joint are enjoying the same level of stresses, which was reduced by 62%, when compared with plain spot-welded joints.

4.3 Stress Distribution of Spot-Welded and Weld-Bonded Dissimilar Thickness T-Peel and Single Lap Joints

Four finite element models were studied a weld-bonded T-peel testing model having similar sheet thicknesses, a T-peel testing weld-bonded model having

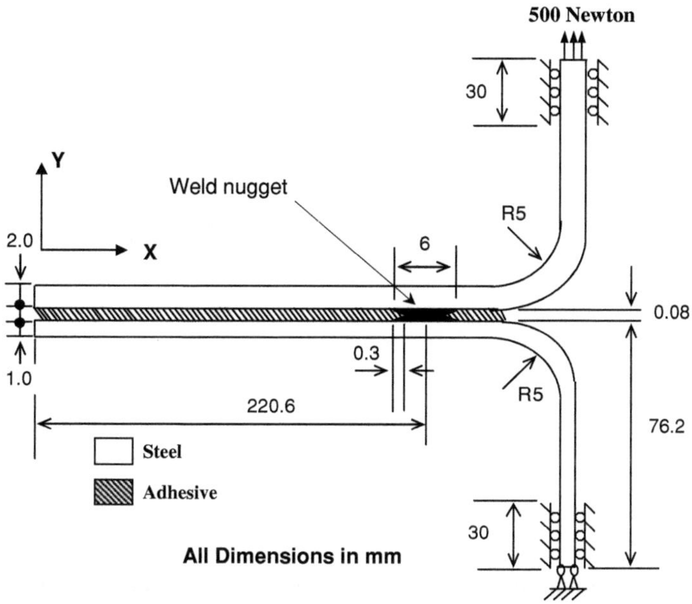

Fig. 28 Structural dissimilar thickness weld-bonded (T-peel strength) model

dissimilar sheet thicknesses, a spot-welded having similar sheet thickness single lap shear testing model, and weld-bonded model (single lap shear testing) having dissimilar sheet thicknesses [3, 17, 18]. Figures 28 and 29 show the configuration,

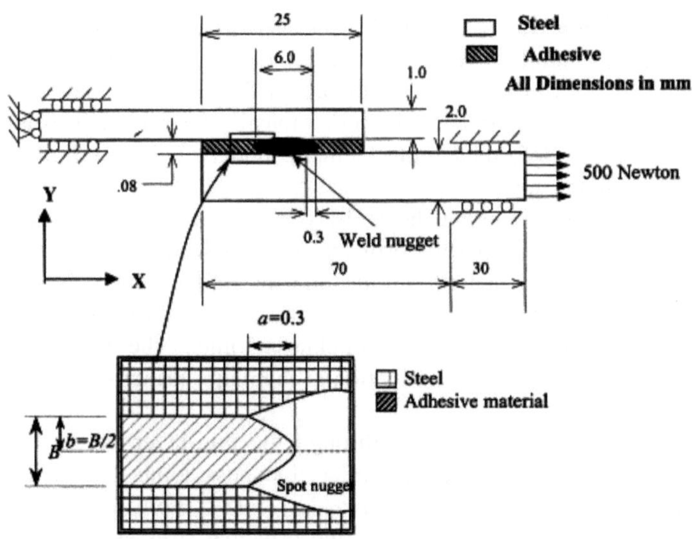

Fig. 29 Structural dissimilar thickness weld-bonded (tensile-shear strength) model

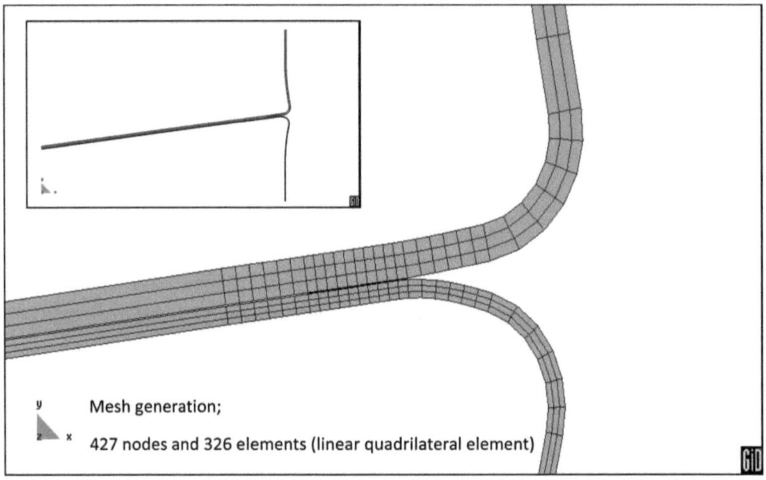

Fig. 30 Deformed finite element model of the dissimilar thickness spot-welded (T-peel) joint

dimensions, constraints and loading conditions of the models considered. In all four models, the joint width was taken to be 25 mm.

In order to simplify the configuration, an elliptical blunt notch Fig. 29 was assumed to approximate the apex of the periphery crab, where *a* and *b* are the lengths of the major and minor axes of the elliptical notch.

The applied load was taken to be a relatively small load (500 N), as the displacement and accordingly the stresses were assumed to be proportional to the load. A gap of 0.08 mm was assumed in the spot-welded specimen in the overlap zone. The meshes of the finite element models were refined until a fair convergence for the solution was obtained. The mesh pattern for the T-peel similar and dissimilar thickness joints models was the same. Figure 30 shows the deformed finite element model of the dissimilar thickness T-peel model. The mesh pattern for the tensile-shear similar and dissimilar tensile-shear models was identical. Figure 31 shows the deformed finite element model of the dissimilar thickness tensile-shear strength model.

The principal stress σ_1 along the mid-layer of T-peel specimen are shown in Fig. 32. This figure shows that the principal stresses were concentrated at the far edge of the weld nugget in both similar and dissimilar joints. However a 55% increase in the major principal stress was realized in case of dissimilar thickness joints, when compared with similar thickness joints. The introduction of an adhesive layer resulted in a 85% reduction in the major principal stress, located at the far end of the joint.

The predicted principal stress σ_1 through the mid-layer of the overlap area of the spot-welded similar and dissimilar thickness joints (per unit load) are shown in Fig. 33 for the single lap joint [3, 17, 18]. From this figure it can be seen that for all the models considered, the principal stresses were concentrated at the far ends of the weld nugget. It can also be seen from Fig. 33. That the

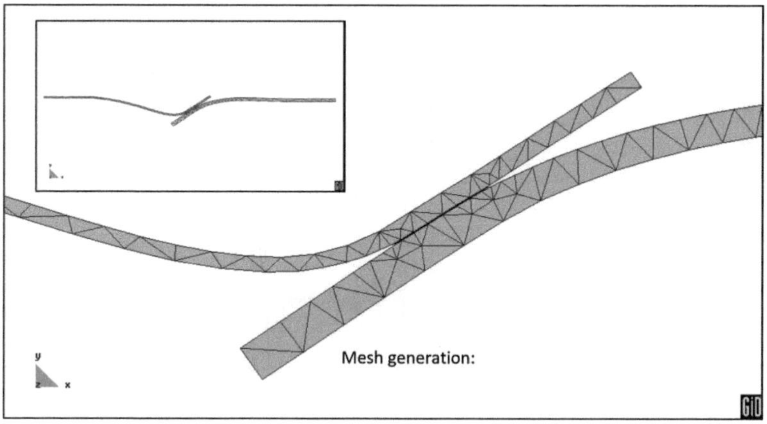

Fig. 31 Deformed finite element model of the dissimilar thickness spot-welded (tensile-shear) joint

stresses were nearly at the same level, for both members of similar thickness spot-welded joints. However, they were highly concentrated towards the thinner part of dissimilar thickness joints, when compared with the thicker part of the joint.

It is worth noting that several similar and dissimilar thickness joints were manufactured and tested experimentally. In almost all cases, joints having similar adherend thicknesses sustained higher failure loads, when compared with those having dissimilar thicknesses. Rupture happened in almost all cases within the thinner part of dissimilar thickness joints around the welding nugget.

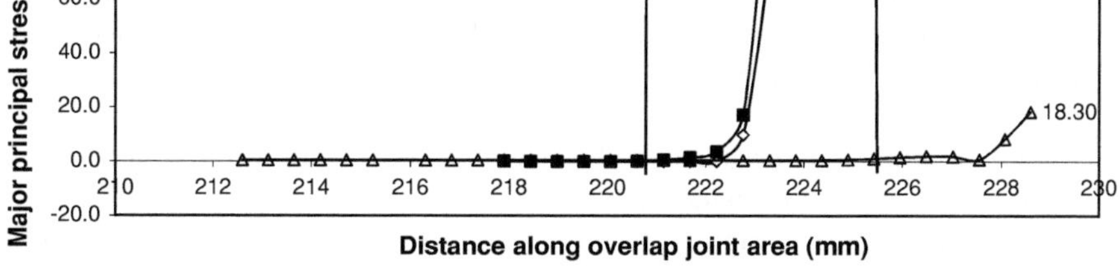

Fig. 32 Predicted major principal stress σ_1 developed in the T-peel model

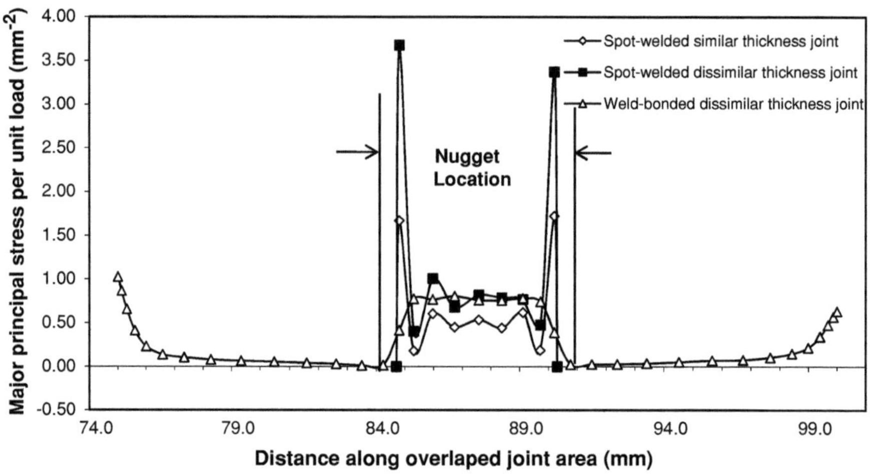

Fig. 33 Predicted major principal stress σ_1 developed in tensile-shear testing model

5 Conclusions

The present work addresses spot welding and weld-bonding techniques from implementation, analysis to characteristics. Based on the present work, the following conclusions can be drawn.

1. A tremendous increase in the tensile-shear strength is noticed in weld-bonded aluminum joints, when compared with the corresponding spot-welded joints and in comparison with adhesively bonded joints.
2. Weld bonded exhibited higher damping capacity with respect to spot-welded joints.
3. The natural frequency seems to be independent of the shear strength as well as the joining technique (spot welding or weld bonding).
4. The stresses are concentrated at the far ends of the overlap area in case of adhesively bonded joints. In case of spot-welded joints, they are concentrated at both ends of the welding nugget. They are concentrated at the overlap region as well as both ends of the welding nugget, in case of weld-bonded joints.
5. The major principal stress developed in the spot-welded joints is nearly five to six times that associated with weld-bonded and adhesively bonded joints, respectively. This demonstrates the role of the adhesive layer in strengthening weld-bonded joints.
6. Single lap joints having adhesive layer with a spew fillet resulted in additional strengthening of weld-bonded joints.
7. The effect of introducing an adhesive layer in conjunction with spot welding resulted in a uniform distribution of stresses along the weld-bonded joint. Both thin and thick members of the joint show the same level of stresses.

References

1. Aidun, D.K., Bennett, R.W.: Effect of resistance welding variables on the strength of spot welded 6061 T6 aluminum alloy. Weld. J. **64**, 15–26 (1985)
2. Alcini, W.V.: A measurement window into spot welding. Weld. J. (1990)
3. Al-Samhan, A., Darwish, S.M.: Finite element modeling of weld-bonded joints. J. Mater. Process. Technol. **142**, 587–598 (2003)
4. American Welding Society. Recommended Practices for Spot Welding Aluminum and Aluminum Alloys. C1.2, 53 T, USA (1953)
5. Apartseva, E.L.: The use of adhesives in mechanical engineering. Vestnik Mashinstroeniya **58**, 47–50 (1978)
6. Bastien, L.J., Gillespie, J.W. Jr.: A non isothermal healing model for strength and toughness of fusion bonded joints of amorphous thermoplastics. Polym. Eng. Sci. **31**(24), 1720–1730 (1991)
7. Bratschun, W.R., Leicht, J.L.: Welding of plated dissimilar metals for RF/EMI shielding. IEEE Trans. Compon. Hybr. Manuf. Technol. **15**(6), 931–937 (1992)
8. Brent-Smith, I.: Development in spot welding technology. Metall. USA **55**, 244 (1988)
9. Brewis, D.M.: Critical assessment of factors affecting bonding of metals. Master Sci. Technol. **2**, 761–767 (1986)
10. Chang, H.S., Cho, H.S.: A study on the shunt effect in resistance spot welding. Weld. J. 308–317 (1990)
11. Cohen, R.L., West, K.W.: Aluminum spot weld strength determined from electrical measurements. Weld. J. 37–41 (1985)
12. Darwish S.M.: Characteristics of weld-bonded commercial aluminum sheets. Int. J. Adhes. Adhes. **23**, 169–176 (2003)
13. Darwish, S.M.: Weld-bonding strengthens & balances the stresses in spot-welded dissimilar thickness points. J. Mater. Process. Technol. **134**, 352–362 (2003)
14. Darwish, S.M.: Analysis of weld-bonded dissimilar materials. Int. J. Adhes. Adhes. **24**, 347–354 (2004)
15. Darwish, S.M., Al-Dekhial, S.D.: Statistical models for spot welding of commercial aluminum sheets. Int. J. Mach. Tools Manuf. **39**, 1589–1610 (1999)
16. Darwish, S.M., Al-Dekhial, S.D.: Micro hardness of spot welded (B.S. 1050) commercial aluminum as correlated with welding variables and strength attributes. Int. J. Mater. Process. Technol. **91**, 43–51 (1999)
17. Darwish, S.M., Al-Samhan, A.: Design of weld-bonded joints. Int. J. Adhes. Adhes. **24**, 367–377 (2004)
18. Darwish, S.M., Al-Samhan, A.: Peel and shear strength of spot-welded and weld-bonded dissimilar thickness joints. **147**, 51–59 (2004)
19. Darwish, S.M., Davies, R.: Adhesive bonding of metal cutting tools. Int. J. Mach. Tools Manuf. **29**(1), 141–152 (1989)
20. Darwish, S.M., Davies, R.: Investigation of heat flow through bonded and brazed metal cutting tools. Int. J. Mach. Tools Manuf. **29**(2), 229–237 (1989)
21. Darwish, S.M., Ghanya, A.: Critical assessment of weld-bonded technologies. J. Mater. Process. Technol. **105**, 221–229 (2000)
22. Darwish, S.M., Niazi, A., Ghania, A., Kassem, M.E.: Formulation effects on some properties of structural epoxy resin adhesives. In: Proceedings of the Third Applied Mechanical Engineering Conference, Military Technical College (1988)
23. Darwish, S.M., Azamy, K.M., Sadek, M.M.: Design rationale of industrial containment joints. Int. J. Adhes. Adhes. 65–70 (1989, 1991)
24. Darwish, S.M., Azamy, K.M., Sadek, M.M.: Dynamic characteristics of industrial double containment joints. In: Proceedings of the International Conference on CAPE Edinburgh (1989)
25. Darwish, S.M., Niazi, A., Gahanya, A., Kassem, M.E.: Improving electrical conductivity of structural epoxy resin adhesives. Int. J. Adhes. Adhes. **11**(1), 37–41 (1990)

26. Darwish, S.M., Azamy, K.M., Sadek, M.M.: Design philosophy of a bonded gear box. Int. J. Mach. Tools Manuf. **31**, 625–631 (1991)
27. Darwish, S.M., Niazi, A., Ghanya, A.: Phase stability of Duralumin machined with bonded and brazed metal cutting tools. Int. J. Mach. Tools Manuf. **32**(4), 593–600 (1992)
28. Darwish, S.M., Soliman, M.S., Al-Fahead, A.M.: Characteristics and variables of spot welding and weld bonding biomaterials. Int. Mater. Manuf. Process. **12**(1), 59–67 (1997)
29. Darwish, S.M., Soliman, M.S., Al-Fahead, A.M.: Manufacturing and characteristics of brass damping sheets. Int. J. Mater. Process. Technol. **79**, 66–71 (1998)
30. Dickinson, D.W.: Characterization of spot welding behavior by dynamic electrical parameter monitoring. Weld. J. **59**, 170–176 (1980)
31. Don, R., Lambing, C.L.T., Gillespie, J.W.: Experimental characterization of processing performance relationships of resistance welded graphite/PEEK composite joints. Polym. Eng. Sci **32**(9), 620 (1992)
32. Gould, J.E.: An examination of nugget development during spot welding, using both experimental and analytical techniques. Weld. J. **66**, 1–10 (1987)
33. Heidi, B., El-Sebakhy, I., Malik, D.: Weld bonding of structural adhesives for body stiffening. In: Worldwide Passenger Car Conference USA (1992)
34. Holmes, S.T., Gillepsie, J.W. Jr.: Thermal analysis for resistance welding of large scale thermoplastic composite joints. In: Proceedings of the American society for Composites, pp. 135–146 (1992)
35. Kaiser, J.G.: The effect of electrical resistance on nugget formation during spot welding. Weld. J. **65**, 167–174 (1982)
36. Karashor, D.A.: The use of adhesives in aircraft construction. Vestnik Mashinostroeniya **58**, 50–53 (1987)
37. Kilik, S., Davies, R., Darwish, S.M.: Thermal conductivity of epoxy resin adhesives. Int. J. Adhes. Adhes. **19**(4), 219–223 (1989)
38. Kimchi, M.: Spot weld properties when welding with expulsion: a comparative study. Weld. J. **63**, 58–63 (1984)
39. Marinelli, J.M., Lambing, C.L.T.: Advancements in welding technology for composite-to-metallic joints. J. Adv. Mater. **25**(3), 20–27 (1994)
40. Mc Knight, S.H., McBride, M., Gillespie, J.W.: Strength of aluminum and glass reinforced polypropylene sandwich single lap joints. In: Proceedings of the American Society for Composites, pp. 419–430 (1994)
41. Nied, H.A.: The finite element modeling of the resistance spot welding process. Weld. J. **61**, 123–132 (1984)
42. Sawhill, J.M., Baker, J.C.: Spot weldability of high-strength sheet steels. Weld. J. **59**, 19–30 (1990)
43. Sawhill, J.M., Watanabe, H., Metchell, JW.: Spot weldability of Mn-Mo-Cb, V-N, SAE 1008 steels. Weld. J. 217–224 (1977)
44. Schwartz, M.M.: Metals Joining Manual Book, pp. 1–32. McGraw Hill, New York (1979)

Technology of Weld-Adhesive Joints

Holger Gaul and Gert Weber

Abstract The application of the weldbonding technique for the lightweight body shell design in the steel car manufacturing is described. For different steel grades of advanced high strength steels and ultra high strength steels and some selected adhesive systems, the influence of the base metal strength and the viscosity of the applied adhesive on the weldbonding process and the weldbonded joints are analyzed. This was done with a focus on the specific requirements for the applications of the automotive industry. In detail, these are discussed regarding process reliability, the mechanical and metallurgical properties of joints, the failure types and the fracture behaviour of the weldbonded joints.

1 Introduction

Already in the mid-1950s, specific process combinations of resistance spot welding and adhesive bonding were applied for joining aluminium alloys in aircraft construction [1, 7]. In such procedures, a spot weld was first produced followed by the application of a low-viscosity adhesive to the joint gap using an injection device. This highly laborious procedure is, however, unsuitable for use in a vehicle mass production sphere. Here, a modified process variant is required consisting in dispensing an adhesive on the parts to be joined prior to resistance

H. Gaul (✉) · G. Weber
BAM Federal Institute for Materials Research and Testing,
Unter den Eichen 87, 12200 Berlin, Germany
e-mail: holger.gaul@bam.de

G. Weber
e-mail: gert.weber@bam.de

Adv Struct Mater (2011) 6: 37–77
DOI: 10.1007/8611_2010_47

Published Online: 2 February 2011

spot welding (RSW) and subsequent curing of the adhesive. This combined mode of area joining by adhesive bonding and punctiform thermal joining by resistance spot welding is referred to as weldbonding (WB) and, thus, represents a hybrid joining technique. Weldbonding of aluminium alloys will not be further considered in this chapter, since the problems known from resistance spot welding become more severe in weldbonding of aluminium alloys. As hybrid joining techniques for aluminium alloys, combinations of adhesive bonding and cold mechanical joining processes such as clinching or self-piercing riveting are employed in modern car body manufacture. Therefore, the subject of this chapter is exclusively weldbonding of steel materials for lightweight body shell manufacture.

The combination of area adhesive bonding and punctiform spot welding counterbalances the advantages and disadvantages of the two joining techniques. For example, the lack of initial strength in adhesive bonding is overcome by the combination with another thermal or mechanical joining procedure, say spot welding, entailing an increased production rate.

In lightweight steel body shell mass production of automobiles, different thermal and mechanical joining techniques are applied. Among the most important thermal and mechanical joining techniques commonly used in the steel car body shell design are resistance spot welding, arc welding, laser welding, brazing and laser brazing. Important mechanical joining procedures are self-piercing riveting and clinching. Additionally, the technique of adhesive bonding and its combinations with conventional thermal or mechanical joining procedures is becoming more and more important. In particular, the combination of adhesive bonding and resistance spot welding called weldbonding is of great importance. The weldbonding procedure affords many advantages over the resistance spot welding procedure, e.g. higher mechanical strengths, improved crash performance and higher fatigue resistances of the joined components. In particular, for newly developed high strength multiphase steels, also called advanced high strength steels (AHSS), these joining techniques have more advantages than other joining procedures for thin steel sheets. The application of AHSS materials in conjunction with economically efficient and reliable joining processes helps saving costs and conserving resources (weight reduction, energy minimization) and provides at the same time consistent or improved safety of the passenger cell (crash optimization). In this context, the Ultra Light Steel Auto Body design (ULSAB) and the New Steel Body (NSB) design are referred to [10, 11]. For more information on AHSS see [9]. There are many contributions on the resistance spot weldability of AHSS, the process reliability of the resistance spot welding process and the mechanical behaviour of spot welded AHSS and some other steel grades [5, 14–16, 18]. Furthermore, a lot of publications have been written about the weldbonding process and the properties of weldbonded joints [1, 4, 7, 8, 12, 13, 19]. In this chapter, statements about the process reliability of both the spot welding process as well as the weldbonding process are given. Furthermore, the problems of joinability for the weldbonding of high strength steels are discussed. Some comments on the mechanical behaviour of both the spot welded and the weldbonded joints for low- and medium-viscosity adhesives and different combinations of a mild and some high strength steels are made. Moreover, failure types of weldbonded joints are described and discussed. Additionally, some remarks on the metallurgical structure and the fracture behaviour of weldbonded joints are given.

2 Specific Requirements in Industry

2.1 Welding Equipment

In industrial mass production of automobiles, different welding equipments are applied for realizing the spot welding and the weldbonding technique. On the one hand, the welding equipment is based on different force generation systems. On the other hand, the welding current form may be an alternating current (AC) or a direct current (DC), realized by a medium frequency inverter equipment (MF-DC). Spot welded and weldbonded joints may be realized by alternating current (AC) and medium frequency direct current (MF-DC) spot welding guns. Figure 1 shows the principle of the resistance spot welding and the weldbonding process.

The main difference is that in the case of weldbonding a specific adhesive is applied between the sheets. The joints can be realized with stiff short arm spot welding guns providing better performance during production. On the other hand, larger components may require welding guns with longer arms resulting in poorer stiffness and therefore lower performance. For both processes, the choice of the electrode force F_e depends on the applied base metals. The electrode forces which are typically used for RSW of high strength steels are 3.5, 4.5 and 5.5 kN. The welding times are 6 Cyc. (120 ms), 10 Cyc. (200 ms) and 20 Cyc. (400 ms) for short, medium and long time welding, respectively [16].

The results for weldbonded joints are given in [2] and described in [19]. The electrode forces and the welding times for the weldbonding processes were chosen

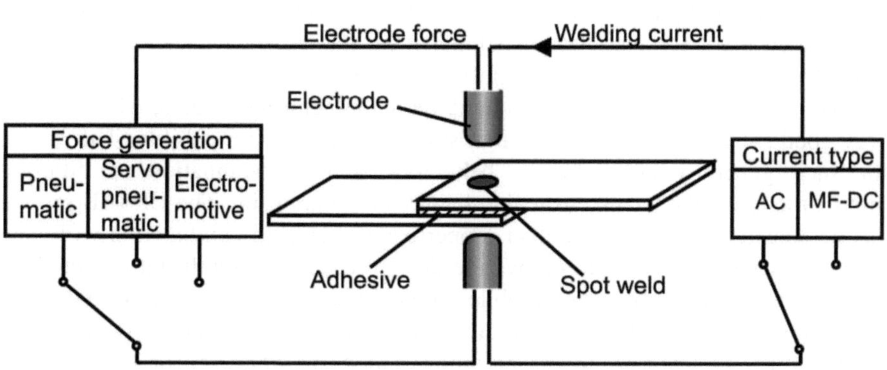

Fig. 1 Principle of resistance spot welding and weldbonding with different electrode force generation systems and with different welding current types

Table 1 Examples of types and forms of electrode caps

Type of electrode caps (ISO 5821)	A16 × 20	B16 × 20	F16 × 20 (flattened)	G16 × 20
Form of caps			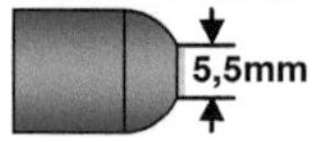	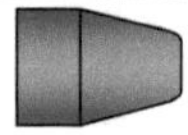

different from those used for RSW. For the high strength steels the electrode forces were 3.5 and 4.5 kN, for the martensitic steel HDT1200M 4.5 and 5.5 kN and for the mild steel 2.5 and 3.5 kN; the welding times were chosen as 260 and 400 ms. The resistance spot welded joints were exclusively realized by an AC spot welding gun with a pneumatic force generation system. In contrast, the weldbonded joints were realized by a MF-DC spot welding gun with an electromotive force generation system.

An important influence on the welding current ranges for both the resistance spot welding and the weldbonding process is given by the type of electrode cap geometry and by the material the electrodes are made of. Commonly used types of electrode caps are given in Table 1.

The results presented in this chapter for the spot welding and weldbonding process were obtained using both the AC and the MF-DC welding equipment by the application of electrode caps of the type F16. In particular, for the AC welding equipment the type of cap was F16 × 20 (flattened, $d = 5.5$ mm) and in the case of MF-DC welding equipment F16 × 22 ($d = 8.0$ mm).

2.2 Base Metals for Weldbonding and Spot Welding

For the body-in-white (BIW) automotive production, materials of different strength are applied (Fig. 2). In particular, for the lightweight steel body shell construction, multiphase steels, martensitic steels and hot stamped manganese boron steels are used. These steels belong to the classes of advanced high strength steels (AHSS) and ultra high strength steels (UHSS).

Due to the chemical composition and the manufacturing process, AHSS offer a very good solution with regard to high strength and good formability. The reason for such generally contradictory behaviour is the combination of different microstructures, like ferrite, austenite, bainite and martensite. The ULSAB project has demonstrated that the excellent properties of AHSS in combination with new manufacturing processes and innovative design leads to a significant reduction in the weight of body shells in automobile production and, accordingly, to a reduction of CO_2 emissions [10]. Furthermore, ultra high strength steels (UHSS) for weldbonding applications in the automotive body-in-white are of great importance. These steel grades are characterized by tensile strength greater than 1,200 MPa.

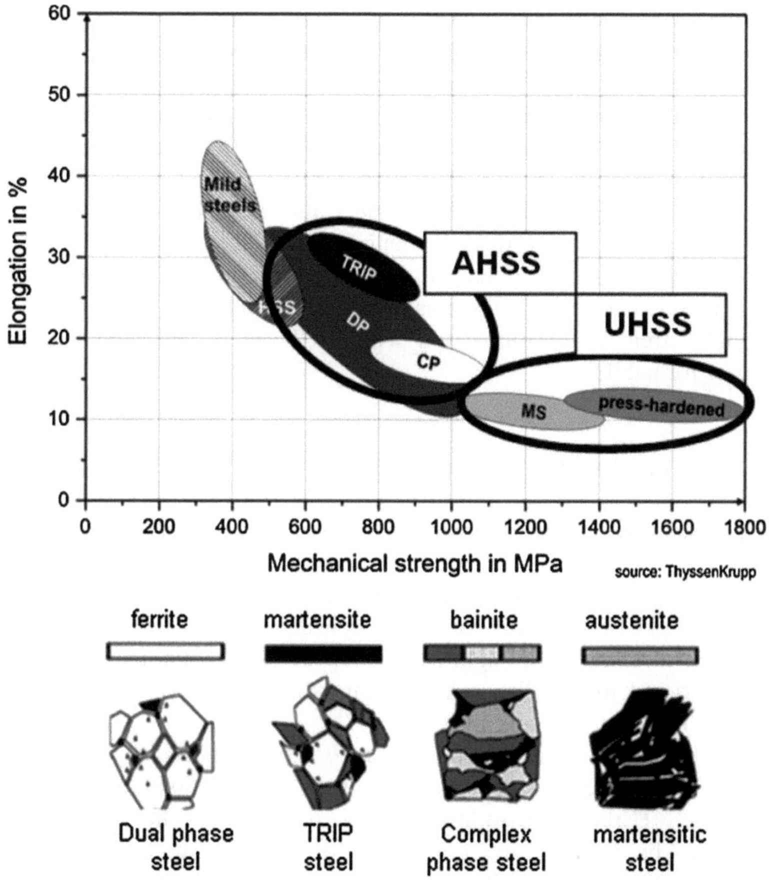

Fig. 2 Formability versus strength for conventional, advanced high strength steels (*AHSS*) and ultra high strength steels (*UHSS*)

2.2.1 Grades of Advanced High Strength Steels

Advanced high strength steels are normally classified into three types, i.e. dual phase steels (DP), transformation induced plasticity steels (TRIP) and complex phase steels (CP). With the aid of a temperature specific manufacturing process and thanks to different chemical compositions the amount as well as the dispersion of different microstructures (ferrite, martensite, etc.) and the mechanical properties of AHSS can considerably be varied according to the user requirements. The different microstructures of AHSS can be characterized as follows [9, 11].

Dual phase steels (*DP*) have a ferritic matrix (ferrite content varying between 85 and 90%) with inclusion of martensite islands. The mechanical properties of DP steels can be influenced by changing the amount of martensite. Generally, increasing the martensite content also increases the mechanical strength.

An advantage of DP steels is the combination of high strength with high elongation when compared to conventional high strength steels, see Fig. 2.

Transformation induced plasticity steels (TRIP) possess a complex microstructure dominated by ferrite (70–85%) with residual austenite (up to 15%) and additions of martensite and bainite. The strength/ductility balance is increased by strain induced austenite to martensite transformation (TRIP effect). Steels with TRIP effect attain high uniform elongation values at high tensile strength levels because of their very strong work hardening.

Complex phase steels (*CP*) have a fine complex microstructure of bainite with islands of retained austenite and inclusions of ferrite and martensite. CP steels offer higher yield strength in comparison with TRIP steels, but with simultaneous decrease of formability.

2.2.2 Grades of Ultra High Strength Steels

Martensitic steels and press hardening manganese-boron steels belong to the class of UHSS.

Martensitic steels (*MS*) offer a predominantly martensitic microstructure with minor quantities of ferrite and bainite. In consequence of the martensitic microstructure, the mechanical strength reaches up to 1,400 MPa and more, but with strongly restricted formability, which is an essential factor to be considered in applications in the automobile industry.

Press hardening manganese-boron steels are produced by different steel companies. Therefore, on the market different realizations of this steel are available (e.g. UsiBor®, Ultraform®, BTR or MBW). The alloying element boron acts as hardening agent during thermo mechanical treatments and provides a material with excellent hardness and high strength. A quenching treatment leads to precipitation of boron carbides at the grain boundaries as well as boron segregations. The suppressing of austenite to ferrite transformation caused by this segregation phenomenon leads to increasing hardenability. Furthermore, substitutional solid solution elements such as Mn influence the strength after quenching.

Mechanical properties, thicknesses and surface conditions of the used AHSS and UHSS are given in Table 2.

3 Adhesive Characteristic Required in the Automotive Body-in-White

For the lightweight vehicle construction certain polymer classes are suitable as basis of the applied adhesives for the weldbonding technique. These polymer classes are mainly acrylates, epoxies, polyurethanes (PUR) and rubber. According to the curing behaviour of the applied adhesives, hot cured and cold cured adhesive

Table 2 Mechanical properties of the used base metals

Steel grade (short term)	Steel grade (standard definition)	Sheet thickness in mm	Surface	Yield strength in MPa	Tensile strength in MPa	Elong. after fract. in %
–	DC01	1.50		185	320	42
TRIP700	HCT690T	1.60		450	740	28
TRIP700	HCT690T	1.60	Z100	470	710	32
TRIP800	HCT780T	1.25	ZE75/75	700	880	29
DP600	HCT600X	1.30	Z100	400	625	26
CP-W800	HDT780C	1.65	–	790	900	13
CP-W800	HDT780C	1.50	Z190	690	910	11
MS-W1200	HDT1200 M	1.55	–	950	1,220	12
UsiBor®	22MnB5	1.50	AlSi	1,280	1,470	11

systems are distinguished. The adhesives must fulfil al lot of requirements regarding the mechanical behaviour. The physical properties of a polymer depend significantly on the temperature. Accordingly, the modulus of elasticity changes from higher to lower values with increasing temperatures. Therefore, the behaviour of the polymer is like glass at lower temperatures, and like rubber at higher temperatures. This has to be taken into account when choosing an adhesive for practical applications of weldbonding. A further important requirement of an applied adhesive is the capability to absorb oil.

The weldbonding processes described in this chapter were carried out based on three one-component hot-curing adhesives. In particular, a rubber-based adhesive [Terostat 5194 (A1)], an epoxy resin-based adhesive [Betamate™ 1480 (A2)] and an epoxy resin PUR-based adhesive [SikaPower®-498 (A3)] were applied, see Table 3.

The viscous behaviour of these adhesives is different. The adhesive Terostat 5194 can be described as an adhesive of low-viscosity, the adhesives Betamate™ 1480 and the SikaPower® 498 can be described as adhesives of medium-viscosity. For instance, Fig. 3 shows the shear viscosity as a function of shear stress.

These results were generated by the application of linear increasing shear stress to the adhesive specimen [2]. Both the viscous behaviour of the adhesives and their yield point in the non-cured condition were determined. It can be seen from Fig. 3 that the SikaPower® 498 adhesive, despite its higher viscosity, exhibits a lower yield point compared to Betamate™ 1480. Terostat 5194 has the lowest viscosity and the lowest yield point. The shear viscosity is an important factor for the squeezing out of the adhesive in order to obtain the electrical contact between the two steel adherends necessary for the welding current flow of the resistance spot welding process. The values of the shear viscosities of the two adhesives Betamate™ 1480 and SikaPower® 498 become similar for the investigated temperatures at shear stresses between 2.3 and 5.0 kPa (see Fig. 3). Therefore, the influence of their squeezability during the application of the electrode force on the electrical resistance between the adherends can be neglected. By application of a force controlled amplitude sweep to

Table 3 Adhesives for the realized weldbonding processes

Type of adhesive		Properties				Elong. after fracture
Name	Abbreviation	Base	Viscosity	Tensile strength in MPa	Young's Mod. in MPa	
Terostat 5194	A1	Rubber	Low	12	NA	NA
Betamate™ 1480	A2	Epoxy resin	Medium	38	1,600	10%
SikaPower®-498	A3	Epoxy resin PUR	Medium	30	NA	5%

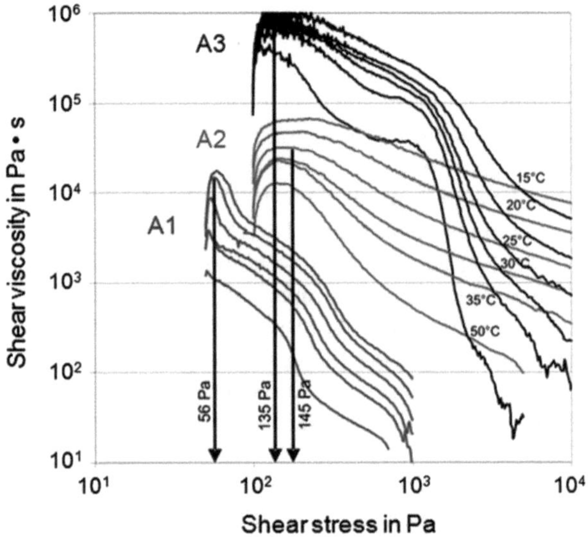

Fig. 3 Shear viscosity versus shear stress—determination of the flow limits

the adhesive specimen, the viscoelastic behaviour of the non-cured adhesives was examined, too. This was accomplished by identifying the characteristics of the complex viscosity [2].

4 Manufacture

4.1 Preliminaries for Car Body Manufacture by the Weldbonding Process

To manufacture lightweight body shell structures of high strength steels by the resistance spot welding procedure and the weldbonding procedure, certain conditions on the quality of the spot welded and weldbonded joints must be fulfilled.

The main quality feature of spot welded and weldbonded joints is given by the spot weld diameter d_p. To ensure sufficiently high strength of the joints, the spot weld diameter must be large enough. The realised spot weld diameter by the spot welding and weldbonding process is a non-linear function of the root mean square value (r.m.s-value) of the welding current I_W. This means that there is a correspondence between the quality feature spot weld diameter d_p and the r.m.s.-value of the welding current I_W. Accordingly, there is a correspondence between ranges of spot weld diameters and ranges of welding currents for the resistance spot welding and the weldbonding process, too. Therefore, the knowledge of the welding current ranges (WCRs) for joining the applied base metal combinations by the resistance spot welding and weldbonding procedure is of fundamental importance. The process reliability of the resistance spot welding process depends on the size of the WCR.

The introduction of the weldbonding process reliability is based on the definition of the resistance spot welding process reliability [15–17, 20]. Thus, to ensure process reliability of the weldbonding process under production conditions, sufficiently large sizes of the welding current ranges must be realized. Different WCRs are obtained depending on the selected welding parameters. The primary welding parameters of resistance spot welding are the r.m.s.-value of the welding current I_W, the stationary electrode force F_e and the welding time t_W. These parameters are the primary joining parameters of the weldbonding process, too.

In resistance spot welding, the WCRs are usually defined on the basis of requirements imposed on the r.m.s.-value of welding current during spot welding. A particular requirement is that the r.m.s.-value of welding current must be kept within certain limits set by quality demands placed on the spot weld diameter d_p.

The lower limit of the r.m.s.-value of welding current is determined by a minimum spot weld diameter, whereas the upper limit of the welding current is given by the physics of the spot welding process and the weldbonding process, respectively. Commonly used lower quality limits are spot weld diameters of $d_p = 3.5\sqrt{t}$ or $d_p = 4\sqrt{t}$, whereby t is the sheet thickness. These lower quality limits are then also referred to as $3.5\sqrt{t}$-limit or $4\sqrt{t}$-limit. The situation is completely different with the upper quality limits. The maximum admissible upper quality limit is usually referred to as splash limit I_{SL}. This limit constitutes a stability limit of the resistance spot welding process and accordingly of the weldbonding process. The splash limit is the very quality limit at which a spot weld can still be performed without the occurrence of a splash. In order to ensure that at and below this limit no splashes do in fact occur, it is necessary in setting this limit to take into account that it varies within a certain scatter band. The variation of the upper quality limit depends on the welding/joining parameters of electrode force and welding time, on the material to be joined and its coating, on the applied adhesive system, on the electrode cap types, on the applied current form as well as on the electrical and mechanical machine properties of the spot welding unit. The spot weld diameter, and hence the upper and lower quality limits depend on the test procedure, and the fracture type must always be indicated in welding current range determinations.

4.2 Description of the Process Reliability by Weldability Lobes

As mentioned in the previous section, the process reliability for both the resistance spot welding and the weldbonding process depends on the size of the WCR. For resistance spot welding and also for weldbonding the WCRs are defined on the basis of special requirements imposed on the weld quality feature characterized by the spot weld diameter d_P. Larger sizes of WCRs imply higher process reliabilities; smaller sizes of WCR imply lower process reliabilities. The dependence of these WCRs on the welding parameter electrode force F_e and welding time t_W are formally described by weldability lobes. The term lobe means the area limited by the welding parameters in electrode force/ welding current diagrams or welding time/welding current diagrams. WCRs for the different steel grades (mild steel, AHSS and UHSS) are formally defined by welding current differences according to

$$\Delta I = I_U - I_L = I_{SL} - I_{4\sqrt{t}}, \tag{1}$$

whereby the lower (I_L) and the upper quality limits (I_U) are set by the $4\sqrt{t}$-limit $I_L = I_{4\sqrt{t}}$ and by the splash limit $I_U = I_{SL}$. The representation of the difference ΔI of the r.m.s.-value of welding current, according to Eq. 1, between the upper and the lower quality limit as a function of the (stationary) electrode force F_e and of the welding time t_W

$$\Delta I = f(F_e, t_W) \tag{2}$$

is referred to as three-dimensional weldability lobe. From the three-dimensional weldability lobe according to this definition, the classical two-dimensional weldability lobes can be derived as special representations [15, 16, 18], i.e.

$$\Delta I = f(F_e = \text{const.}, t_W) \tag{3}$$

$$\Delta I = f(F_e, t_W = \text{const.}) \tag{4}$$

Furthermore, WCRs can be described by extended weldability lobes [18]. The idea of extended weldability lobes is to describe several WCRs depending on one of the welding parameters of electrode force or welding time plus a further influence on the welding current range. In this chapter only a special form for an extended weldability lobe is used. This form is given by

$$\Delta I = f_E(F_e, t_W = \text{const.}, A), \tag{5}$$

where the variable parameter A describes the applied adhesives for the weldbonding process.

4.3 Joinability of AHSS and UHSS for Weldbonding

Depending on the choice of the welding parameters electrode force F_e and welding time t_W, as discussed in the previous section, different process reliabilities can be realized. Based on different welding parameter combinations, different WCRs occur. With an optimal choice of welding parameter combinations, in particular, for the case of medium- and long-time welding, sufficiently high process reliabilities can be generated. But the existence of process reliability is not a sufficient condition for the weldability of the steel sheets. According to ISO 18278, both the possibility of spot welding and the reproducibility must be given at the same time. This demand must be fulfilled for the weldbonding process, too.

The possibility of spot welding as well as of weldbonding is guaranteed by the existence of determined welding parameter combinations for which sufficiently high process reliabilities are realized. The possibility of reproducibility for spot welding of a given material is only guaranteed by sufficiently little electrode wear. This is to say that the electrode life must be sufficiently long. The same is demanded for the weldbonding process. It was shown that for hot dip zinc coated, and hence for uncoated, AHSS sheets the electrode life is sufficiently high [16]. Therefore, for the weldbond joinability of AHSS and UHSS sheets both sufficiently large WCRs and sufficiently long electrode lifes are needed. Due to the influences of the adhesives and, in particular, of their viscosities on the conductance of the electrode areas, it can be expected in general that the degree of joinability will be lower for the weldbonding process than for the spot welding process.

4.4 Joinability of a Special AHSS

In the following, the joinability of TRIP steels is discussed. This is done first for the spot welding process. For a given uncoated and hot dip zinc coated TRIP steel, the process reliability is discussed. For various welding parameters, the lower and upper limits of the welding current and the WCRs for the base metals HCT690T and HCT690T+Z are given in Table 4 and depicted in Figs. 4, 5.

It can clearly be seen in Table 4 that the WCRs for the chosen welding times in the case of hot dip zinc coated sheets are appreciably smaller than those in the case of sheets without zinc coat. Due to the zinc influence on the electrical process parameters and on the quality feature of weld diameter, short-time welding is not sufficient for the hot dip zinc coated TRIP steel HCT690T+Z in terms of the process reliability, since the WCRs in Fig. 5 for 6 Cyc. welding time become too small. It can be expected for weldbonding, that the similar WCRs become smaller. But this is, from a practical point of view, not a problem when short-time welding is not applied. Figure 5 shows further, that the choice of the welding parameter

Table 4 Lower and upper limits of the welding current and WCRs for uncoated and hot dip zinc coated HCT690T

Joining technique		RSW								
		$t_W = 120$ ms			$t_W = 200$ ms			$t_W = 400$ ms		
Base metal	F_e in kN	Welding current limits/ranges in kA								
		I_L	I_U	ΔI	I_L	I_U	ΔI	I_L	I_U	ΔI
HCT690T (1.6 mm)	2.5	5.4	6.2	0.6	4.7	6.1	1.4	4.2	5.8	1.6
	3.5	5.7	7.0	1.3	5.0	6.7	1.7	4.4	6.5	2.1
	4.5	6.0	7.0	1.0	5.1	6.9	1.8	4.9	7.5	2.6
HCT690T+Z (1.6 mm)	2.5	8.2	8.2	0.0	6.1	6.8	0.7	5.1	6.1	1.0
	3.5	8.7	9.1	0.4	6.5	7.3	0.8	5.4	6.2	0.8
	4.5	9.2	9.4	0.2	6.6	7.2	0.6	5.5	7.3	1.8

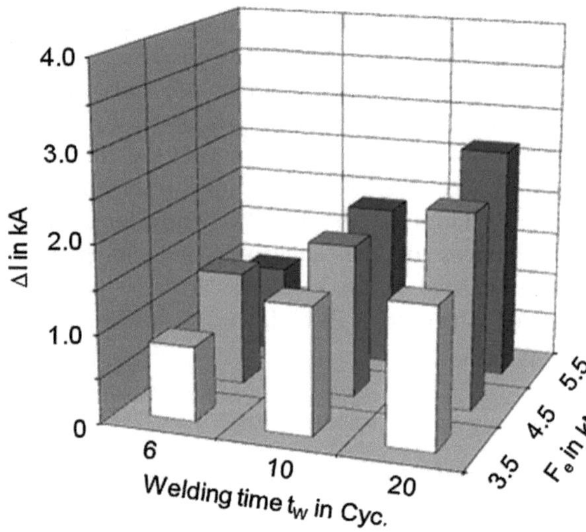

Fig. 4 Weldability lobe for an uncoated AHSS (HCT690T)

electrode force must be high enough for weldbonding of hot dip zinc coated AHSS. Therefore, the choice of $F_e = 4.5$ kN is convenient with regard to a sufficiently high process reliability.

Figure 6 shows the electrode life of the hot dip zinc coated steel HCT690T+Z for different welding current forms. It can be seen that for both cases the number of spot welds is approximately 1,900 when the weld diameter falls below the lower quality limit. This means that the electrode life for the hot dip zinc coated HCT690T+Z is sufficiently high. The results show, that the above mentioned spot weldability in terms of sufficiently wide WCRs and sufficiently long electrode life is given in this case. The same procedure must be applied for showing the weldbond joinability of this AHSS and other AHSS or UHSS, too.

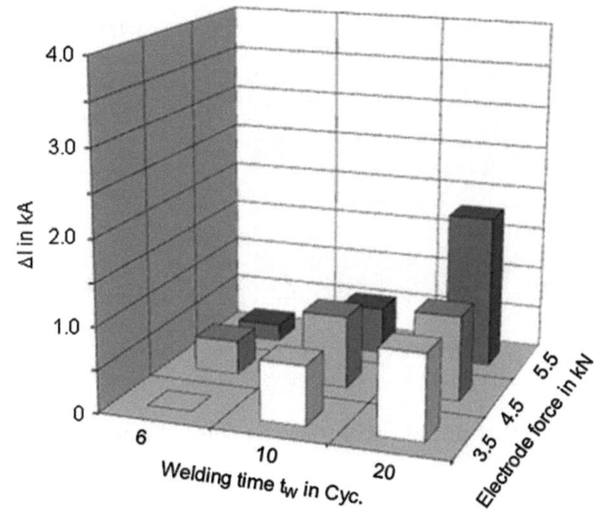

Fig. 5 Weldability lobe for a hot dip zinc coated AHSS (HCT690T+Z)

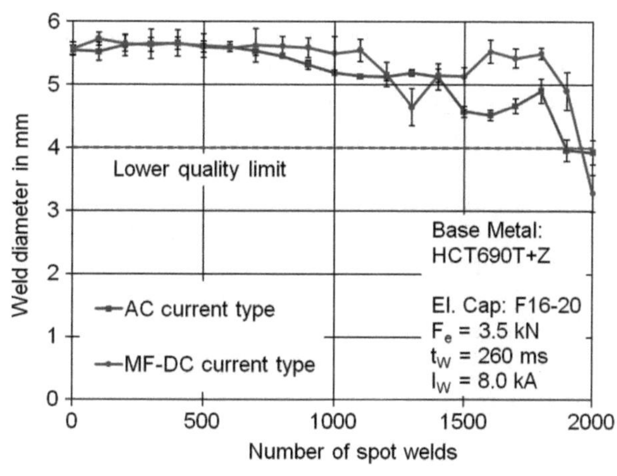

Fig. 6 Electrode life represented by the weld diameter versus number of spot welds for a hot dip zinc coated AHSS (HCT690T+Z)

4.5 Influence of Electrode Cap Type on the Process Reliability for Spot Welding and Weldbonding

In order to show the influence of different electrode caps on the process reliability for resistance spot welding and weldbonding of AHSS, the welding current ranges of an uncoated and a hot dip zinc coated TRIP steel are presented as an example. The welding time was chosen as 20 cycles (400 ms). The electrode force was changed in three steps according to 3.5 kN/4.5 kN/5.5 kN. The WCRs are depicted in Fig. 7 as three-dimensional extended weldability lobes of the form

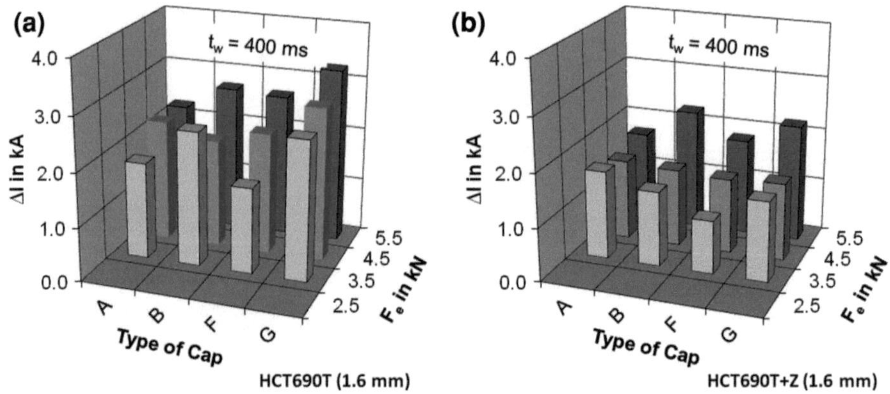

Fig. 7 Extended weldability lobe for different types of electrode caps **a** uncoated and **b** hot dip zinc coated steel

$$\Delta I = f_{\mathrm{E}}(F_{\mathrm{e}}, t_{\mathrm{W}} = \text{const.}, C) \tag{6}$$

where C is a parameter for the type of electrode cap.

Depending on the choice of the welding parameter combinations for each type of cap, different values of welding current limits and WCRs are produced by the resistance spot welding process [17] as well as by the weldbonding process. As already shown in the previous section, the sizes of the WCRs for the zinc coated sheets are significantly smaller than those for the uncoated case. From Fig. 7a it follows that for spot welded uncoated TRIP steels HCT690T the maximum WCRs are realized by the electrode caps of Type G. With the maximum value of $\Delta I = 3.25$ kA for the welding parameters $F_{\mathrm{e}} = 5.5$ kN and $t_{\mathrm{W}} = 20$ cycles/400 ms, this cap dominates the others. For the same welding parameters, the caps of Type B and Type F (flattened) produce sufficiently large WCRs ($\Delta I = 2.7$ and 2.65 kA), too [17].

According to the extended weldability lobes of the WB process (see also the following section), similar results are expected for the WCRs in the case of the WB process when the same different electrode cap types are used. Due to the comparison of the WCRs for the resistance spot welding and the weldbonding process, similar changes of the WCRs and hence the process reliability are expected if the electrode cap is changed. This is important with regard to the weldbond joinability of mild steels, AHSS and UHSS for different applied adhesives. Results on the joinability for a special type of electrode cap are not the same for another type of electrode cap. The different conditions of the electric conductivity between the adherends/steel sheets depend on the base metals, its surface conditions (e.g. zinc coating) and the applied adhesives. Additionally, these conductive conditions are mainly influenced by the geometry of the electrode cap type. Therefore, the choice of the electrode cap must be optimized with respect to maximum process reliability and a minimum electrode wear for the realized special weldbonding process.

4.6 *Welding Current Ranges and Process Reliability for the Weldbonding Process*

4.6.1 Weldbonding of Same Base Metal Combinations

As mentioned above, the size of the welding current range is a measure for the process reliability of the weldbonding process. In order to show the weldbonding process reliability for different weldbonded AHSS, the corresponding weldability lobes were first established (Table 5).

The applied adhesives are given in Table 3. The mechanical properties of the tested base metals are given in Table 2. For the tested base metals DC04ZE-KSP, HCT600X, HCT780Z+ZE50/50, HDT1200M and 22MnB5 (UsiBor®) and the applied adhesives Terostat 5194 (A1), Betamate™ 1480 (A2) and SikaPower®-498 (A3) the lower and upper quality limits I_L and I_U of the welding current and the welding current ranges are given in Table 5.

In the following, three-dimensional weldability lobes in accordance with Eq. 5 are used. Depending on the tested base metal, different constant welding times (from 260 up to 400 ms) were used. Therefore, according to Eq. 5, the welding current ranges for the weldbonding process are represented by an extended weldability lobe of the form

Table 5 Lower and upper limits of the weldbonded quality and welding current ranges ΔI for mild steel, AHSS and UHSS and different weldbonding processes (WB)

Base metal combinations and thickness		Joining technique											
		Weldbonding									Resistance spot welding		
		WB1			WB2			WB3			RSW		
Welding parameters	F_e in kN	Welding current limits/ranges in kA											
		I_L	I_U	ΔI	I_L	I_U	ΔI	I_L	I_U	ΔI	I_L	I_U	ΔI
DC04ZE-KSP (1 mm) $t_W = 300$ ms	2.5	6.6	7.2	0.6	6.8	9.3	2.5	6.7	8.6	1.9	6.4	8.6	2.2
	3.5	7.1	8.7	1.6	7.3	9.8	2.5	7.3	9.4	2.1	6.8	9.0	2.2
HCT600X (1.3 mm) $t_W = 260$ ms	3.5	6.6	8.5	1.9	7.1	8.3	1.2	7.1	9.6	2.5	6.8	8.0	1.2
	4.5	7.1	9.6	2.5	7.5	9.5	2.0	8.1	10.1	2.0	7.2	9.8	2.6
HCT780T (1.3 mm) $t_W = 260$ ms	3.5	5.4	7.3	1.9	5.3	7.0	1.7	5.5	7.3	1.8	5.2	7.0	1.8
	4.5	5.6	7.5	1.9	5.8	8.1	2.3	6.2	7.8	1.6	5.7	8.2	2.5
MS-W1200 (1.5 mm) $t_W = 300$ ms	4.5	6.4	7.9	1.5	6.3	8.6	2.3	6.2	8.5	2.3	6.4	8.8	2.4
	5.5	6.6	8.3	1.7	6.6	8.8	2.2	6.7	9.6	2.9	6.8	9.8	3.0
22MnB5 (2.0 mm) $t_W = 400$ ms	3.5	5.8	8.5	2.7	5.7	7.5	1.9	5.9	7.8	1.9	5.6	8.6	3.0
	4.5	6.4	8.7	2.3	6.0	9.0	3.0	6.1	8.5	2.4	6.1	9.2	3.1

Table 6 Abbreviations for the adhesives and the related joining techniques

Joining technique	Index i in Eq. 7	Type of adhesive	
		Name	Abbreviation
WB1	1	Terostat 5194	A1
WB2	2	Betamate™ 1480	A2
WB3	3	SikaPower®-498	A3
RSW	0	No adhesive	–

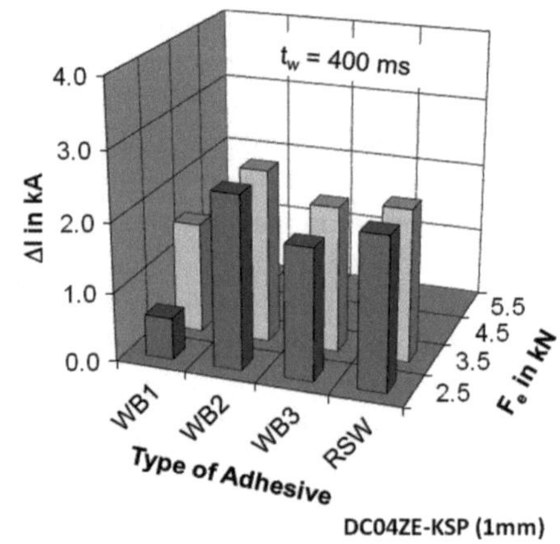

Fig. 8 Extended weldability lobe for weldbonded mild steel

$$\Delta I = f_{\mathrm{E}}(F_{\mathrm{e}}, t_{\mathrm{W}} = \text{const.}, A_i), \quad i = 0, 1, 2, 3, \tag{7}$$

where the index i characterize the applied adhesive according to Table 6.

Equation 7 expresses that only the influences of the welding parameter electrode force F_e on the WCRs was investigated for the different combinations of base metals and for the three applied adhesives, as represented in Table 6. The corresponding weldability lobes for these combinations are given in Figs. 8, 9, 10.

Table 5 shows that in the case of weldbonding WB1 (=RSW+A1), WB2 (=RSW+A2) and WB3 (=RSW+A3) both quality limits I_L and I_U increase with increasing electrode force F_e. Furthermore, it can be seen that for all tested base metals and each applied adhesive these quality limits shift towards higher current values. This can be easily seen from two-dimensional weldability lobes of the form in Eq. 4 given in [19]. This means that the energy input of the WB process is higher than in the case of RSW. Clearly, roughly speaking, the electrical energy needed for spot welding and weldbonding is proportional to the second power I_W^2 of the welding current. At the beginning of the WB process, the energy input is higher compared to the RSW process due to the influence of the applied adhesive. The reason for this fact is the shunt behaviour of the electrical resistance, which is strongly increased by the insulating effect of the adhesive between the two metal sheets at the beginning of the WB process [2].

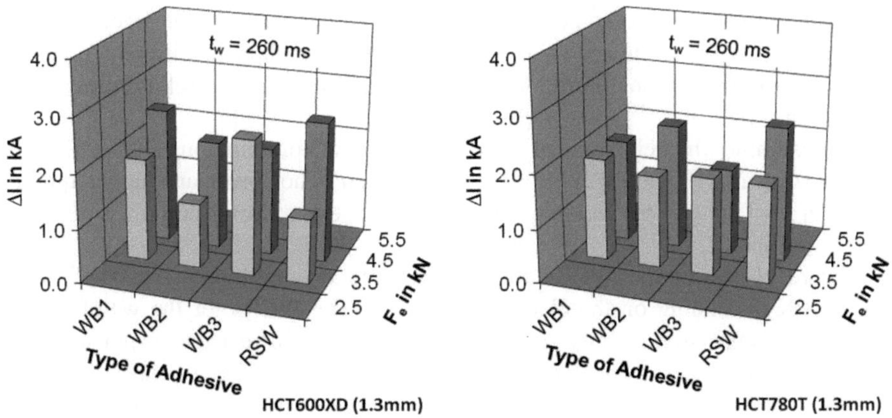

Fig. 9 Extended weldability lobes for weldbonded AHSS

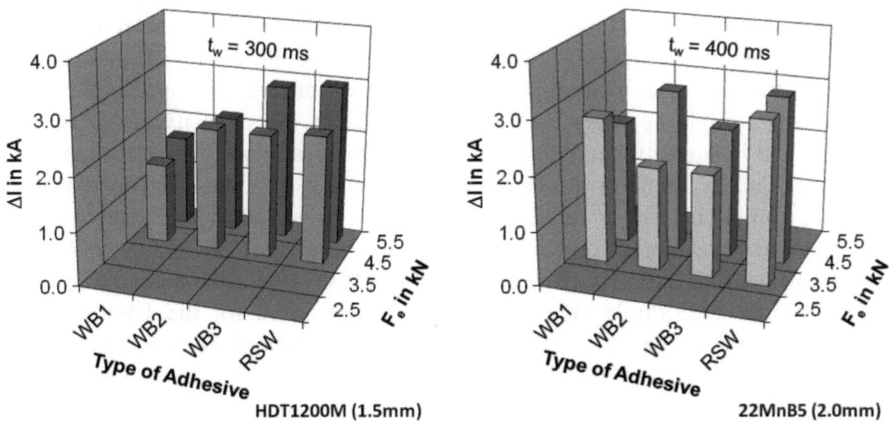

Fig. 10 Extended weldability lobes for weldbonded UHSS

An influence of the viscosity of the applied adhesives can be seen in Fig. 8 for the low strength base metal DC04ZE-KSP and for the ultra high strength martensitic steel HDT1200M in Fig. 10 (left). In these cases, the low-viscosity adhesive A1 produces smaller WCRs than the medium-viscosity adhesives A2 and A3. But for the other three types of tested high strength steel grades such an influence of the viscosity on the WCRs and consequently on the process reliability cannot be observed.

4.6.2 Weldbonding of Different Base Metal Combinations

In summary, one can say that the influence of the tested low- and medium-viscosity adhesives for all weldbonded base metals of the same type on the welding current ranges is not significant in comparison to the spot welding process. Thus, the influence of the same adhesive applied to different types of base metal combinations on the WCRs and hence the process reliabilities are of further interest.

For example, the behaviour of the two base metal combinations DC04ZE-KSP/HCT600X and HCT600X/22MnB5 are discussed. The lower and upper quality limits I_L and I_U for the welding current I_W and the welding current ranges are given in Table 7.

Figure 11 shows the WCRs for the base metal combination DC04ZE-KSP/HCT600X depending on the electrode force F_e. The WCRs for the weldbonding process becomes significantly smaller than in the case of resistance spot welding. It can be easily seen from two-dimensional weldability lobes of the form given in Eq. 4 [19], that a different behaviour of lower and upper welding current limits for the low-viscosity and the two medium-viscosity adhesives occurs (see Table 7).

For the low-viscosity adhesive, the lower quality limits are nearly the same and the upper limit changes towards lower welding currents. The behaviour for the medium-viscosity adhesives is completely different: while the lower quality limit changes towards higher welding currents, the upper limit remains nearly the same. These results might be interpreted as an influence of the viscosity of the applied adhesives. It is interesting to note the significantly lower energy input for realizing the upper quality limits in the case of the low-viscosity adhesive Terostat 5194 (A1). A similar behaviour of the WCRs cannot be stated in the case of weld-bonding the base metal combination of dual phase steel HCT600X and press hardening steel 22MnB5 (UsiBor®) (see Table 7 and [19]). But analogously to the results of the AHSS/mild steel-base metal combination, for the UHSS/AHSS-combination the behaviour of the WCRs for the weldbonding process with the low-viscosity adhesive Terostat 5194 (A1) is quite different from that one of the weldbonding processes with the medium-viscosity adhesives Betamate™1480

Table 7 Lower and upper limits of the weldbonded quality and welding current ranges ΔI for the combination of mild steel/AHSS and AHSS/UHSS and different weldbonding processes (WB)

Base metal combinations and thickness		Joining technique											
		Weldbonding									Resistance spot welding		
		WB1			WB2			WB3			RSW		
Welding parameters	F_e in kN	Welding current limits/ranges in kA											
		I_L	I_U	ΔI	I_L	I_U	ΔI	I_L	I_U	ΔI	I_L	I_U	ΔI
DC04ZE-KSP (1.0 mm)/ HCT600X (1.3 mm)	2.5	7.2	8.7	1.5	8.3	10.1	1.8	8.3	10.0	1.7	7.4	9.7	2.3
$t_W = 300$ ms	3.5	8.0	9.5	1.5	8.9	10.5	1.6	9.0	10.8	1.8	8.0	10.5	2.5
HCT600X (1.3 mm)/ 22MnB5 (2.0 mm)	3.5	6.7	7.9	1.2	6.6	8.4	1.8	6.8	8.4	1.6	6.5	8.4	1.9
$t_W = 400$ ms	4.5	7.1	8.2	1.1	7.4	8.7	1.3	7.2	8.7	1.5	6.9	8.7	1.8

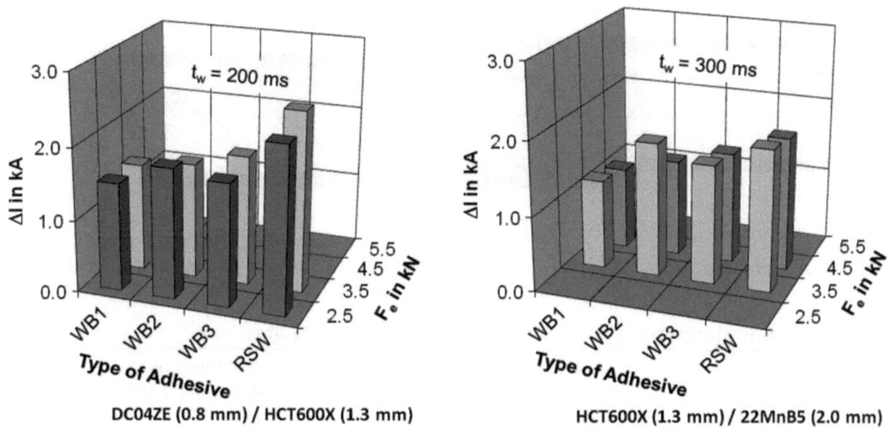

Fig. 11 Extended weldability lobes for combinations of weldbonded mild steel (DC04KSP), AHSS (HCT600X) and UHSS (22MnB5)

(A2) and SikaPower®-498 (A3). This fact may also be interpreted as an influence of the viscosity of the used adhesive when combinations of base metals are weldbonded.

5 Strength and Durability

In this section, a comparison of some selected mechanical properties of weldbonded and spot welded sheets of AHSS and UHSS and low strength reference sheets of mild steel is carried out. For the characteristic data of the tested steel grades, see Table 2. The most important mechanical properties of weldbonded and spot welded joints are the behaviour under mechanical quasi-static and impact loads. Furthermore, the fatigue behaviour of weldbonded joints in comparison with spot welded joints is of major importance.

5.1 Tensile Shear Strength of Spot Welded and Weldbonded Joints

In the following, some selected results of the tensile shear strength under quasi-static and impact load for weldbonded and spot welded joints of some different high and low strength steels are given and commented. Furthermore, the absorbed energy of the materials and their combinations under quasi-static and impact load is given, too. The used specimen geometry is depicted in Fig. 12.

All results are based on the same types of adhesives, as presented in Sect. 4.6 for the establishment of weldability lobes showing the process reliability of the weldbonding process. The results of the tensile shear forces based on quasi-static tests are summarized in Table 8.

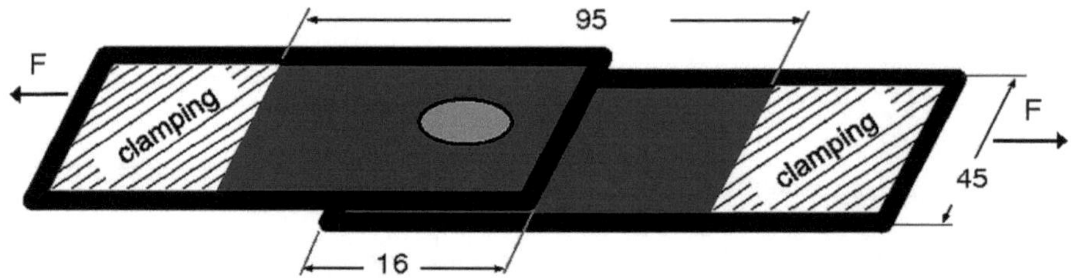

Fig. 12 Specimen geometry for the mechanical tests (all dimensions in mm, not to scale)

Table 8 Tensile shear forces for weldbonded and spot welded joints

Joining technique	Weldbonding Betamate 1480 (A2)	RSW –	Weldbondinq Betamate 1480 (A2)	RSW –
Base metal combinations and thickness	Same base metal		Base metal combinations with HCT600X(1.3 mm)	
	F_s in kN	F_s in kN	F_s in kN	F_s in kN
DC04(1.5/0.8 mm)	14.3	9.7	10.9	5.1
HCT600X(1.3 mm)	22.8	13.8	22.7	13.9
HCT780T(1.25 mm)	26.2	19.4	26.1	16.6
HCT980X(1.5 mm)	33.9	23.7	25.3	16.7
HDT1200 M (1.55 mm)	33.9	27.5	27.3	15.9
22MnB5 (1.5/2.0 mm, Usibor)	29.8	26.8	24.6	18.8

It can be seen that the tensile shear force of the weldbonded joints is higher than of the resistance spot welded joints in all cases. The influence of the base metal strength on tensile shear forces for spot welded and weldbonded joints of same base metal combinations and different base metal combinations with the same adhesive A2 is shown in Figs. 13, 14.

These figures show that the increase of tensile shear force between the weldbonded and the spot welded joints given by

$$\Delta T_S = T_{S,WB} - T_{S,RSW} \tag{8}$$

is significantly higher for different base metal combinations than in the case of the same base metal combination. Figure 15 shows the shear force differences ΔT_S for the investigated base metals.

The influence of the different adhesives on the tensile shear force and the absorbed energy is given in Table 9.

The tensile shear forces for the weldbonded and spot welded AHSS/AHSS-base metal combination HCT780T/HCT600X under quasi-static shear load (a) and impact shear load (b) for different adhesives A1, A2 and A3 are given in Fig. 16. This figure shows a significant increase of the tensile shear force values for the medium-viscosity adhesives A2 and A3 compared with the low viscosity adhesive A1. The influence of the viscosity is stronger in the case of impact load.

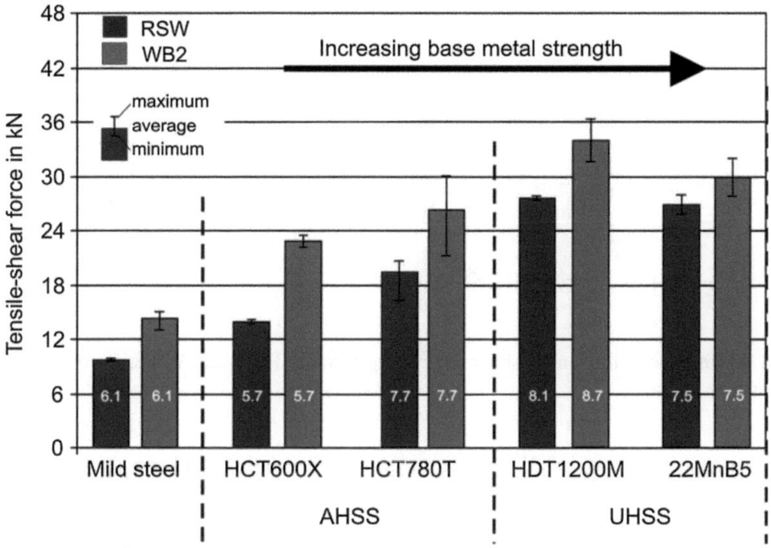

Fig. 13 Influence of the base metal strength on the strength of weldbonded (WB2) and spot welded (RSW) joints for same base metals, with weld diameters (in columns)

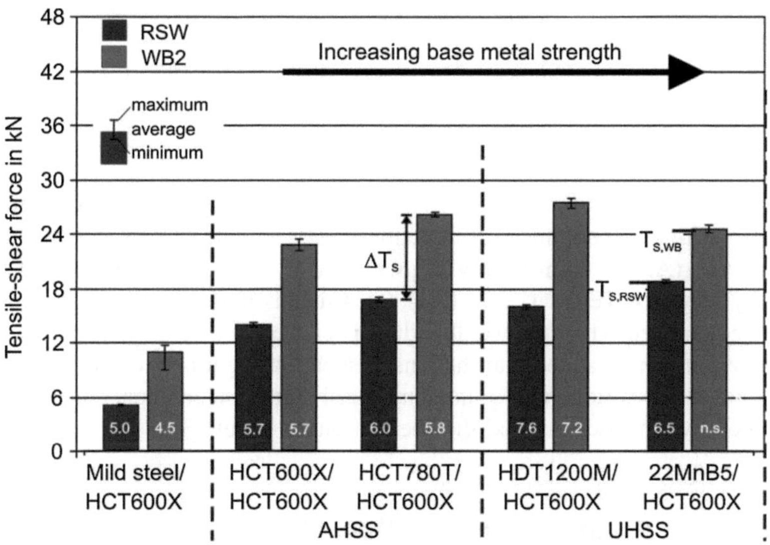

Fig. 14 Influence of the base metal strength on the strength of weldbonded (WB2) and spot welded (RSW) joints for base metal combinations, with weld diameters (in columns)

The absorbed energy of the spot welded and the weldbonded AHSS/AHSS-base metal combination HCT780T/HCT600X with the applied adhesives Terostat 5194 (A1), Betamate™ 1480 (A2) and SikaPower®-498 (A3) under quasi-static shear load (a) and impact shear load (b) is illustrated in Fig. 17.

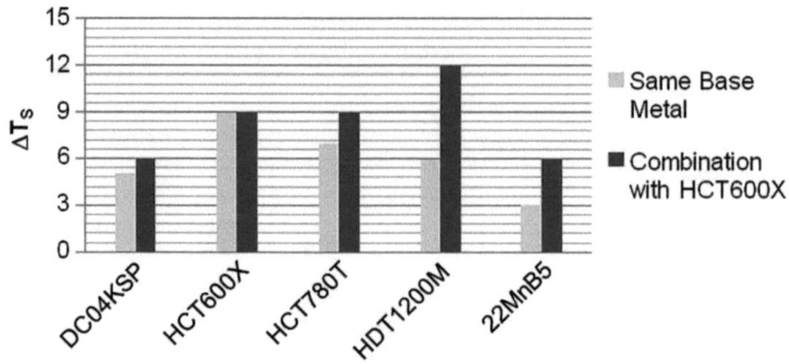

Fig. 15 Difference of the tensile shear forces ΔT_S between RSW and WB joints for the same and different base metals

Table 9 Tensile shear forces and absorbed energies for weldbonded and spot welded joints

Joining technique		Weldbonding						RSW	
		Terostat 5194 (A1)		Betamate 1480 (A2)		SikaPower 498 (A3)		–	
Base metal combination and welding parameter	Type of load	F_s in kN	W_{ABS} in J	F_s in kN	W_{ABS} in J	F_s in kN	W_{ABS} in J	F_s in kN	W_{ABS} in J
HCT780T (1.25 mm)/ HCT600X(1.3 mm)	Quasi-static	16.4	53.2	26.2	86.2	25.3	80.5	16.7	60.2
$F_e = 4.42$ kN $t_W = 300$ ms $I_w = 9.0$ kA	Impact	22.2	62.8	38.9	177.9	35.3	154.4	21.9	39.6

Here, an analogous effect occurs. The absorbed energy of the weldbonded joints produced with the medium-viscosity adhesives Betamate™ 1480 (A2) and SikaPower®-498 (A3) is significantly higher than of those produced with the low-viscosity adhesive Terostat 5194 (A1). This means that a better crash behaviour might be achieved when the weldbonding is carried out with adhesives of higher viscosities. It can be stated that the tensile shear strength behaviour and the crash behaviour of the weldbonded joints and quasi-static under impact shear load are influenced by the viscosity and the shear strength (Table 3).

5.2 *Fatigue Behaviour of Spot Welded and Weldbonded Joints*

In the following, some selected results of the fatigue strength under shear load for weldbonded and spot welded joints of some different high and low strength steels are given and commented. In order to analyze the fatigue behaviour of weldbonded joints, Wöhler tests were carried out in a dynamic testing machine under a force-controlled regime with a load ratio of $R = 0.1$. This was done for shear specimens

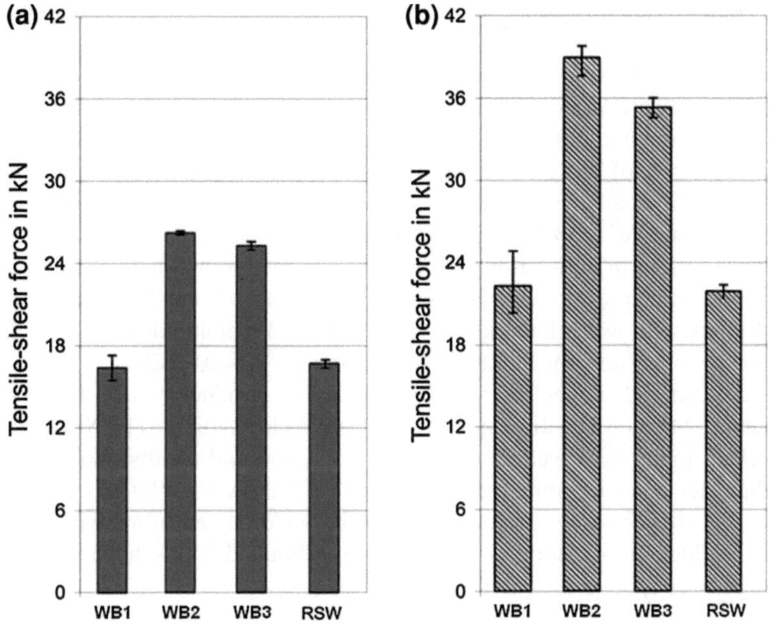

Fig. 16 Tensile shear force of the base metal combination HCT780T/HCT600X spot welded (RSW) and weldbonded with different adhesives (WB1, WB2, WB3) under **a** quasi-static shear load and **b** impact shear load

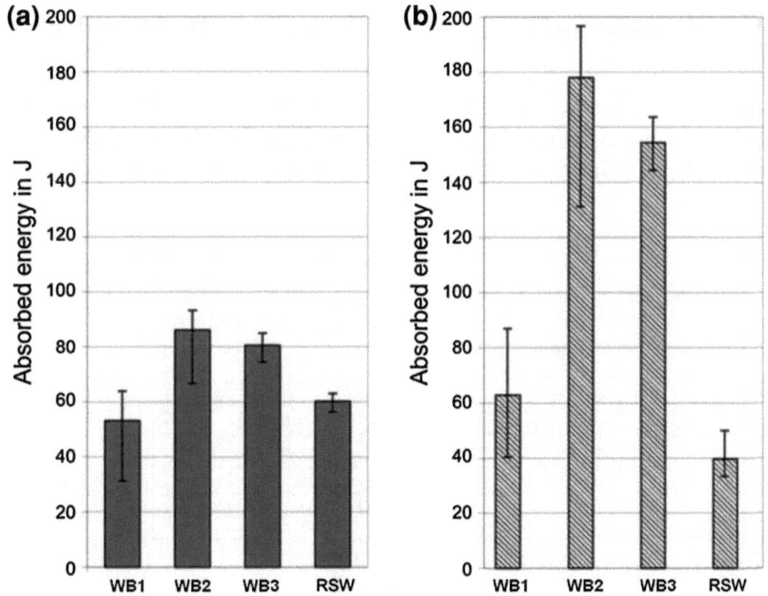

Fig. 17 Absorbed energy of the base metal combination HCT780T/HCT600X spot welded (RSW) and weldbonded with different adhesives (WB1, WB2, WB3) under quasi-static (**a**) and impact shear load (**b**)

using tensile-fatigue test procedures according to EN ISO 14324. The applied test stop criterion was the total fracture of the tested specimens. It should be mentioned in this connection that plug failures were the dominating fracture types in the materials. For example, the fatigue behaviour under tensile shear load for the AHSS/UHSS-base metal combination HCT600X/22MnB5 is discussed (Figs. 18, 19, 20).

A comparison of the *S/N*-curves for spot welded joints with weld diameter $d_p = 6.7$ mm and adhesive bonded joints produced with the medium-viscosity adhesives Betamate™ 1480 (A2) and SikaPower®-498 (A3) is given in Fig. 18. It can be seen that in both cases of adhesively bonded joints there is an advantage with respect to the spot welded joints. Figure 19 shows the *S/N*-curves for spot welded joints with weld diameter $d_p = 6.7$ mm and weldbonded joints with weld diameter $d_p = 6.3$ mm for the adhesive Betamate™ 1480 (A2).

There is an advantage of weldbonded joints produced with the adhesive Betamate™ 1480 (A2) with respect to the spot welded joints. The *S/N*-curves for spot welded joints with weld diameter $d_p = 6.7$ mm and weldbonded joints with weld diameter $d_p = 6.5$ mm for the adhesive SikaPower®-498(A3) are given in Fig. 20. This figure shows an advantage of the weldbonded joints with respect to the spot welded joints, too. In summary, weldbonded joints have a significant advantage over spot welded joints.

6 Causes of Weld Failures

6.1 Weld Discontinuities in Spot Welded and Weldbonded Joints

By definition, weld discontinuities in spot welded and weldbonded joints are interruptions of the typical structure of the joined materials. Weld discontinuities are imperfections within or adjacent to the weld nugget of a spot welded or

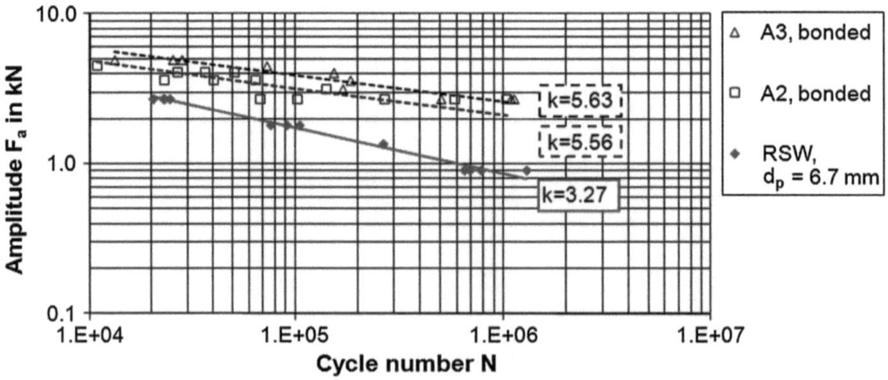

Fig. 18 Fatigue behaviour (*S/N*-curves) of the AHSS/UHSS-base metal combination HCT600X/22MnB5 for spot welded (RSW, $d_p = 6.7$ mm) and adhesive bonded joints (A2, A3)

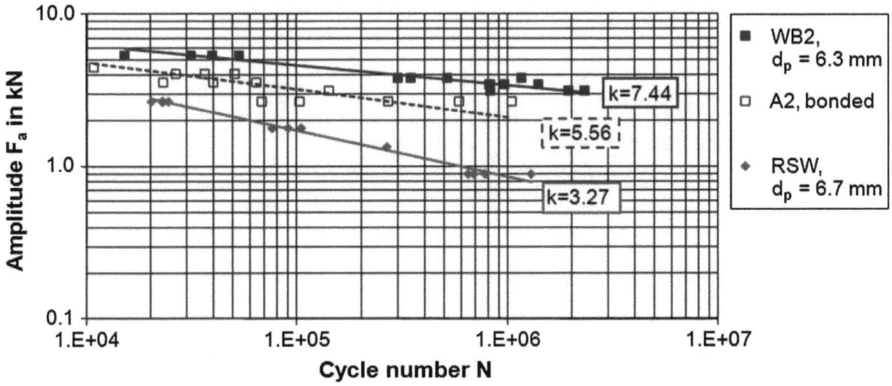

Fig. 19 Fatigue behaviour (*S/N*-curves) of the AHSS/UHSS-base metal combination HCT600X/22MnB5 for spot welded (RSW, $d_p = 6.7$ mm), adhesive bonded (A2) and weldbonded (WB2, $d_p = 6.3$ mm) joints

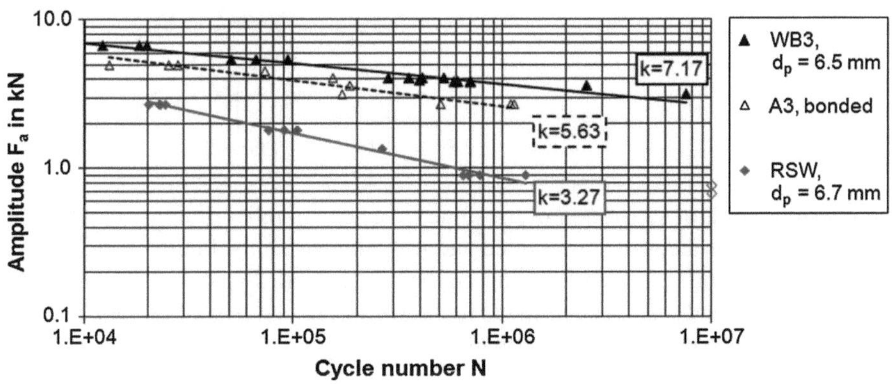

Fig. 20 Fatigue behaviour (*S/N*-curves) of the AHSS/UHSS-base metal combination HCT600X/22MnB5 for spot welded (RSW, $d_p = 6.7$ mm), adhesive bonded (A3) and weldbonded (WB3, $d_p = 6.5$ mm) joints

weldbonded joint. These imperfections may depend on their sizes and/or locations. Depending on unacceptable sizes and/or locations these discontinuities can cause weld failures. Section 2.2 describes innovative high strength steels for the body-in-white automotive engineering exhibiting partly complex microstructures which arise from special temperature-specific production processes and entail the particular mechanical-technological properties. Yet, welding fabrication may involve major problems regarding the formation of weld discontinuities in the joining area. Specifically, in resistance spot welding and weldbonding, due to the localized heat input combined with the comparatively high cooling rates in the joining area, discontinuities may evolve, as depicted in Fig. 21. Apart from inclusions and shrinkage cavities, minor surface cracks as well as excessive deformation of the

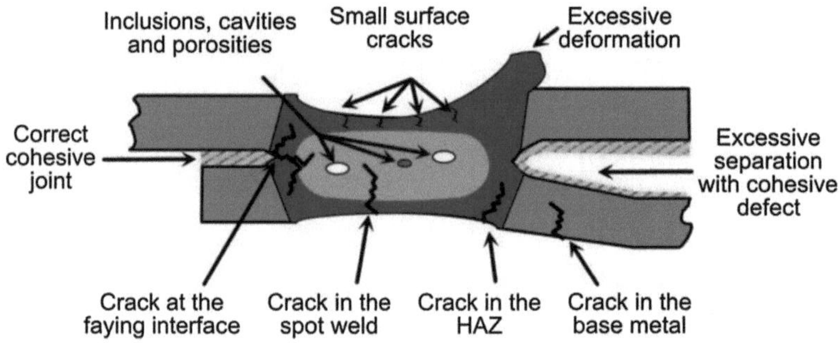

Fig. 21 Overview of weld discontinuities for spot welded and weldbonded joints

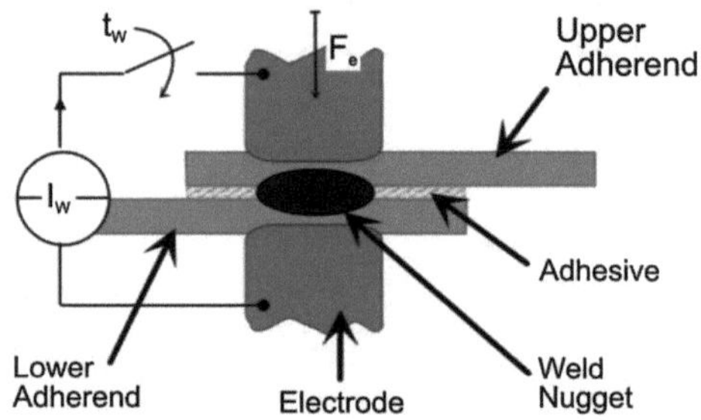

Fig. 22 Joining parameters and welding arrangement

spot welded joint or separation of the joined sheets, a number of larger cracks may appear in various areas of spot welded and weldbonded joints. In this section, some of such cracks are examined focusing on their formation. Typical causes of cracks in spot welded and weldbonded joints as shown in Fig. 21 as well as their features are therefore discussed in the following.

6.1.1 Possible Causes of Weld Discontinuities in Weldbonded Joints

Weldbonding in a highly automated production environment is influenced by many factors. The primary causes of weld discontinuities are the welding/joining parameters themselves. As introduced in Sect. 2.1, the welding/joining parameters are the electrode force F_e, the welding current I_W, the welding time t_W, and, in the case of weldbonding, the adhesive properties (see Fig. 22).

Among those four parameters, the influence of the electrode force and of the adhesive properties is up until now unclear, but these two parameters are not suspected of causing crack-like weld discontinuities. A known impact on the occurrence of weld discontinuities, in particular in the form of cracks, is given by the welding current and the welding time. From the experience on the production floor, it is known that currents at the upper limit of the weldbond quality (i.e. splash limit) as well as very long welding times lead to an increasing number of cracks in joints made of microalloyed steel, AHSS and UHSS.

Fig. 23 Cross section of a weld nugget with shrinkage cavity (base metal: UHSS 22MnB5)

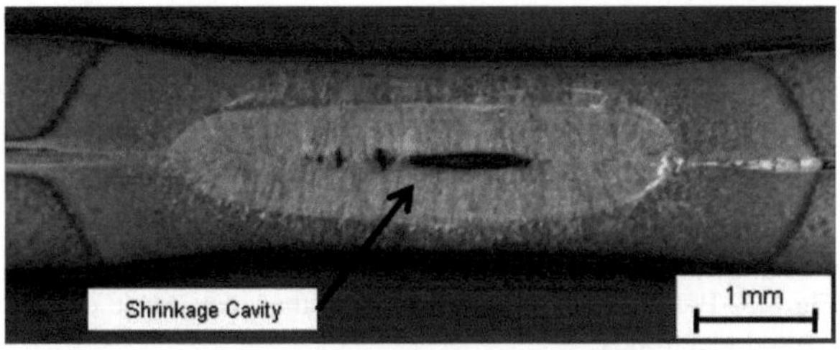

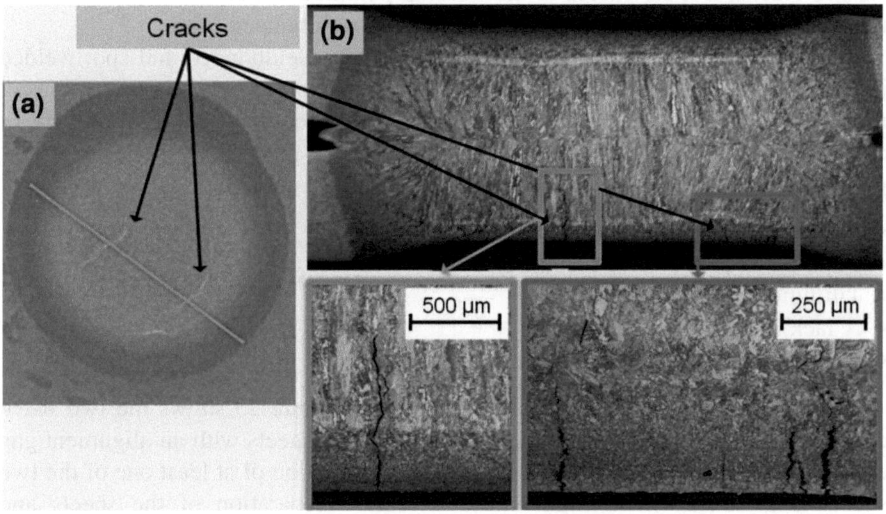

Fig. 24 Cracks in spot welded/weldbonded joints **a** X-ray image, **b** cross section, the *yellow line* represents the plane of the cross section

The occurrence of shrinkage cavities is shown in Fig. 23 for the spot welded UHSS 22MnB5. Similar results occur for weldbonded specimens made of the same steel grade. It is well known that for other examples of the combinations of the same and different high strength steel grades by the choice of the joining parameters shrinkage cavities are generated. This means that shrinkage cavities can be avoided both in the case of weldbonded and spot welded joints by the suitable choice of the joining paramaters given above.

Figure 24 shows a typical example of cracks in a spot welded joint. In part (a) of the figure, the cracks were detected by means of an X-ray image (non-destructive). The visible weld discontinuities are short in comparison to the size of the spot welded joint. In order to analyse how deep the cracks penetrate into the weld nugget, part (b) of Fig. 24 shows cross sectional images (destructive).

It can be seen that the crack in the middle of the spot welded joint (magnification on the left hand side) even penetrates the weld nugget. The other cracks in the magnification on the right hand side are, in the plane of the cross-section, only

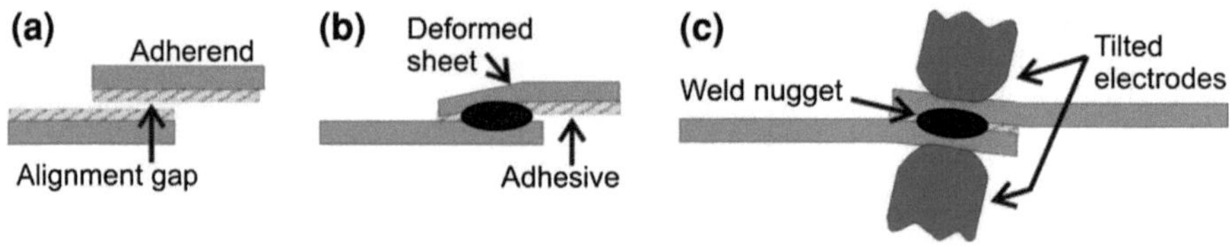

Fig. 25 Production-specific influences that may cause discontinuities: **a**, **b** gap between sheets, **c** tilted electrodes

surface cracks. But the choice of the welding parameters alone do not lead to such kinds of weld discontinuities of spot welded or weldbonded joints.

The secondary causes of weld discontinuities in weldbonded and spot welded joints arise from the conditions of the production environment, i.e. they result mainly from a possible malfunction of the welding equipment, misalignment of components or welding guns prior to the welding process or mistakes in the programmed welding schedule. From these circumstances,

- Electrode wear
- Gaps between the steel sheets/adherends
- Tilting of the electrodes
- A lack of electrode cooling
- Two-times welding of the same spot welded joint

may result and affect the quality of the weld. Figure 25 shows the two listed consequences of misalignment. In part (a), two steel sheets with an alignment gap are depicted. The welding process leads to a deformation of at least one of the two sheets (Fig. 25b), which might cause excessive separation of the sheets and therefore a lack of adhesive bonding (Fig. 21).

Figure 25c shows the effect of electrode tilting. This kind of misalignment does not lead to a separation of the spot welded/weldbonded sheets since they are deformed parallel to each other. But due to the tilting of the electrode faces, excessive deformation of the spot welded joint might be observed. In addition to accidentally applied extreme welding parameters as discussed before, further mechanical stress is imposed on the spot welded structure which might abet the occurrence of cracks.

The consequence of a gap between two components before joining due to an alignment deviation can be influenced by the welding equipment. When simple equipment is used, one electrode is typically fixed. This results in an asymmetrical deformation of one sheet, and an excessive separation on one side of the overlap joint is the result (Fig. 26a). If the separation is too large, the layer of the applied adhesive might not be thick enough to fill the gap, resulting in a lack of bonding. When modern equipment with jaw balancing is used, the gap is compensated from both sides of the components. Therefore, the separation is distributed equally on both sides and is hence less severe (Fig. 26b).

A crack on the surface of a spot welded joint caused by an electrode tilting of 5° in combination with extreme welding parameters can be seen in Fig. 27a). The crack is circumferential and surrounds almost half of the spot welded joint. Due to

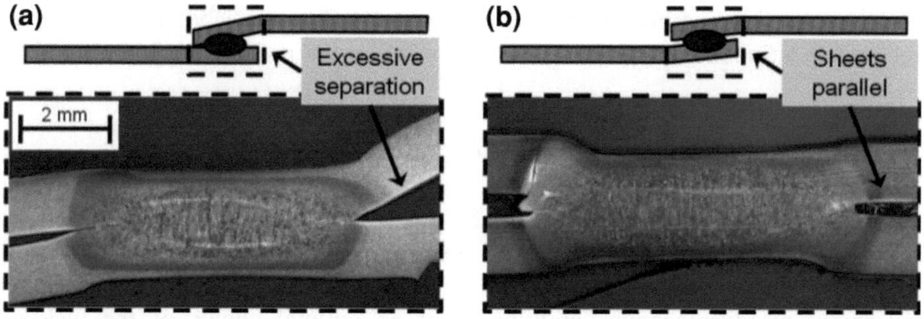

Fig. 26 Influence of an alignment gap on the weldbond quality **a** without jaw balancing, **b** with jaw balancing

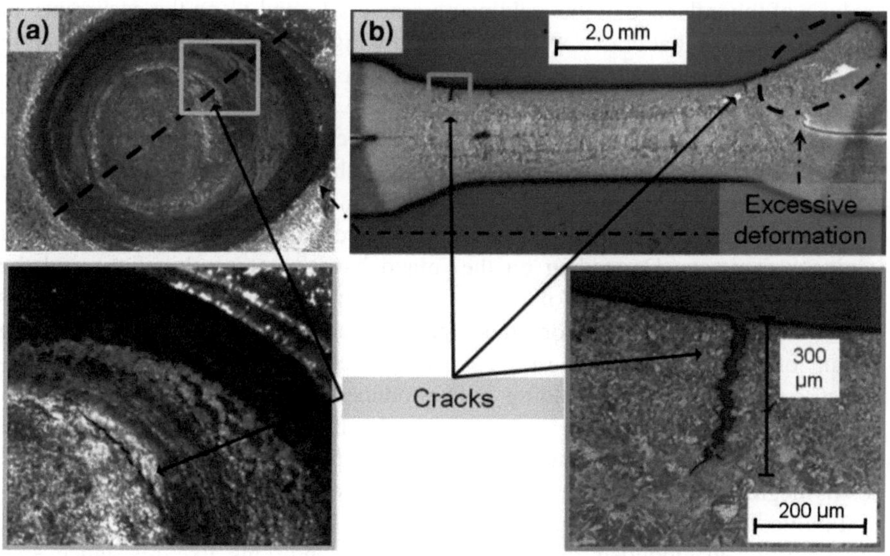

Fig. 27 Cracks on surfaces of a spot welded/weldbonded joint caused by electrode tilting **a** surface image, **b** cross-sectional image

the wide opening of the crack, it can be easily seen with an optical microscope and some experience. Although the cracks look large in the optical surface image, the cross-section (Fig. 27b) reveals that those cracks are only surface cracks which do not penetrate the weld nugget. Furthermore, the tilted electrode leads to an excessive deformation of the spot welded joint, which can be seen in the optical surface image shown in Fig. 27a as well as in the cross section shown in Fig. 27b (on the right hand side). Since some material is squeezed out of the weld nugget, the profile of the joint and therefore its strength is weakened.

An overview of the possible weld discontinuities in weldbonded joints made of steel sheets and their possible causes has been discussed in this section.

The next section summarizes the experience gained in the laboratory and gives recommendations to avoid weld discontinuities on the production floor.

6.1.2 How to Prevent Weld Discontinuities in Weldbonded Joints

An important fact that was discovered by investigating the causes of occasionally occurring weld discontinuities in spot welded joints under laboratory conditions is that weld discontinuities of potentially critical size are observed only under extremely disadvantageous conditions. For an optimal choice of the welding parameters sufficiently large sizes of WCRs can be realized for spot welded AHSS. Cracks can thus be avoided conveniently. But the influence of cracks on the mechanical strength of the joined components is up until now unknown. The analysis of the influence of the welding parameters I_W and t_W on the occurrence of cracks has shown that cracks were provoked by high currents near the splash limit and long welding times. In combination with a lack of electrode cooling, the discontinuities occurred more frequently and were larger in size. This means that a maximized total heat input into the spot welded/weldbonded joint with at the same time minimized heat output is one scenario which should be avoided in order to prevent cracks.

Shrinkage cavities can be avoided, as mentioned above, by the correct choice of joining parameters. Depending on the joined base metal combinations and the applied adhesives, the kind of the energy input into the weldbonded joint is the determining factor. Therefore, the interdependency of the choice of the welding current and welding time is of great importance. This means a higher welding current and a shorter welding time produces other results such as a lower welding current and a longer welding time for the same state of energy input. Furthermore, regarding weldbonded joints, the influence of the type of adhesive, i.e. its degree of viscosity (low, medium, high) on the occurrence of weld discontinuities has not been investigated until now. In relation to the viscosity, the behaviour of weldbonded joints under similar conditions like resistance spot welded joints might be different.

The additional influences resulting from the conditions on the production floor can combine with disadvantageous welding parameters in an unfavourable way. Worn out electrodes, a disturbance of the electrode cooling or two-times welding of the same spot welded joint can increase the total number of spot welded joints exhibiting cracks when high welding currents and long welding times are used at the same time. In order to obtain discontinuity-free weldbonded joints, electrodes should be changed early enough before the number of discontinuities grows, cooling circuits should be maintained regular and welding schedules should be programmed very accurately to avoid two-times welding of the same spot welded joint. But Gaps between the sheets of up to 2 mm and electrodes tilted by up to 5° are typical alignment deviations which are not avoidable in a highly automated mass production sphere at the present state of the art. With regard to weldbonding, electrode tilting does not harm the cohesive strength of the joint since the sheets are not separated from each other, as it is the case when gaps occur. With an excessive separation between the sheets, a delamination of the adhesive layer reduces the adhesively joined area which would decrease the strength of the weldbond. The use of a welding gun with jaw balancing is recommended here.

6.2 Metallurgical Structure of Weldbonded Joints

For the AHSS HCT780T and the UHSS 22MnB5 (UsiBor®) the metallurgical structures of the weld nuggets for resistance spot welded and weldbonded joints with the medium-viscosity adhesive Betamate™ 1480 (A2) are shown in Figs. 28 and 29.

It can be seen that there is no difference between the spot welded and weldbonded nugget structures. These results are consistent with the hardness curves, also given in the figures. There are no significant differences between the hardness curves for the weldbonded steels mentioned above and the applied adhesive Betamate™ 1480 (A2). Similar results are given for other base metals and their combinations with the adhesives A1 and A3 [2].

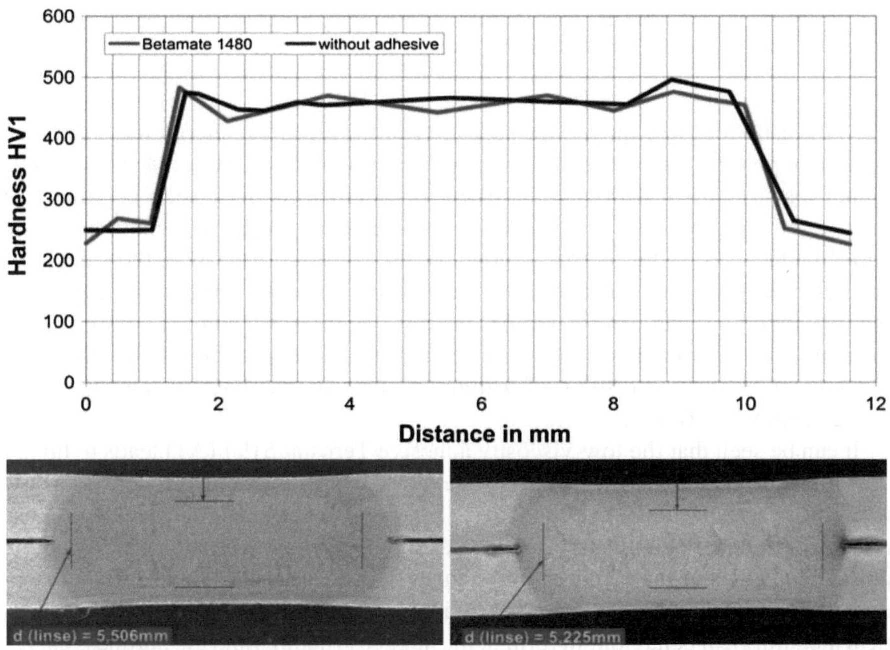

Fig. 28 Weld nugget structures and hardness curves of spot welded (*left*) and weldbonded (*right*) AHSS-base metal HCT780T for WB2 (A2)

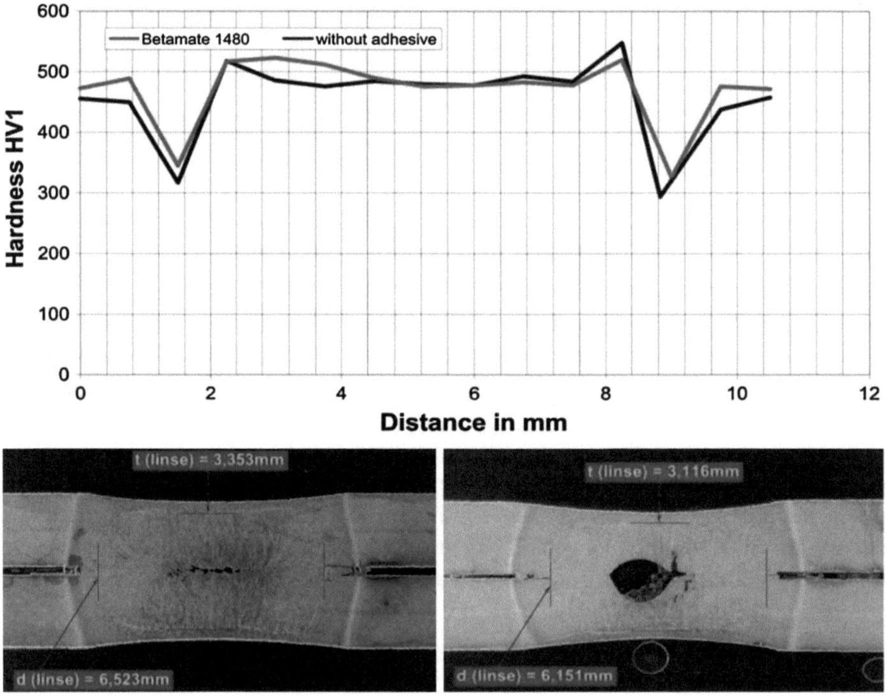

Fig. 29 Weld nugget structures and hardness curves of spot welded (*left*) and weldbonded (*right*) UHSS-base metal 22MnB5 for WB2 (A2)

The reason for these facts is the similarity of the electrical energy input after the beginning phases of the weldbonding and the resistance spot welding process. The higher total electrical resistance between the steel sheets during the beginning of the weldbonding process is attributed to the electrical insulating behaviour of the applied adhesives. After the break down of the constant resistance, the calculated total electrical resistance between the steel sheets achieves comparable values for the weldbonding and spot welding process. Figures 30 and 31 show, as an example, the electrical resistance between the steel sheets during the welding time for the weldbonding and the spot welding process for the base metals HCT780T and 22MnB5 (UsiBor®).

It can be seen that the low-viscosity adhesive Terostat 5194 (A1) leads to higher deviations from the electrical resistance for the spot welding process than the medium-viscosity adhesive Betamate™ 1480 (A2) for the above mentioned base metals during the beginning phase (process time < 80 ms). But after the beginning phase, the values of the electrical resistances between the steel sheets are independent of the applied adhesives, i.e. nearly identical in all cases. Therefore, the identical metallurgical behaviour in form of the nugget structures and the hardness curves given above can be interpreted as a result of the similar electrical behaviour of the weldbonding and the resistance spot welding process after the beginning phase.

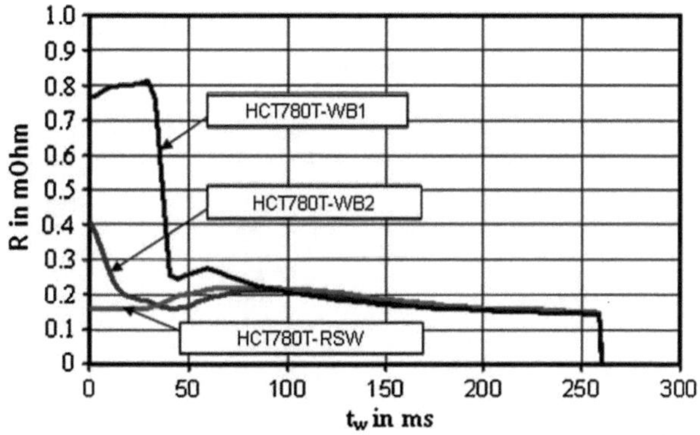

Fig. 30 Electrical resistance R between the AHSS sheets during t_W for RSW, WB1 and WB2 for HCT780T+Z

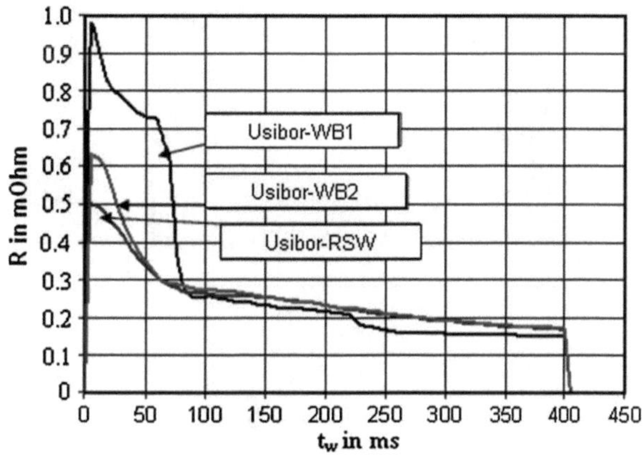

Fig. 31 Electrical resistance R between the UHSS sheets during t_W for RSW, WB1 and WB2 for 22MnB5 (UsiBor®)

6.3 Fracture Behaviour of Weldbonded Joints

A comparison of the fracture behaviour of weldbonded and spot welded joints of the AHSS/UHSS-base metals HCT780T (TRIP Steel) and 22MnB5 (UsiBor®) for different electrode forces ($F_e = 3.5/4.5$ kN) and for the lower and upper quality limits is given in Figs. 32, 33, 34 and 35. The applied adhesives in these cases are the low- and medium-viscosity adhesives A1, A2 and A3 (see Table 3). It can be seen that there is no influence of the applied adhesives and, consequently, of the viscosity on the failure type. The fracture behaviour of the UHSS 22MnB5 (UsiBor®) is very irregular (Figs. 34, 35). But the fracture behaviour of the weldbonded TRIP steel HCT780T shows, compared with the spot welded joints, a tendency towards a plug failure independently of the applied adhesives (Figs. 32, 33). Furthermore, the Figures show the influence of different electrode forces F_e for the upper and lower quality limits of the weldbonded and spotwelded joints. It can be seen that there is no significant difference in the fracture behaviour for the different joining parameter electrode force for both the lower and the upper limit of weldbonded quality.

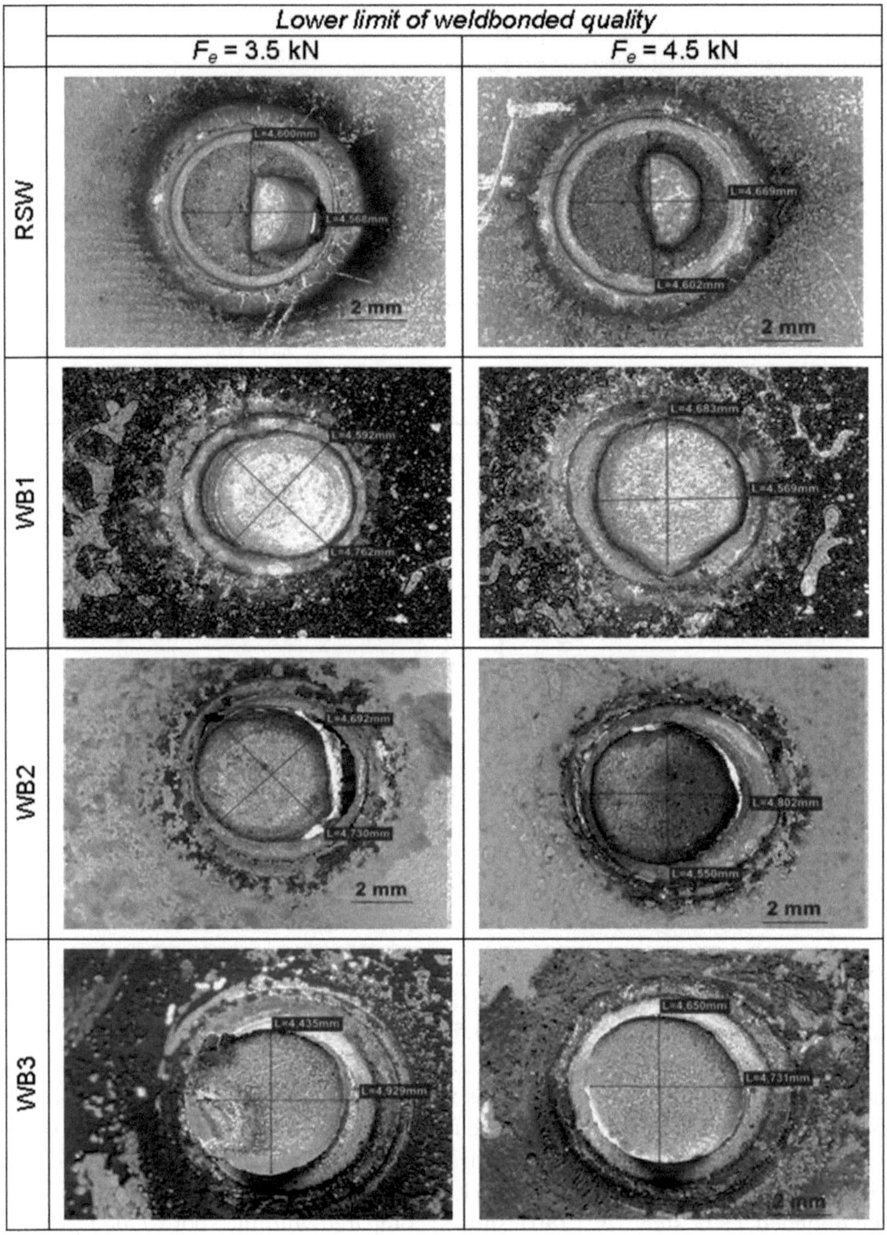

Fig. 32 Fracture behaviour at the lower limit of weldbonded quality (RSW and WB1-WB3, AHSS HCT780T)

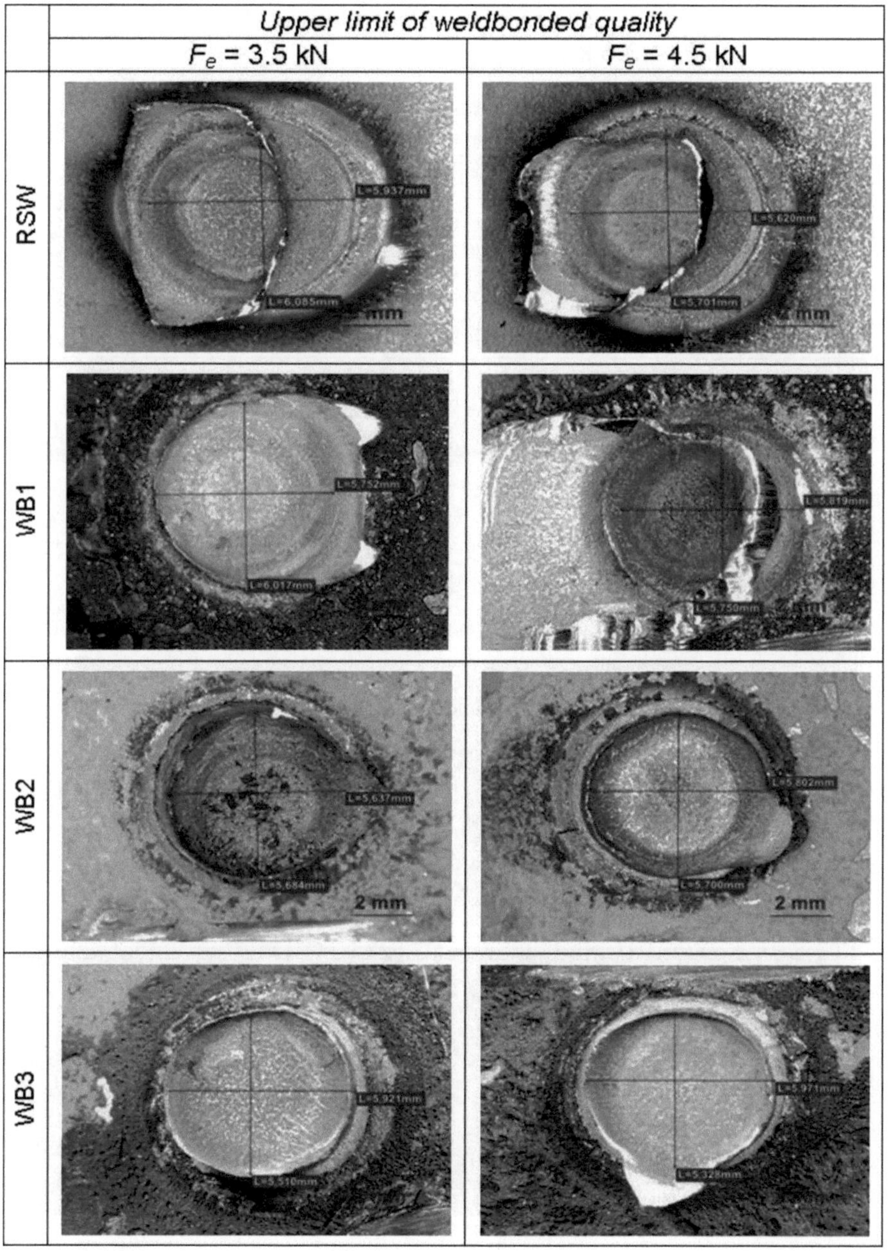

Fig. 33 Fracture behaviour at the upper limit of weldbonded quality (RSW and WB1-WB3, AHSS HCT780T)

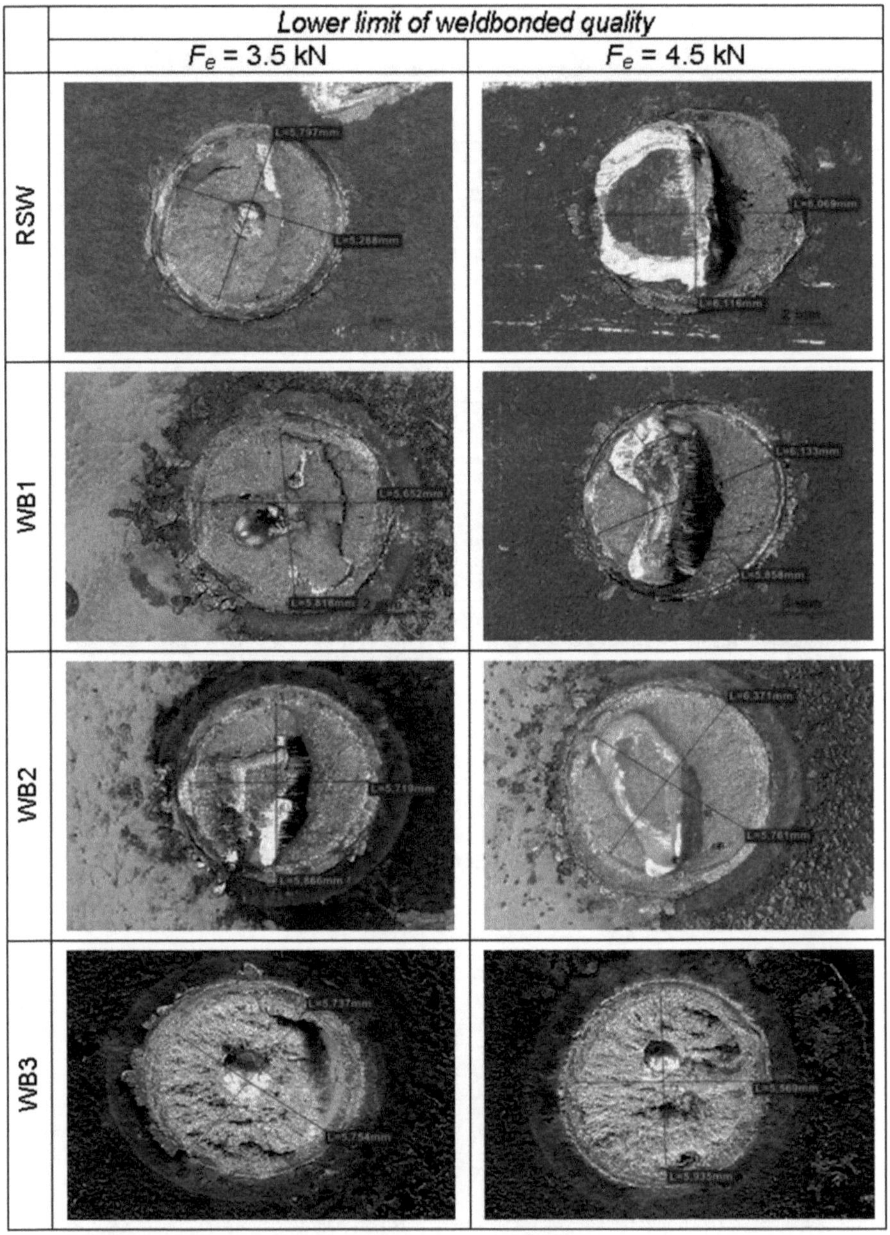

Fig. 34 Fracture behaviour at the lower limit of weldbonded quality (RSW and WB1-WB3, UHSS 22MnB5)

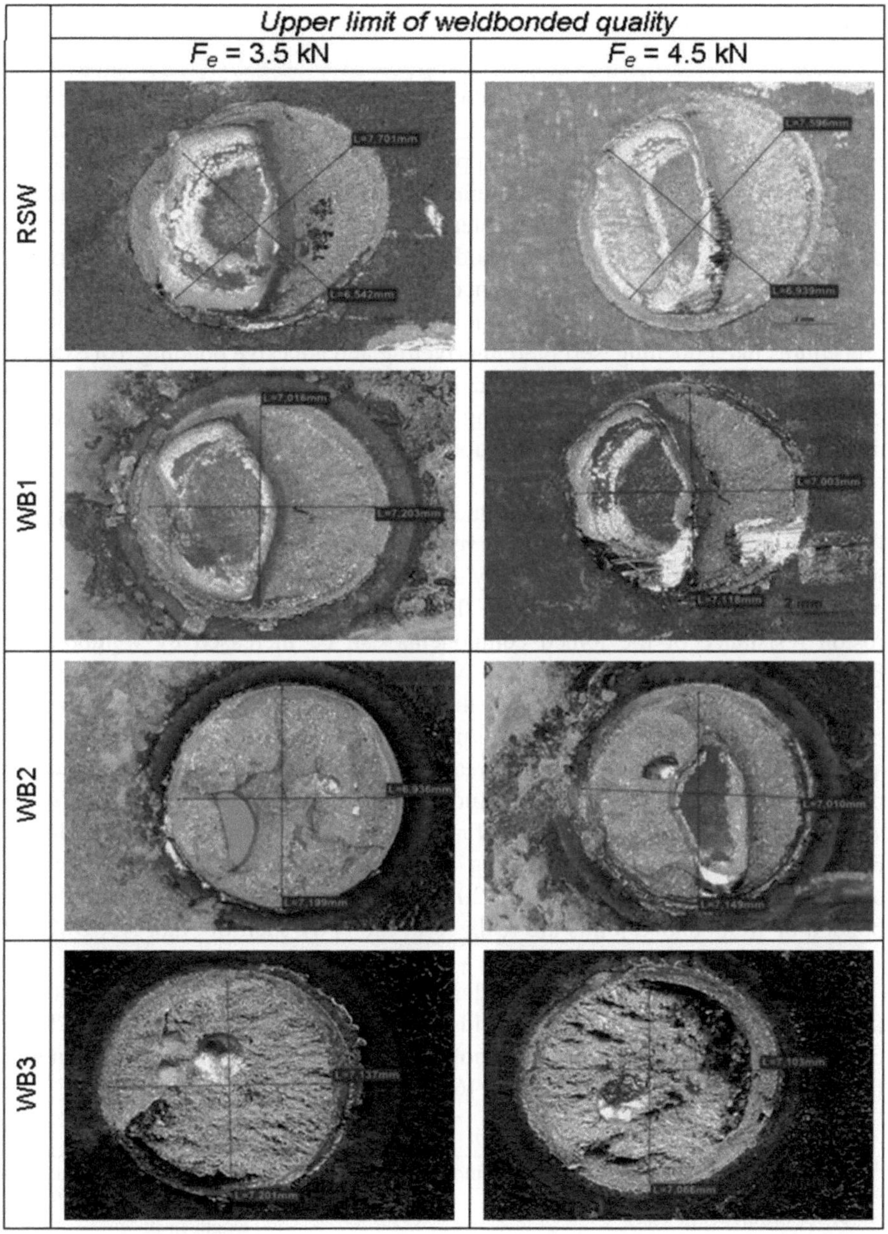

Fig. 35 Fracture behaviour at the upper limit of weldbonded quality (RSW and WB1-WB3, UHSS 22MnB5)

7 Examples of Use

The most important range of applications of weldbonding is located in car body manufacturing and in collision repair. In the body-in-white automotive applications, several joining techniques are applied. Depending on the requirements of the lightweight body-shell design, different selected materials and material combinations of the body-in-white require selected joining processes. In this connection, different warm and cold joining methods are applied. The important warm joining methods are: laser welding, resistance spot welding, MIG/MAG welding and brazing/laser brazing. Important cold joining methods are the mechanical fastening procedures like clinching, self piercing riveting and adhesive bonding. An example for the distribution of these selected joining methods is published in [3]. The combination of resistance spot welding and adhesive bonding, is reported there, is mainly applied for joining the wheel-house and the floor panel in the vicinity of the doors.

As already mentioned in the introduction, the weldbonding technique is applied in car body manufacturing to reduce vehicle weight and fuel consumption and to improve the durability of the car body as well as the safety and the comfort of the passenger cell. It is well known that the stiffness of the body shell and the corrosion resistance of the joined components can be increased by weldbonding. Furthermore, the application of adhesive bonding in combination with resistance spot welding enables a reduction of the number of spot welds needed for a car body shell. Therefore, weldbonding implies cost reduction for the body-in-white automotive applications. In recent years, weldbonding has been introduced by many automotive companies, e.g. Audi, BMW, Daimler, Mazda, Opel, Volkswagen, and Volvo. Many examples of applications of the combination of adhesive bonding and resistance spot welding with different concepts of adhesives are described in [6]. In particular, from 20 to 90 m or more than 100 m of adhesive bonded or weldbonded joints are realized in the body shell of Audi A5, BMW 7, C-class Mercedes, Opel Insignia and Volvo V70 and XC60 (see Table 10).

In Table 10 some examples of the length of adhesively bonded components and the number of the needed spot welds for a car body shell are given. These examples are only a few of the whole range of applications of weldbonding. Further examples are summarized in [6].

Table 10 Length of applied adhesives and number of spot welds for different car body shell construction according to [6]

	Car model	Structural adhesive length in m	No. of spot welds
I	Audi A5 (2008)	89.7	4,975
II	BMW 7 (2009)	90.0	NA
III	Mercedes C-Class(2007)	62.4	5,394
IV	Opel Insignia (2008)	21.1	6,321
V	Volvo V70 (2008)	24.9	4,170
VI	Volvo XC60 (2008)	25.4	4,337

8 Conclusions

The application of the weldbonding technique for the body-in-white automotive manufacturing of lightweight steel cars was described. This was done for different steel grades and some selected adhesive systems.

The influence of the base metal strength and the viscosity of the applied adhesive on the weldbonding process and the weldbonded joints were analyzed. Depending on the applied low-viscosity and medium-viscosity adhesives for mild steels, AHSS and UHSS, different weldbonding processes and the corresponding weldbonded joints were realized. The welding current ranges and the process reliability were discussed and compared.

For the different weldbonding processes, the welding current ranges were determined and described by extended weldability lobes. Based on these representations, implications on the process reliability were stated in comparison to the well known resistance spot welding process. It was shown that the process reliability for the different weldbonding processes is sufficiently large to be applied on the production floor. The important influence of electrode caps on the weldbonding and resistance spot welding processes was discussed. The weldbond joinability was introduced based on the components of process reliability and electrode wear. This was demonstrated for the case of a special AHSS.

Another requirement regarding the quality of weldbonded joints is their strength and durability. In this connection, weldbonded joints were tested by quasi-static, impact and fatigue loads. For all types of load, weldbonded joints show a higher mechanical strength than the resistance spot welded joints. In particular, when different base metal combinations are joined, the increase of tensile shear strength is even extended. One important result regarding the viscosity of the applied adhesive was that weldbonded joints with medium-viscosity adhesives show higher tensile strengths in quasi-static and impact load tests. Moreover, the crash behaviour of weldbonded joints was analyzed by the determination of the absorbed energy. In this case, the medium-viscosity adhesives show higher absorbed energy values, corresponding to the above mentioned tensile shear strength. Overall, weldbonded joints show a higher mechanical performance than resistance spot welded joints, as expected.

The types of weld discontinuities for weldbonded and spot welded joints were discussed. The important types of weld discontinuities discussed were shrinkage cavities, cracks caused by the weldbonding process and the influence of alignment gaps and electrode tilting on the quality of the joints. It was found that these weld discontinuities only appear under extremely disadvantageous conditions, and methods to prevent weld discontinuities in weldbonded and spot welded joints were given. The metallurgical behaviour for special weldbonded base metal combinations of AHSS and UHSS was analyzed. There is no difference between the spot welded joints and weldbonded joints regarding the hardness and the metallurgical structure of the weld nuggets. The fracture behaviour of weldbonded and spot welded joints for different quality limits was discussed. This was done for an AHSS and an UHSS base metal and different applied adhesive systems. The failure mode of the weldbonded UHSS was very irregular. But in comparison to the spot welded joints, for the weldbonded joints of AHSS, plug failures occurred. Furthermore, some applications of weldbonding in the body-in-white automotive were demonstrated.

In summary, different base metal combinations of mild steels, AHSS and UHSS combined with several adhesive systems are applicable in the modern lightweight body shell car manufacturing.

References

1. Budde, L., Hahn, O.: Adhesive bonding in combination with spot welding or clinching. In: Welding in the World, vol. 30, no. ½, pp. 26–32. (1992)
2. Cramer, H., Bschorr, T., Hahn, O., Thommes, H., Zech, F.: Final report for FOSTA-No. P704/10, AiF-No. 14476 N. Forschungsvereinigung Stahlanwendung e. V., Düsseldorf (2009)
3. Cordes, R., Germann, V., Tunger, R.: EuroCarBody, the body in white of the new Volkswagen Passat B6, Bad Nauheim (2005)
4. Darwish, S. M., Al-Samhan, A.: Design rationale of weldbonded joints. Int. J. Adhes. Adhes. **24**, 367–377 (2004)
5. Fritzsche, C., Laurenz, R.: Widerstandspunktschweißen kaltgewalzter Mehrphasenstähle—DVS 2935-2. Treffpunkt Widerstandsschweißen, vol. 20, pp. 33–40. (2007)
6. Giovanoli, J.: Dow Betamate™ Adhesives Overview. URL: http://www.dowautomotive.com/pdfs/AdhesiveInformationPowerPoint.pdf. Letzter Zugriff, 28 June 2010
7. Hahn, O., Peetz, A.: Eigenschaften und Wirtschaftlichkeit kombiniert gefügter Blechverbindungen. In: Konferenz-Einzelbericht: DVS-Berichte, vol. 1, pp. 272–277. Düsseldorf, Germany (1995)
8. Kötting, G., Schmid, G.: Widerstandspunktschweißkleben geklebter und beschichteter Karosseriebleche: Verbindungseigenschaften, Technologie und Gefahrstoffemission. In: Konferenz-Einzelbericht: DVS-Berichte, vol. 165, pp. 17–20. Düsseldorf, Germany (1995)
9. Maggi, S., Murgia, M.: Introduction to the metallurgic characteristics of advanced high-strength steels for automobile applications. In: Welding International, vol. 22, no. 9, pp. 610–618. (2008)
10. N.N.: Ultra light steel auto body final report. American Iron and Steel Institute, Washington, DC (1998)
11. Osburg, B., Patberg, L., Grüneklee, A., Flöth, Th., Große-Gehling, M., Hinz, M., Mebus, H.: New steel body—Sicherer und wirtschaftlicher Karosserieleichtbau mit Stahl. In: *Automobiltechnische Zeitschrift ATZ*, vol. 106, no. 3, pp. 190. (2004)
12. Santosa, I.O., Zhang, W., Goncalvesa, V. M., Bayb, N., Martins, P. A. F.: Weldbonding of stainless steel. In: Int. J. Mach. Tools Manuf. **44**(14), 1431–1439 (2004)
13. Schmid, G., Korte, M., Walther, U.: Punktschweißkleben im Automobilbau. In: Konferenz-Einzelbericht: DVS-Berichte, vol. 213, pp. 53–58. Düsseldorf, Germany (2001)
14. Sierlinger, R., Szinyur, J., Ritsche S.: Schweißeignung und Gebrauchseigenschaften von punktgeschweißten hochfesten Multiphasenstählen. In: Proceedings of the Eurojoin, vol. 5, pp. V10. Wien, Austria (2004)
15. Weber, G., Göklü, S.: Resistance spot welding of advanced high strength steels—influence of welding parameters and electrode cap type. In: Proceedings of the International Conference of the International Institute of Welding (IIW), pp. 135–155. Prague, Czech Republic (2005)
16. Weber, G., Göklü, S.: Resistance spot welding of uncoated and zinc coated advanced high-strength steels (AHSS)—weldability and process reliability—influence of welding

parameters. In: Welding in the World, Journal of the International Institute of Welding (IIW), vol. 50, no. 4/4. Rossy, France (2006)

17. Weber, G., Göklü, S., Rethmeier, M.: Influence of the type of electrode caps on the welding current ranges and the process reliability in resistance spot welding. In: Annual Assembly of IIW, vol. 61. Graz, Austria (2008)
18. Weber, G., Brauser, S., Rethmeier, M.: Extended weldability lobes in resistance spot welding of advanced high strength steels (AHSS). In: Proceedings of EUROJOIN, vol. 7. Lido di Venezia, Italy (2009)
19. Weber, G., Thommes, H., Bschorr, T., Cramer, H., Hahn, O., Rethmeier, M.: Hybrid bonding of advanced high strength steels in the lightweight body shell design for the automobile manufacturing. In: International Conference on Advanced Computer Engineering and Experimenting, Rome, Italy (2009). Also published in Materials with Complex Behaviour. Series: Advanced Structured Materials, Vol. 3, L.F.M. da Silva; H. Altenbach (Eds.) Springer-verlag Berlin Heidelberg (2010)
20. Weber, G., Brauser, S., Pepke, L.-A., Rethmeier, M.: Description of welding current ranges by extended weldablity lobes in resistance spot welding. In: Annual Meeting SC-Auto of the International Institute of Welding (IIW). Wolfsburg, Germany (2009)

Technology of Rivet: Adhesive Joints

Fabrizio Moroni and Alessandro Pirondi

Abstract In general hybrid bonding allows to match together the properties of two different joining techniques, with the aim to obtain a joint with a better set of properties (that could involve strength, energy absorption at failure, cost, etc.) with respect to the simple joints. In the literature, several works dealing with the evaluation of the strength for this kind of joints and a global enhancement of the properties, in terms of static strength, fatigue strength and energy absorption are found. The first hybrid technique "discovered" was the weld bonding and therefore in the literature it is the most deeply analyzed. In terms of mechanical performance, weld-bonding is the hybrid joining technique where better result are obtained with respect to other hybrid joining techniques such as rivet bonding, clinch bonding, etc., but at the same time it is the most troublesome in terms of joint manufacturing (influence of the presence of the adhesive on the welding performance, weld compliant adhesive requirements, etc.). On the opposite, the hybrid techniques which combine adhesive with mechanical fastening (like rivet, clinching, self piercing riveting) give in general a lower enhancement of performance if compared with weld bonded joint, but certainly they are characterized by an easier and trouble less manufacturing. In this chapter, a brief introduction to the rivet bonding technology is given, dealing with why the hybrid joints are useful and which kind of industrial requirements this kind of joints are able to satisfy. Later on, the adhesive requirements are discussed with the aim to find the best adhesive for different purposes. One of the most important benefits of the rivet bonded joints is the ease of manufacture, therefore the way to produce rivet

F. Moroni (✉) and A. Pirondi
Dipartimento di Ingegneria Industriale, Università degli Studi di Parma, viale G.P. Usberti, 181/A, 43124 Parma, Italy
e-mail: moroni@ied.unipr.it

A. Pirondi
e-mail: alessandro.pirondi@unipr.it

Adv Struct Mater (2011) 6: 79–108
DOI: 10.1007/8611_2010_37

Published Online: 30 July 2010

bonded joints is discussed, and a literature example is shown with the aim to emphasize the different manufacture procedure of rivet bonded joints when compared with traditional weld bonding techniques. In this case, a noticeable benefit in terms of manufacture cost is also shown with respect to traditional joining techniques. An example of the evaluation of the strength of rivet bonded joints (pop rivet and self piercing rivet) in comparison with the performances of simple riveted and simple bonded joint, is then given. It is shown how the performance of hybrid joints depend mostly on the strength of the adhesive bond, and that the rivet becomes relevant only when the adhesive performance decreases (i.e. when the service temperature is higher than the adhesive glass transition temperature for example) or in general when the adhesive fails. Finally, the failure mode is discussed for both pop rivet bonded and self piercing rivet bonded joints. The failure of a pop rivet bonded joint is simulated using appropriate damage models for the rivet and for the adhesive. In particular the parameters of the damage models are tuned by comparison with experiments performed on simple joints and they seem to be adequate for the simulation of the failure behaviour of rivet bonded joints.

1 Specific Requirements in Industry

Prior to define the requirement in industry of hybrid Pop Riveted (PR) and Self Piecing Riveted (SPR) joints, a quick look to the two simple riveting techniques is given. Both are joining techniques for sheets, where the junction takes place by means of an external body called rivet. The PR is a fastening method that involves the plastic deformation of the rivet only. This allows to use this kind of joint for almost all the materials: the only need is to make a hole in the parts. Moreover, a pop riveted joint can be done even in the case the access is limited to one side of the joints.

On the other hand, SPR is a joining technique similar to clinching but a rivet is placed between the plates. Here, during the rivet setting, both the rivet and the parts are plastically deformed, therefore, brittle material cannot be joined with this technique. Furthermore, for the manufacturing of the joint, the access to both the sides is needed and a relatively powerful and expensive equipment (if compared with that of PR) is required.

Despite these advantages of PR joints, the SPR joints show typically a higher strength and durability, therefore they are used where the mechanical performances are more relevant.

The possibility to combine adhesive bonding with riveting is addressed to reduce the drawbacks of each single joining technique. For example, using the rivet-bonded joint, the strength of a bonded joint (produced with an epoxy structural adhesive) can be coupled with the rapidity of manufacturing of a riveted joint. Indeed, the presence of the rivet allows to hold the parts together during the polymerization without any other tool.

Looking at the industry, certainly one of the most relevant requirements is the joint strength. Depending on the industrial field, this requirement has to be coupled with some other features: for example in aerospace applications other relevant requirements involve the stiffness, the damage tolerance and the weight of the joint. All of these features are characteristics of adhesive bonding. If the adhesive is coupled with riveting, a quicker and easier joint production will result and at the same time the joint will also have a higher damage tolerance. In aerospace, the research trend is to tailor the choice of the material for the specific application with the aim to obtain a lighter structure [1], and it is rather common to meet joints where a strong and stiff steel is coupled with a lighter aluminium. Moreover, nowadays composite materials and polymers in general are also widely used. This leads to have in the same product several different kinds of materials: steel, aluminum alloys, polymers, composites, which can be difficult to be joined together with traditional joining techniques like welding. A solution could be bolting but it brings in a significant weight increase (absolutely to be avoided in the modern lightweight structures) and its performances are not comparable to welding. The hybrid rivet bonding technique seems to be a solution since it usually satisfies all the requirements in terms of strength, cost and weight. In this industrial field, hybrid joining in general, but especially rivet bonding is used also for patch repairs [4].

Considering the automotive, bus, truck or railway coach construction, the rivet bonding technology could be a good solution for the bodywork assembly (especially if made out of aluminum alloy) as enhancement with respect to the simple riveted joints. In fact, when parts are held using structural pop rivets, the intrinsic strength of the rivets is low, therefore butt-straps are generally needed to extend the joining area (an example is shown in Fig. 1). In this case, the number of rivets can be reduced (reducing therefore the assembly costs) adding an adhesive layer between the straps and the profile, producing at the same time a stiffer and stronger joint.

In the nautical industry, the coupling of adhesive and rivet attracts attention especially for the possibility to obtain a strong (by the rivet) and sealed (by the adhesive, or better the sealant) joint. Here, the materials involved are brittle (typically fiberglass), therefore pop riveting in the only possible choice.

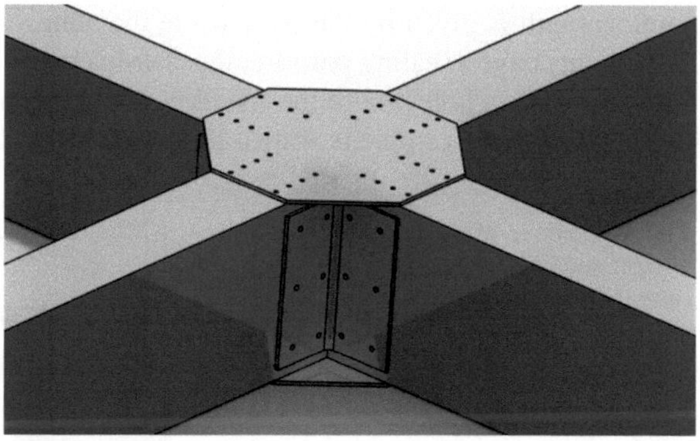

Fig. 1 Example of butt-strap riveted joint for vehicle frame applications

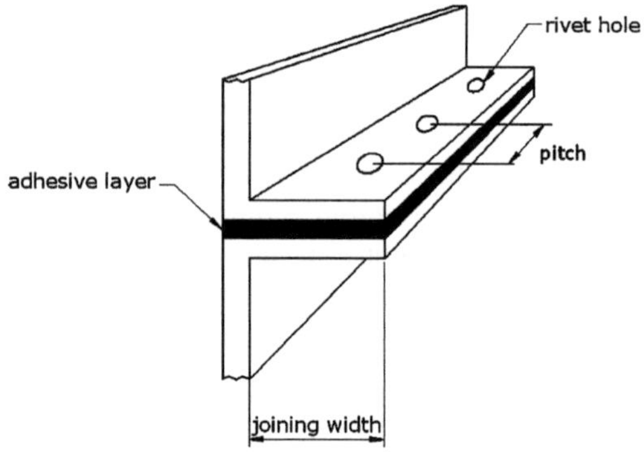

Fig. 2 Example of hull/vessel joint in naval application

The typical geometry in this kind of application is shown in Fig. 2. In the majority of cases the joint is designed to sustain the load only by the rivets, while the adhesive has only sealing purpose.

If the adhesive used is not only a sealant, but it has also good strength, the joint performance can be significantly increased.

However, regardless the specific industrial field, the most important industrial requirements involve strength and costs. Apart from this two factors, other relevant factors to be considered from the industries when a joining techniques is selected concern the materials to be joined, the energy absorption at failure (especially in the automotive crash tests) and aesthetic characteristics. All of these requirements seem to be potentially satisfied by hybrid joints.

2 Adhesive Characteristics Required

It is obvious that the mechanical performance requested for the joint affects directly the adhesive choice. If the industrial needs involve the joint strength, a stiff and strong adhesive (typically an epoxy structural adhesive) is the only possible choice. On the other hand, if the joint has a sealing purpose, a flexible but also weak adhesive (typically a sealant like silicone) has to be coupled with the rivet: in this manner the joint strength is given by the rivet while the adhesive has only sealing properties. If the joints have a sealing purpose, the classic pop rivets cannot properly be used, in fact they have a hole where the mandrel is placed. In this case SPR or a specially designed PR (an example is shown in Fig. 3) has to be used.

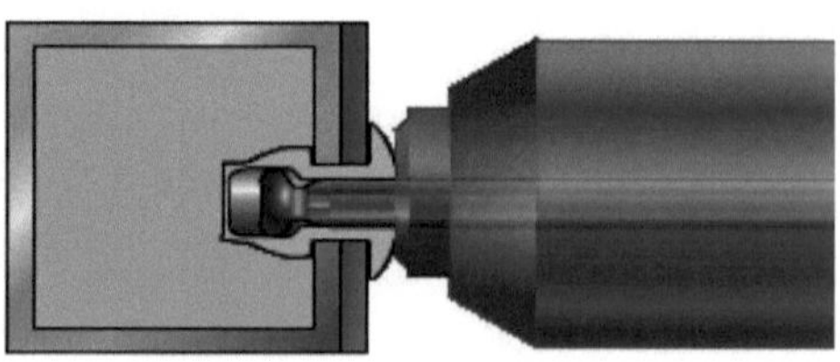

Fig. 3 Example of rivets for airtight applications (http://www.far.bo.it)

In this manner, a complete sealed rivet bonded joint can be obtained: this can be useful in nautical application and in the building of low-pressure tanks.

Except for a particular manufacturing procedure that will be discussed in the next paragraph, it can be said that the requirements dealing with the mechanical performance of rivet bonded joints are the only relevant requirements that have to be taken into account for the adhesive selection. This is a clearly visible benefit in comparison with other kind of hybrid joints, especially the weld bonded joints ones, where the adhesive must be properly formulated in order to permit the welding process, without an excessive burn-out or damage in general.

3 Manufacture

The manufacture of a hybrid joint consists essentially of three phases, two regarding the adhesive (adhesive deposition and adhesive polymerization) and one regarding the rivet (rivet setting): by changing the order of this three phases, different manufacturing techniques can be obtained. Weld bonded joints can be produced using a weld-through or a flow-in techniques [6]. The first is the most severe for the adhesive but is the most popular industrially, because of its rapid application in comparison with the second one which does not produce any adhesive burning, but needs time for the adhesive infiltration between the welded parts. Concerning rivet bonded joints, three manufacturing techniques can be ideally defined:

- Flow-in: the rivet is set, then the adhesive is placed and cured. In this case, a low viscosity adhesive is needed, and similarly to the case of weld bonding a long time is necessary to allow the infiltration of the adhesive between the parts by capillarity. This manufacturing technique can only be useful with the aim to increase the strength of an existing riveted joint.
- Rivet-through uncured adhesive: this procedure is essentially identical with respect to that of weld bonded joints. The main difference is that the rivet setting does not bring any adhesive degradation by heat. This is certainly the most used technique because it allows to obtain good mechanical properties (provided by the adhesive) with an easy and quick manufacturing (provided by the rivet) without any equipment for holding the parts during the adhesive polymerization.
- Rivet-through cured adhesive: the last procedure consists in the application of the rivet after the adhesive polymerization. Without any doubt, this is the less useful way to manufacture hybrid joints. In fact, all the drawbacks produced by the adhesive polymerization are similar to those shown by a simple bonded joint. The only purpose which make useful this procedure is to repair or increase the strength of pre-existing bonded joints.

From the manufacturing point of view, an interesting and comprehensive work was done by Haraga et al. [9]. This work deals with the advantages brought in by the use of pop rivet bonding. In particular, the rivet bonded joints are compared from the mechanical and economical point of view with resistance spot welded joints in case of panels assembly.

In order to understand the benefit of rivet bonded joints, their strength and manufacturing processes were compared with those of simple joints, like arc welded and simple riveted. Concerning the strength, it appears that the joints produced by the combined use of adhesive and rivets have almost the same strength as arc welding. Moreover, the strength of the hybrid joints (adhesive + rivets) is rather similar to that of simple adhesively bonded joints. This seems to be because the strength of the rivet itself is much smaller than that of the adhesive, and the force is not transmitted to the rivet since it is located in the middle of the overlap. The use of rivet bonding techniques allows to noticeably simplify the manufacturing process in comparison with the traditional arc welding, since reshaping after the distortion caused by welding and puttying to improve the aesthetic properties become unnecessary. Moreover the painting process can also be omitted when pre-coated steel is used in conjunction with rivet bonding technique.

Hybrid joining leads therefore to a significant reduction of work time (until 40% in comparison to arc welding) and costs (until 30% in comparison to arc welding), and moreover a greater reduction can be obtained if pre-painted steel is used in conjunction with rivet bonding (pre-painted steel cannot be used with arc welding).

Whit the environment in mind the energy consumption in the assembly process was also analyzed. Combined joining allows a reduction of electrical power consumption up to 84% with respect to welding.

It can be therefore stated that rivet bonded joints are cheap from the manufacturing point of view, and at the same time they give mechanical performances at least equal to the performance of traditional joining technique like resistance arc and spot welding.

4 Strength and Durability

4.1 Literature Analysis

As previously discussed, the strength and the durability of rivet bonded joints depend on what kind of adhesive is used. A first thought about the strength of a rivet bonded joint could be: "*in the same joint the rivet and the adhesive coexist: therefore the strength of the rivet bonded joint will be the sum of the strength of the two simple joints*". Unfortunately this sentence is wrong. If the joint designer is smart enough, a rivet bonded joint with a strength higher than those of simple joints (but however lower than their sum) can be obtained. But sometimes, if the joint is designed in the wrong way, the performances of the hybrid joint can also be lower than those of the simple joints.

A prediction of the strength of rivet bonded joints was extensively discussed by Gomez et al. [7]: in particular a mechanical model combining springs and dampers

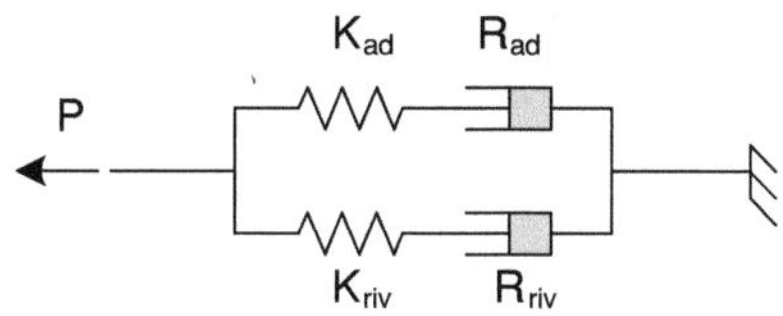

Fig. 4 Schematic representation of the model proposed by Gomez et al. [7]

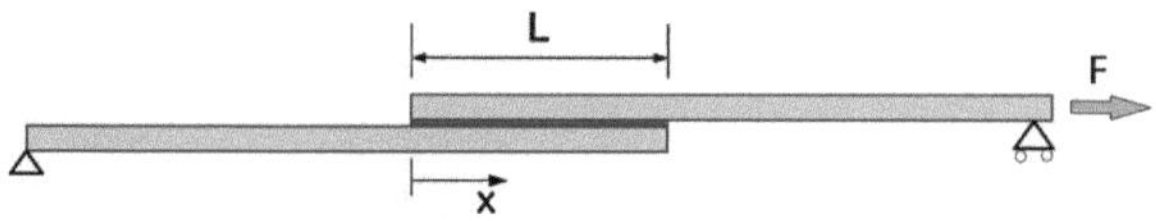

Fig. 5 Example of geometry of a single lap bonded joint

for the representation of the behaviour of a single lap hybrid joint was developed (Fig. 4). The model is divided in two branches: one representing the adhesive and the other one representing the rivet. Each branch is made by a spring (that represents the elastic behaviour) and a damper (that represents the inelastic behaviour). This model assumes that the two parts (rivet and adhesive) work simultaneously, therefore the most rigid component will carry most part of the load, and it will be likely the first to fail, while the most compliant and ductile part will carry the load until final fracture. This model well represents the behaviour of a rivet bonded joint when the two joining techniques cooperate and they do it in the right way. With the aim to understand when an hybrid joint gives better mechanical properties than a simple joint, the joint geometry has to be analyzed.

The shear stresses in the adhesive layer of a single lap bonded joint are generated by the differential deformation in the adherends [20]. Taking for example the geometry shown in Fig. 5, where L represents the overlap length, the shear stresses generated by the load P (assumed to be 100 N by mm of joint width) normalized with respect to the average shear stress along the overlap length are shown in Fig. 6 for three different values of L, as a function of the normalized coordinate x/L.

It can be seen that the increase of the overlap length leads to a progressive concentration of the stresses at the ends, and at the same time a progressive

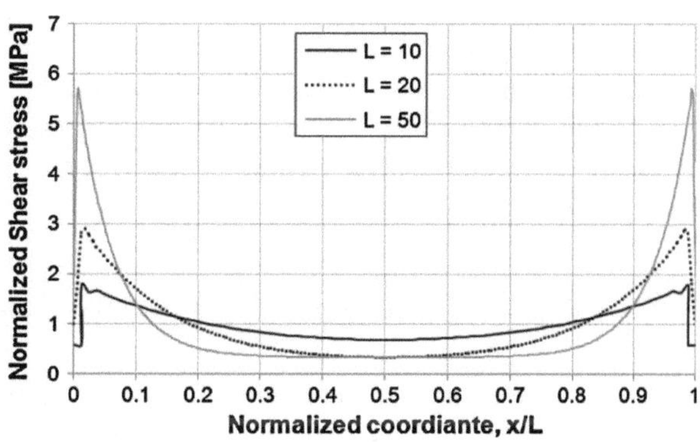

Fig. 6 Normalized shear stress versus normalized coordinate, for three different overlap length (L) for a single lap bonded joint

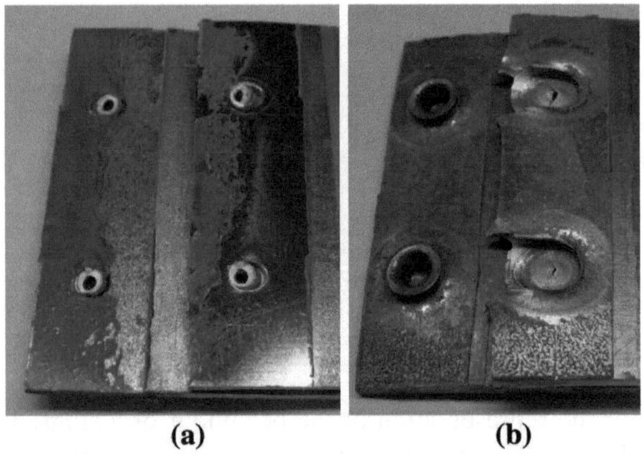

Fig. 7 Fracture surfaces of **a** hybrid PR and **b** hybrid SPR joints

unloading of the zone in the middle of the overlap. This means that, in the middle of the overlap, the differential deformation is really low.

In the case of rivet bonded joints, the rivet is usually placed in the middle of the overlap. Although the rivet generate a stiffness that locally is higher than that generated by the adhesive (especially in the case of SPR joints), in the case of "long" bonded joints the differential deformations in the middle of the overlap are low and therefore the rivet contributes to the load bearing only in a negligible manner. This leads to the assumption that in a structural application, especially in the case of thin and stiff adhesives and long overlap lengths, the strength of the rivet bonded joint depends on the adhesive.

On the contrary, when the overlap is short, the mechanical fastening contributes significantly to the load bearing. Sometimes, the fastened dot brings in a reduction of bonding area in the zone where the adhesive participates to the load bearing. If this reduction is not compensated by the strength of the rivet, the hybrid joints can result in a joint weaker than the simple bonded joint. This phenomenon is not so pronounced in case of pop rivet bonded joints, where the reduction of area is relatively low, but it is significant for self piercing riveted joints, as for example shown in Fig. 7 [14].

It can be seen that in rivet bonded joints (Fig. 7a) the unbonded area is essentially equal to the transverse section of the rivet, while in case of self piercing riveted joints (Fig. 7b) the contact pressure generated during the forming process spread the adhesive away, resulting in a vast unbounded area. In a recent study [13] it was demonstrated that this reduction of bonding area leads to a hybrid joint strength lower than that of a simple bonded joint.

4.2 Strength of PR and SPR Adhesive Joints

Recently, an exhaustive experimental campaign addressed to the evaluation of the strength of hybrid PR and SPR bonded joints in comparison with those of simple

Fig. 8 Specimens with one or two rivets, respectively, representing a pitch of 60 mm (**a**) and 30 mm (**b**)

riveted and simple bonded joints for different environmental and geometrical conditions was done [14]. In particular, the performances were evaluated for different temperature levels, from −30°C up to 90°C, with or without an ageing treatment, for different adherend thickness and for different pitches between the rivet dots. The joint geometry is shown in Fig. 8. Two plates of galvanized steel (S 275) and aluminum alloy (AA 5052) with dimension 100 mm × 60 mm were joined together with an overlap length of 15 mm. This value was chosen taking into account the strength of the substrates in order to avoid the failure of the aluminum plates away from the overlap. The pitch of the rivet was evaluated maintaining the same joint geometry and changing the number of rivets: in order to represent different value of pitch using a joint width of 60 mm, a single rivet was placed to represent a pitch of 60 mm, while two rivets were placed to represent a pitch of 30 mm, as shown in Fig. 8.

The ageing was done following a German standard (VDA 621-415) typically used in the automotive field to test the resistance to corrosion.

The factors evaluated and their levels are summarized in Table 1. The analysis was done using a Design of Experiments (DoE) methodology, and in particular a ¼ factorial plan, with 3 repetitions for each defined treatment, was done. Since a factorial analysis was adopted, the temperature factor was split in two two-level factors as indicated in Table 2. The factor Temp 1 indicates a large

Table 1 Factors evaluated and their levels in the experimental campaign

Symbol	Factors	Level (−)	Level (+)
A	Thickness (mm)	1.5	2.5
B	Pitch (mm)	30	60
C	Ageing	No	Yes
T	Temperature (°C)	−30/23/60/90	

Table 2 Definition of temperature levels

T	D-Temp 1	E-Temp 2
−30	−	−
23	−	+
60	+	−
90	+	+

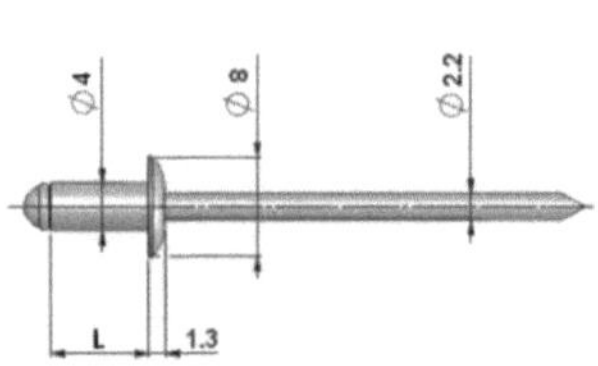

Sheet thickness [mm]	L [mm]
1.5	8
2.5	10

Fig. 9 Dimension of the rivet (http://www.far.bo.it)

Table 3 Self Piercing Rivet joining parameters as a function of plate thickness (http://www.bollhoff.it)

	1.5 mm thickness	2.5 mm thickness
Machine Size	MBF	MPF
Rivet	R 5 × 7	K 5 × 7

variation of temperature, while the factor Temp 2 indicates a small variation of temperature.

The pop rivets used here were made of aluminum alloy (AlMg 3.5) and the geometry is schematically shown in Fig. 9. Figure 9 also shows a small table indicating the length of the rivet as a function of the thickness of the plates.

Concerning the self piercing rivets, the machine and rivet size were changed depending on the sheet thickness, as shown in Table 3. For both the plates thicknesses, the same rivet diameter (5) and the same rivet length were used (7) and only the shape was different.

For the experimental campaign, the Henkel Terokal 5077 adhesive was used: this adhesive is a pasty, heat curing one-part adhesive based on an epoxy resin. One of its main feature is to guarantee a good adhesion also to oily metal surfaces. This property makes this adhesive very attractive for several industrial fields since it allows to reduce the surface preparation that sometimes could be expensive and difficult in the manufacturing process [2, 19, 21]. Another feature of this adhesive is the possibility to be combined with other joining techniques like for example resistance spot welding. Since this adhesive was developed to guarantee high peel and impact peel resistance (load condition typically met in case of torsional and crash impact), it is particularly useful for bonding in the automotive industry.

Being a paste adhesive, its deposition at room temperature is rather troublesome; therefore, in order to reduce the adhesive viscosity, prior to deposition, the adhesive and the adherends were heated up until 50°C. This also improve the adhesion performance since the reduction of the viscosity allows the adhesive to better penetrate in the surface roughness. Immediately after the deposition of the adhesive, the parts were joined in order to avoid an increasing of the viscosity of the adhesive and a consequent reduction of the adhesion performance. The adhesive was then cured in an oven at a temperature of 160°C for 30 min.

In order to understand the influence of the temperature over the mechanical behaviour of the adhesive, tensile tests on bulk adhesive specimens were performed following the ASTM standard D 638 (Type I geometry was selected).

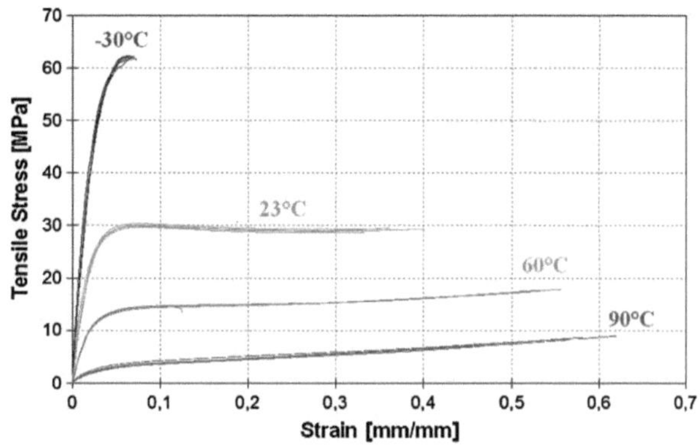

Fig. 10 Stress–strain plots of the bulk adhesive tests

The results of the tensile tests are shown in terms of tensile stress versus strain plots in Fig. 10. It can be seen that the adhesive performance is deeply affected by the temperature. In particular, taking as reference the strength at room temperature, at −30°C the adhesive becomes much stiffer, and it can withstand a tensile stress approximately two times that at room temperature. On the opposite, at high temperature the adhesive shows a viscoelastoplastic behaviour.

The maximum strength is significantly lower than that at room temperature, although the adhesive can withstand an high elongation (until 60%) under this condition. This performance in terms of tensile test can be later compared with the strength shown by the bonded joints for the different temperatures.

Figure 11 shows the load versus displacement behaviour of tests carried out at 90°C: after failure, the elastic elongation is immediately recovered, and from the knowledge of the stress level and of the elastic modulus, the elastic elongation is supposed to be approximately 10%. Specimens were measured few minutes after the test and a residual elongation of 25–27% appears. Therefore an amount of 15–25% of viscous deformation is recovered after the failure (when the material is unloaded).

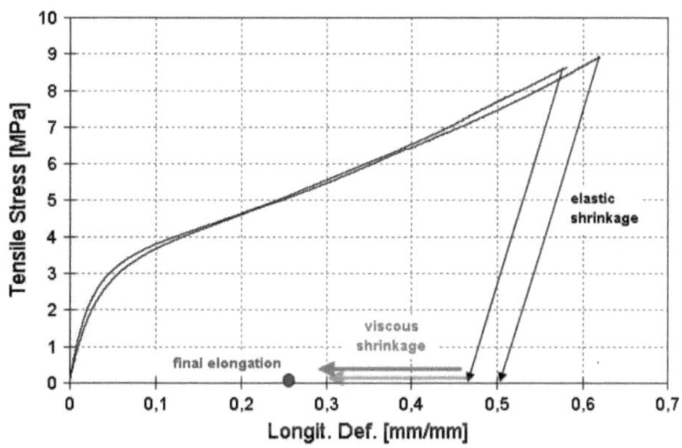

Fig. 11 Load versus displacement plot of tensile test at 90°C

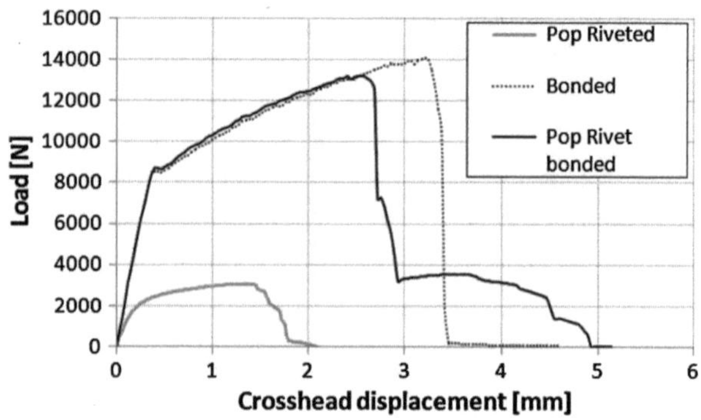

Fig. 12 Comparison of load versus displacement curves of simple bonded, simple pop riveted and hybrid joints

Information about the glass transition temperature (T_g) of the adhesive was not given by the producer, however it can be assumed from the reduction of the Young modulus to be quite higher than 60°C.

Coming back to the performance of rivet bonded, bonded and simple riveted joints, an example of the results obtained with the single lap joint is given in Fig. 12 in case of pop rivets. Comparing the simple pop riveted and simple bonded joints, it can be seen that the latter is the strongest and the stiffest. The riveted joint shows a linear behaviour until 1,500 N, and this is followed by a significant amount of plastic deformation. This plastic deformation is completely due to the deformation of the rivet without any significant plastic deformation of the substrates, as can be shown in Fig. 13, and also demonstrated by the authors in a previous work [15]. The bonded joint shows instead a linear behaviour until 8,500 N. Here, a strong variation of the slope of the curve can be seen, corresponding to the yielding of the aluminum plate. The aluminum adherend begins to accumulate plastic deformation until failure is reached in the adhesive layer.

The hybrid joint resulting from the combination of adhesive and rivet shows a load versus displacement behaviour that is in practice the same as that shown by bonded joint until the maximum load (only a small difference can be noticed in terms of maximum load). Immediately after the maximum load, the bonded joint suddenly fails and loses completely any load carrying capability. On the contrary, the hybrid joint shows a sudden load reduction, but it continues to withstand a load which is similar to the load carried by the rivet.

Fig. 13 Fracture surface of a simple riveted joint

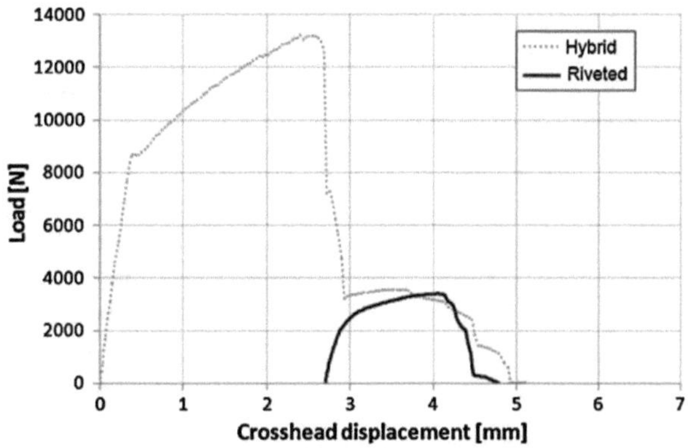

Fig. 14 Comparison of load versus displacement plot of a hybrid pop riveted joint and a simple pop riveted joint shifted of the value of the displacement at the strength of the hybrid riveted joint

The failure mechanism can be better understood looking at Fig. 14. Here the load versus displacement plot of the riveted joint is shifted on the displacement axis of a value of displacement corresponding to the maximum load of the hybrid riveted joint. This curve is compared with that of the hybrid riveted joint. It appears that the behaviour of the riveted joint is similar to that of hybrid joint after the failure of the adhesive.

Therefore, before the adhesive fails, the rivet is only slightly deformed and loaded. In the hybrid joint, the initial plastic deformation is concentrated in the plates and in the adhesive at the end of the overlap, while it is in practice negligible in the middle of the overlap where the rivet is placed. The rivet takes the load only when the adhesive fails.

The same comparison shown in Fig. 12 in the case of pop riveted joints, is shown in Fig. 15 in the case of SPR joints. Although the strength of SPR joints is noticeably higher than that of PR joints, the trend in rather similar. Alike in the previous case, the strength of the adhesive leads to the yielding of the aluminum plates. Some differences can only be noticed concerning the failure of the adhesive layer. In the case of SPR bonded joints, the fracture does not involve immediately the entire bonded region, but the presence of the SPR dot (stronger and stiffer than PR dot) contributes to a partial fracture stop: this can be noticed from the sudden

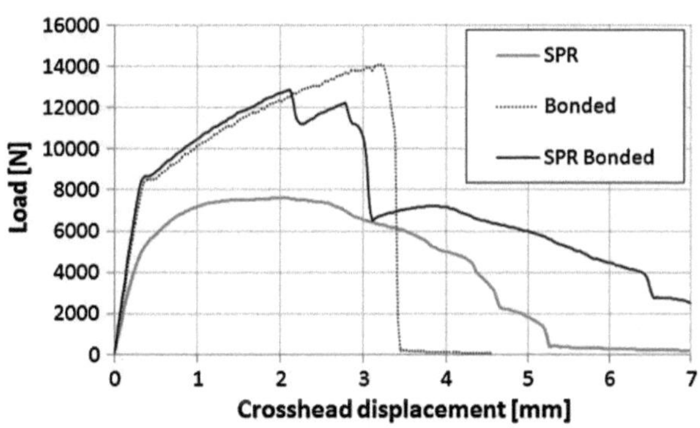

Fig. 15 Comparison of load versus displacement curves of simple bonded, simple self piercing riveted and hybrid joints

	Coefficient estimate [N]		
Factor	Bonded	PR	PR Bonded
Intercept	12334	2328	11470
A - Thickness	504		
B - Pitch		-772	
C - Ageing	-987		-1583
D - Temp 1	-2189		-1481
E - Temp 2	-698		-1254
AC	-423		-928
AD	-447		598
DE	-685		-638

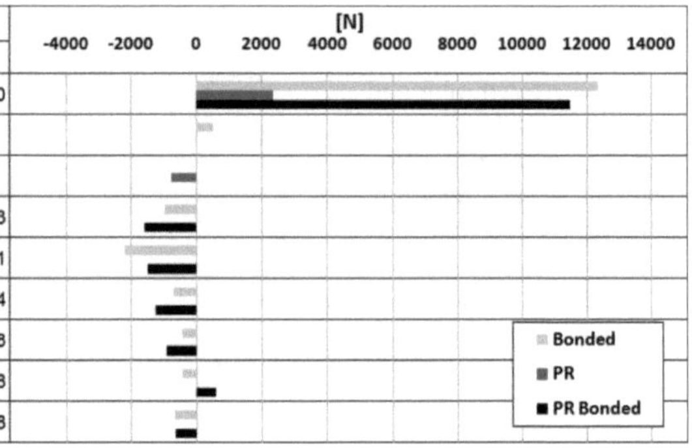

Fig. 16 Average value and influence of factors for simple bonded, pop riveted and pop rivet bonded joints

drop of load close to a displacement of 2 mm (initial failure of adhesive), followed by a further load increase until complete adhesive fracture. Here the SPR dot carries the entire load and the joint behaves as a simple SPR joint until complete fracture.

A better and comprehensive view of the performances of simple and hybrid joints can be seen in Figs. 16 and 17 where the results of the analysis of variance are given in terms of the influence of the analyzed factors together with the average value of the tests.

In particular, Fig. 16 deals with the result of simple bonded and pop riveted joints and their combination. It can be seen that in average, the bonded joints are much stronger than simple riveted joints, and that the hybrid joints show an average strength rather similar to that of simple bonded. Moreover, the strength of simple riveted joints does not depend on the thickness of the plates but is obviously affected by the pitch. This is because the load is sustained simply by the rivet section and the diameter of the rivet does not depend on the plates thickness (in the rivet, only the body length is a function of the plates thickness). Since the loads

	Coefficient estimate [N]		
Factor	Bonded	SPR	SPR Bonded
Intercept	12334	7898	11061
A - Thickness	504	2209	687
B - Pitch		-2666	-688
C - Ageing	-987		-1197
D - Temp 1	-2189		-1749
E - Temp 2	-698		-763
AC	-423		
AD	-447		
DE	-685		-591

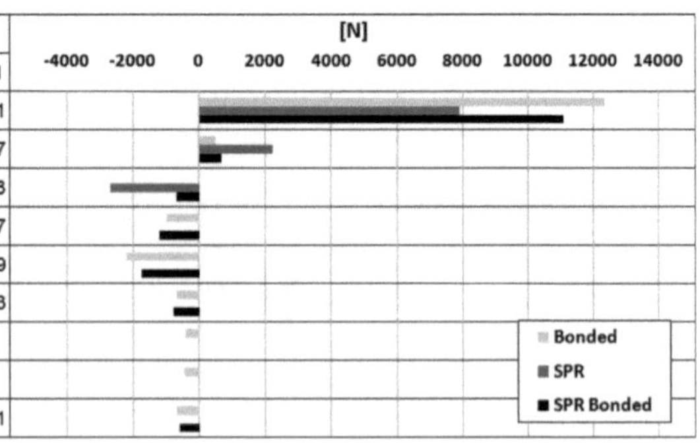

Fig. 17 Average value and influence of factors for simple bonded, self piercing riveted and self piercing rivet bonded joints

reached by the simple riveted joint are relatively low, no bending of the plates is met, therefore the rotation of the overlap is negligible. This leads to a pure shear failure of the rivet that does not depend on the plate thickness. Concerning the pitch, it is obvious that a joint with two rivets will sustain a load that is close to the double of the load sustained by a joint with a single rivet. Concerning the other types of joints, the strength of the simple bonded joint is positively influenced by the plate thickness, and in a more significant manner negatively affected by the temperature and the ageing. Moving to the hybrid joints, it can be seen that the general trend is rather similar to that of the simple bonded joint. Indeed, except for the plate thickness, the relevant factors for the hybrid joints are exactly the same of those of the simple bonded joints.

The result of the analysis of variance for the simple bonded joints, SPR joints and their combination is given in Fig. 17. Again the simple bonded joint is stronger than the SPR one, although this latter is in turn significantly stronger than simple PR. In average it appears that the SPR bonded joints are weaker than PR bonded joints.

Differently from the PR bonded joints, for the SPR bonded joints the pitch assumes a significant influence. This means that in the hybrid SPR bonded joints the mechanical fastening plays a significant role in carrying the load. This fact can be confirmed looking at the influence of the temperature: the strength of the hybrid joints is less affected by the temperature than that of simple bonded joints, and this can be explained by the presence of the mechanical fastening. The same conclusion can be drawn looking at the influence of the thickness: the coefficient estimate for the hybrid joints is quite higher than that of simple bonded joints, and this is because of the presence of the SPR dot (for the simple SPR joints the thickness influence in rather high).

Using the analysis of variance (ANOVA), the strength for different geometries and environmental conditions can be predicted. Starting with the geometrical factors, Fig. 18 shows a comparison between the strength of simple bonded, simple riveted (PR) and hybrid joints (PR + adhesive) tested at room temperature

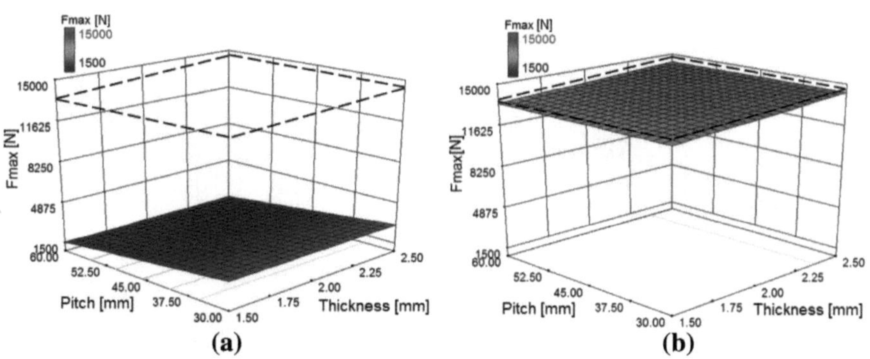

Fig. 18 Strength prediction of simple PR (**a**) and hybrid PR (**b**) joints, in comparison with that of simple bonded joints (*dotted lines*)

as a function of the pitch and the plates thickness. It appears immediately that bonded and hybrid joints are much stronger than pop riveted joints. Moreover, it can also be noticed that the surfaces of hybrid and bonded joints are relatively flat: this means that the variation of geometrical factors slightly affects the strength. However, it can be seen that the bonded joints strength is positively affected by the thickness: in fact the higher is the thickness, the lower are the peel and the shear stress concentration at the end of the overlap, while the pitch is a parameter that cannot be evaluated for simple bonded joints. The hybrid joints show a behaviour that is rather similar to that shown by simple bonded joints: therefore, a small, but significant influence of the thickness and a negligible influence of the rivet pitch is found. This is confirmed also for different geometries: in the case of structural adhesives, the strength of hybrid pop rivet bonded joints is predominately given by the adhesive.

Figure 19 refers to SPR joints. In the same manner of PR joints, the pitch is relevant because the higher is the number of rivets per unit length, the higher is the joint strength, but here the thickness becomes also relevant. For this kind of joint, the failure corresponds to the failure of the adherends and therefore the higher is the thickness the higher is the strength (further details of the failure mode will be later given in the Sect. 5). Moving to the hybrid joints strength, it can be noticed that their strength is lower than that of simple bonded joints. This can be explained by the significant reduction of the bonded area in SPR bonded joints, in comparison with PR joints. However, it can be seen that the hybrid joints strength is mainly given by the adhesive (the hybrid joints strength values are closer to the bonded joints strength values). Differently from PR bonded joints, here a significant influence of the SPR dot pitch can be found. This means that, although the strength of SPR bonded joints is mainly given by the adhesive, the rivet also contributes to carry the load. There is therefore a kind of balance: the presence of the SPR dot reduces the bonded area hence the adhesive bond performance, but at the same time the SPR dots are strong and stiff (on the contrary of PR) and therefore they contribute to the strength.

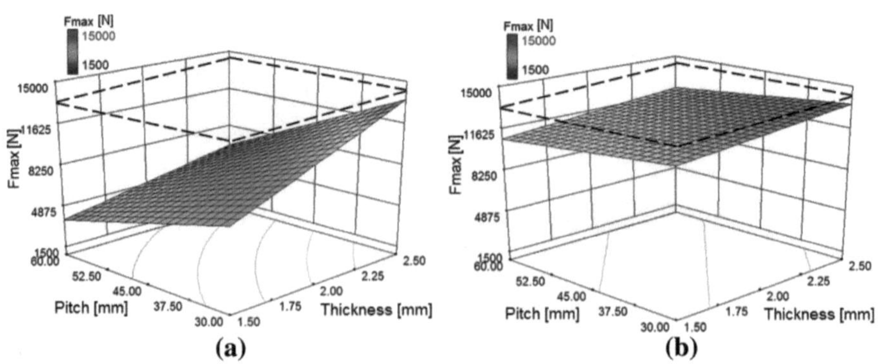

Fig. 19 Strength prediction of simple SPR (**a**) and hybrid SPR (**b**) joints, in comparison with that of simple bonded joints (*dotted lines*)

Table 4 Tensile stress of bulk adhesive specimens and predicted strength of bonded joint as a function of temperature (the variation refers to the strength at room temperature)

Temperature	Bulk adhesive tensile strength (MPa) (%)	Bonded joint strength (N) (%)
−30	62.1 (+107)	13,379 (−10)
23	30	14,960
60	16.5 (−45)	12,605 (−15)
90	8.8 (−71)	8,605 (−42)

So far, the behavior at room temperature has been discussed, but one of the declared benefits of hybrid joints is to compensate the loss of performance of bonded joints at high temperature. It was previously discussed how the adhesive performances decrease as the temperature increases in the case of the bulk adhesive and the same behaviour can be expected for the bonded joint. Taking as reference the room temperature performances, Table 4 shows the difference in terms of bulk tensile strength and lap shear strength as a function of temperature.

It can be seen that at low temperature (−30°C), the tensile strength increases significantly, but at the same time the strength of bonded joints decreases. At high temperature, the loss of strength of bonded joints is noticeably lower than the loss of adhesive tensile stress. This different behaviour can be explained looking at the properties of the adhesive. In fact, at low temperature, the adhesive is stronger than at room temperature, but at the same time it is also much stiffer and brittle. Therefore, the stress concentration at the end of the overlap is more pronounced and the failure occurs at a lower level of average shear stress. At high temperature, the more ductile behaviour of the adhesive allows a better stress distribution and therefore the loss of strength in bonded joints is lower that the loss of bulk adhesive tensile strength.

Concerning hybrid joints, their strength as a function of temperature is shown in Fig. 20 for PR bonded joints and in Fig. 21 for SPR bonded joints. It seems that hybrid and simple bonded joints have similar strengths at room temperature and 60°C, while differences emerge at −30° and 90°C. At high temperature, the

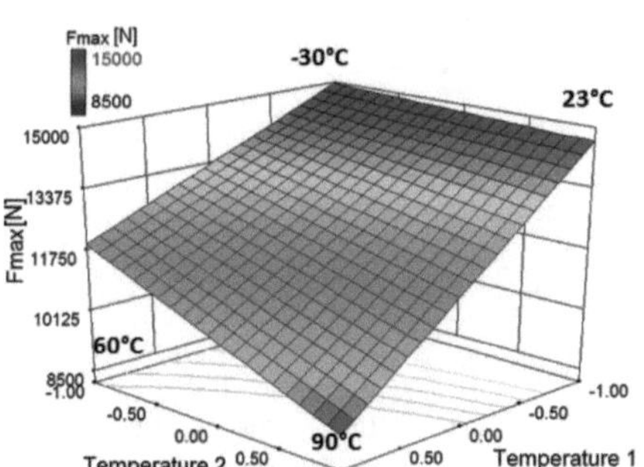

Fig. 20 Strength prediction of PR bonded (**a**) and simple bonded (**b**) joints as a function of temperature (plates thickness 2.5 mm)

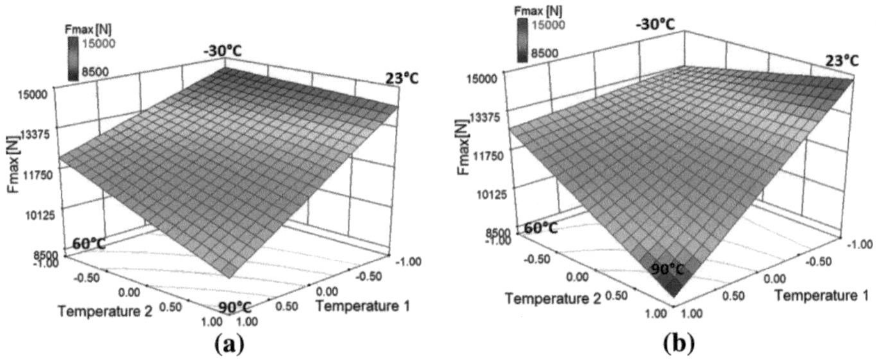

Fig. 21 Strength prediction of SPR bonded (**a**) and simple bonded (**b**) joints as a function of temperature (plates thickness 2.5 mm)

adhesive becomes "soft" and its load bearing capability is reduced. The presence of the rivet allows to reduce the strength reduction, and in particular it allows to ensure a minimum level of load in case of complete adhesive failure. For the materials and geometry studied (adhesive rather stiff, relatively weak rivets, large pitch between rivets), the contribution of the rivet is not so evident, however the strength goes from 8,570 N for simple bonded joints to 9,260 N for hybrid PR joints. A significant improvement of the performance can also be noticed at low temperature. In this case, the presence of the rivet reduces slightly the peel stress within the joint and, therefore, the peel stress concentration that are critical when the adhesive is brittle.

In the case of SPR bonded joints (Fig. 21), the strength still shows the loss of performance alike PR bonded joints, but the reduction is quite lower since the SPR is stronger than the PR.

Figure 22 shows the influence of the ageing for simple bonded and SPR bonded joints. In both cases the ageing affects negatively the joint strength and in

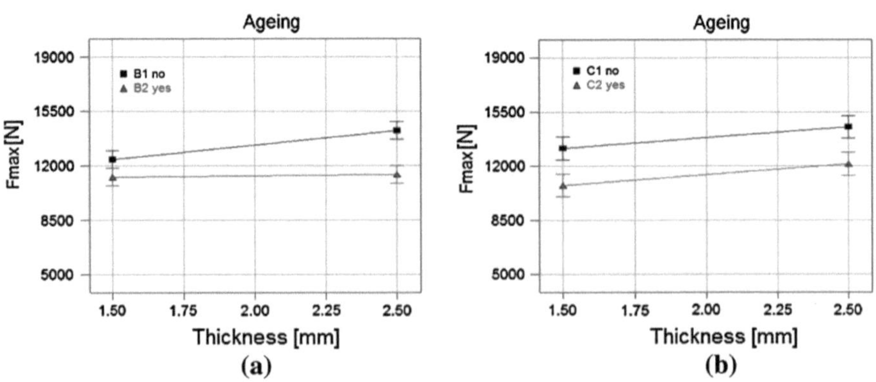

Fig. 22 Strength prediction of bonded (**a**) and SPR bonded (**b**) joints as a function of plate thickness with and without ageing

particular in simple bonded joints the influence of the ageing increases when the plate thickness increases. On the opposite, it can be noticed that for hybrid joints the combined effect of thickness and ageing can be neglected, since the two lines are almost parallel.

5 Types of Failures

5.1 Fractographic Examination

As widely shown in literature and practical experience, when weak materials but also composite laminates are riveted together failure is likely to occur in the substrates [3, 12]. The most typical kinds of failure are schematically shown in Fig. 23.

In this case, the failure occurs in the adherends because the load is transferred locally. This leads to the failure of the joint at a level of load that can be significantly lower than the strength of the mechanical fastening and of the adherends. Using the rivet bonded joints, the adhesive allows to transfer the load uniformly along the bonded width and, therefore, the possible kinds of failure are the failure of the adhesive and the rivet, or the failure of the entire section of the adherends. This better stress distribution permits to reach the critical condition at higher loads, and therefore the materials can be used at the best of their performances.

In case the of metal adherends, as previously shown, the failure of the hybrid riveted joints initially involves the failure of the adhesive (eventually preceded by the yielding of the adherends), and later the failure of the riveted dot. Since the two failure mechanisms occur in sequence, they are independent with respect to each other. In the rivet bonded joint, the adhesive fails as in a simple bonded joint, and the riveted dot fails as in a simple riveted joint.

Looking at Fig. 24, it can be seen that the simple bonded joint fails with a mixed adhesive fracture that propagates at the interfaces.

On the other hand, the pop rivet joint fails due to the shear failure of the rivet (Fig. 25): in particular the body of the rivet initially ovalizes, then it is sheared by the plates which act as a blade.

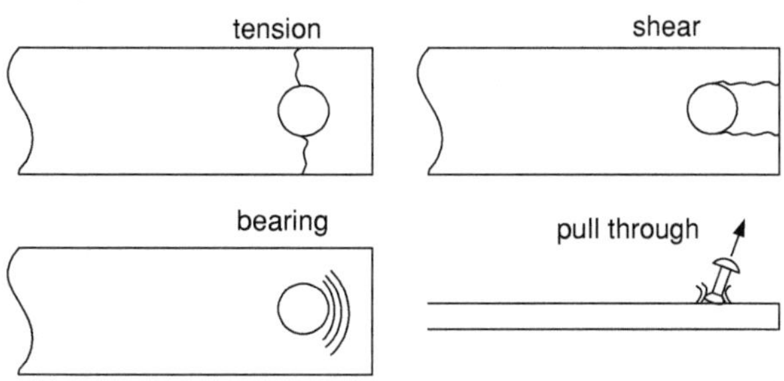

Fig. 23 Typical fracture modes for fiber laminated riveted joints [3]

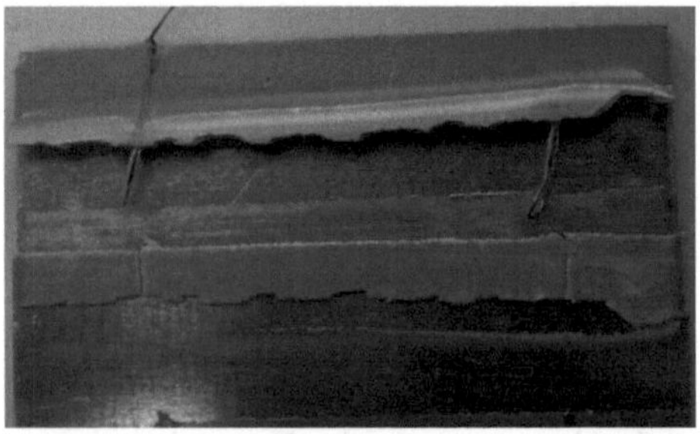

Fig. 24 Fracture surface of a simple bonded joint (the visible wires are those used for the control of adhesive thickness)

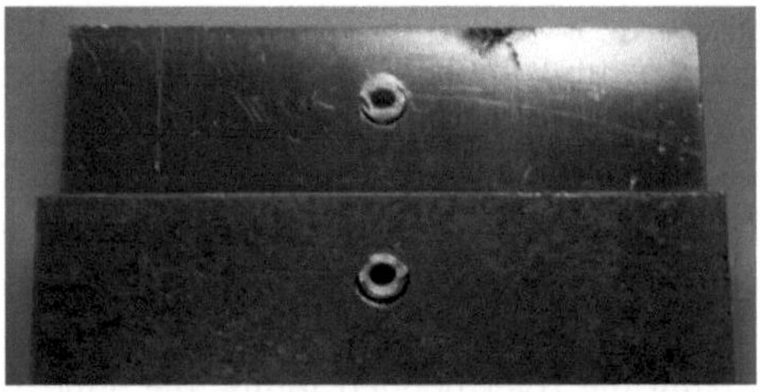

Fig. 25 Fracture surface of a simple PR joint

Fig. 26 Fracture of a SPR joint

A completely different failure is shown by the SPR joints (Fig. 26). Here the rivet is made of high strength steel, therefore the failure occurs due to shearing of the plates. In this way, the strength potential of this joining technique in not completely exploited, even though this kind of failure involves a noticeable volume of material leading to a high energy absorption at failure.

Coming to the hybrid joints, an example of fracture surface is given in Fig. 27. Some differences can be noticed in comparison with simple bonded joints regarding the adhesive fracture. In bonded joints, the adhesive thickness was controlled and set to 0.25 mm by using wires. On the other hand, in riveted joints, the adhesive thickness was not controlled in order to avoid the introduction of an artificial thickness variation: if during the rivet set up wires or spheres were used, a complete lack of control of the adhesive thickness would result as shown schematically in Fig. 28. Therefore, in hybrid joints, the adhesive thickness is quite smaller than in simple bonded joints, resulting in a more cohesive fracture surface.

Fig. 27 Fracture of a PR bonded (**a**) and SPR bonded joint (**b**)

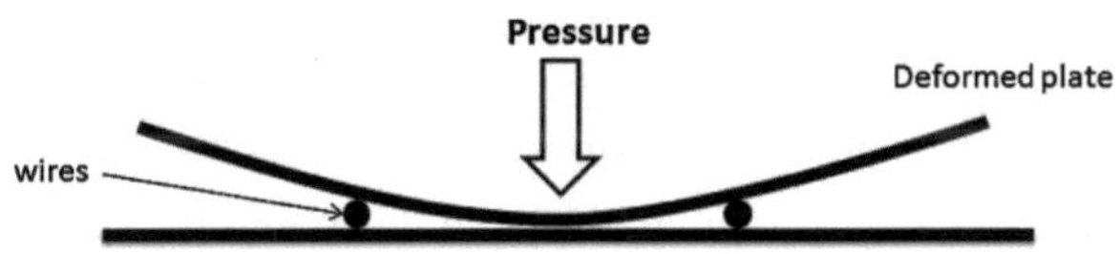

Fig. 28 Gap between the plates in the case of the presence of wires during the rivet bonded joint manufacturing

5.2 Simulation of Damage and Failure

The failure of pop rivet bonded joints was simulated in detail by the authors using damage models for both the adhesive and the rivet [15]. In particular, a cohesive zone model was used for the adhesive, while for the rivet (again made with aluminum alloy AlMg 3.5) the Ductile Damage (DD) model or the Gurson-Tveergard-Needleman (GTN) models were used alternatively. The main aim of the work was to identify the damage models parameters for the rivet and for the adhesive using simple bonded and riveted joints, respectively, and later on use these parameters for the simulation of rivet bonded joints. This work is not intended to give details about the damage models, but it is focused on the hybrid joint, therefore only a brief introduction to the models is given while more details are given in the [15]. The cohesive model is a micromechanical model widely used for the simulation of fracture propagation in bonded joints or composite materials. In particular, it combines the stress and the opening at the crack tip in an energetic criterion. The parameters for this model are usually tuned by comparison with experimental tests and then used for the simulation. Assuming a trapezoidal law in terms of stress-opening behaviour at the crack tip, as shown in Fig. 29, the set of parameters is represented by:

- The fracture energy Γ, i.e. the area underlying the curve.
- The maximum stress σ_{max},
- The shape factors λ_1 and λ_2, which are the values of the displacement corresponding to the vertex normalized with respect to the maximum opening.

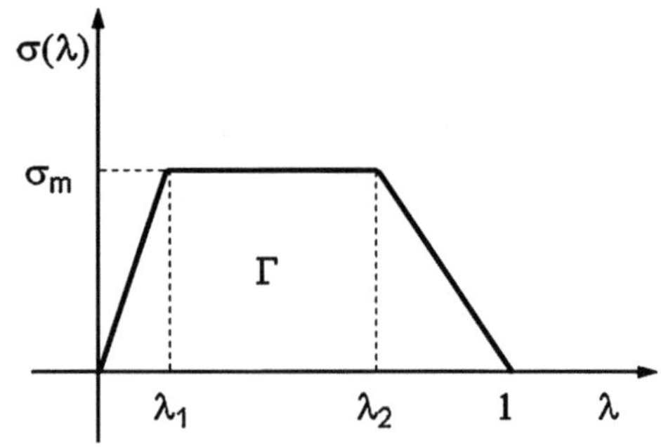

Fig. 29 Trapezoidal stress–displacement relation of the cohesive law used for the adhesive

The ductile damage model [10, 11] is widely used for the fracture simulation of ductile materials failure: using appropriate relations it links the stress with the level of elongation and the stress triaxiality. The damage is represented as the degradation of the elasto-plastic behaviour, after the threshold plastic deformation for damage initiation $\left(\bar{\varepsilon}_0^{pl}\right)$ is attained. Beyond this point, the stress in the material is computed as

$$\sigma = (1 - D)\bar{\sigma} \tag{1}$$

where D is the scalar damage (Fig. 30).

The damage variable D is described as a function of the plastic displacement $\bar{u}^{pl}$ (obtained from the plastic strain by means of a characteristic length of the element L) and is related to the fracture energy, that represent the integral of the stress over the plastic displacement after the damage onset. This trend is schematically shown in Fig. 31, where an exponential law was adopted in order to reduce strain localization and instability of the analysis.

The GTN model is also used for the simulation of the fracture behaviour of ductile metals. The original model [8] describes the material yielding surfaces (defined by the yielding stress σ_0) of a sphere with a void inside, in terms of stress state (in terms of average stress σ_m and equivalent stress σ_{eq}) and of the void volume fraction (f) using the equation.

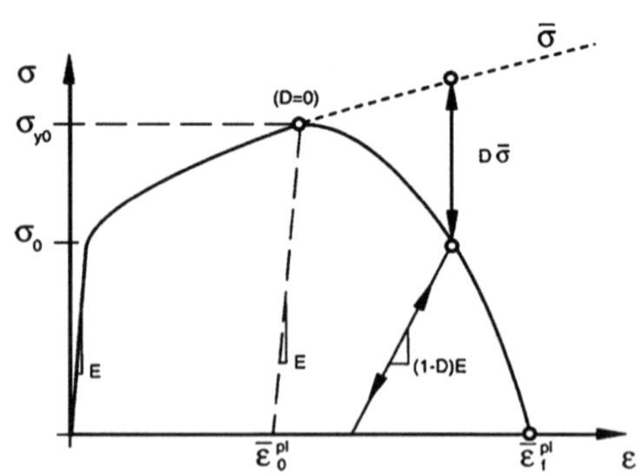

Fig. 30 Representation according to the ductile damage (DD) model of the stress–strain behaviour of a ductile material

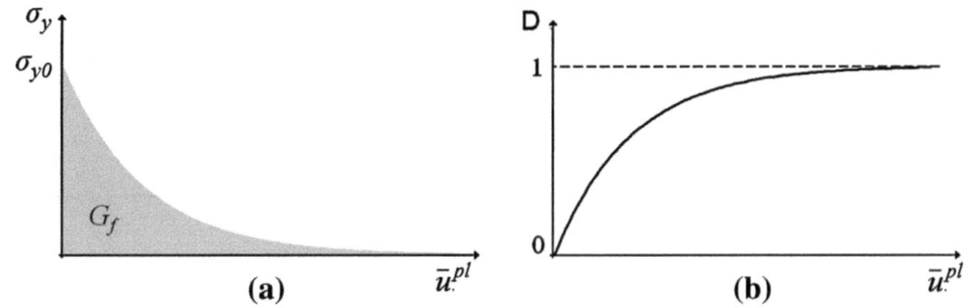

Fig. 31 Yield stress σ_y (**a**) and damage parameters D (**b**) as a function of the plastic displacement $\bar{u}^{pl}$

$$\frac{\sigma_{eq}^2}{\sigma_0^2} + 2q_1 f \cosh\left(\frac{3}{2}q_2\frac{\sigma_m}{\sigma_0}\right) - 1 - (q_3 f)^2 = 0 \tag{2}$$

where q_1, q_2 and q_3 are parameters used to better fit the experimental results. The voids volume growth ($\dot{f}_{growth}$) is obtained from the conservation of mass by means of the plastic flow rate $\dot{\varepsilon}_{kk}^{pl}$,

$$\dot{f}_{growth} = (1 - f)\dot{\varepsilon}_{kk}^{pl}. \tag{3}$$

In this manner, the model takes only into account the growth of the voids. The model was successfully improved by [5, 16–18] in order to take into account the nucleation, growth and coalescence of voids. More details of this model can be found in Chap. 5 of this book regarding the simulation of clinch bonded joints.

The experimental tests for the identification of damage parameters were carried out on adherends and adhesive different from those of the experimental phase described before, taking into account only pop riveted joints and steel sheet plates with the geometry shown in Fig. 32, and the rivet used is that shown in Fig. 9.

The adhesive used for the hybrid joints for tuning models is the Hysol 9466, a two component structural epoxy resin supplied by Henkel. This adhesive has a curing cycle of 24 h at room temperature.

For the characterization of the cohesive zone model, mode I fracture tests on Double Cantilever Beam (DCB) joints were used and the set of parameters identified is shown in Table 5. As a first approximation, a mode mixity

Fig. 32 Geometry of the riveted joint (dimensions in mm) used to tune the parameters of the metal damage models

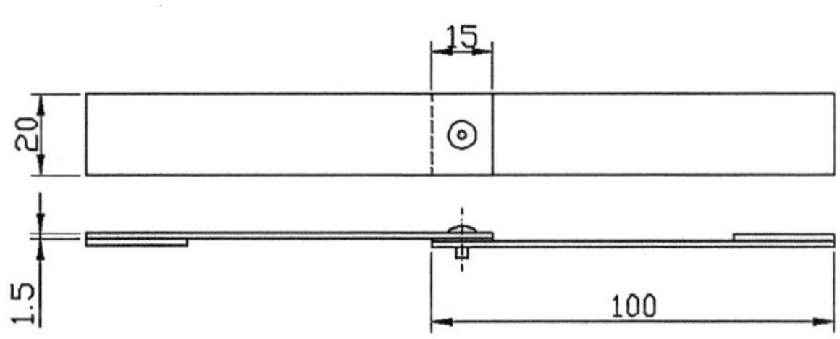

Table 5 CZ parameters for the adhesive Hysol 9466

CZ parameters	
Γ_0 (J/m^2)	725
σ_m (MPa)	60
λ_1	0.2
λ_2	0.5

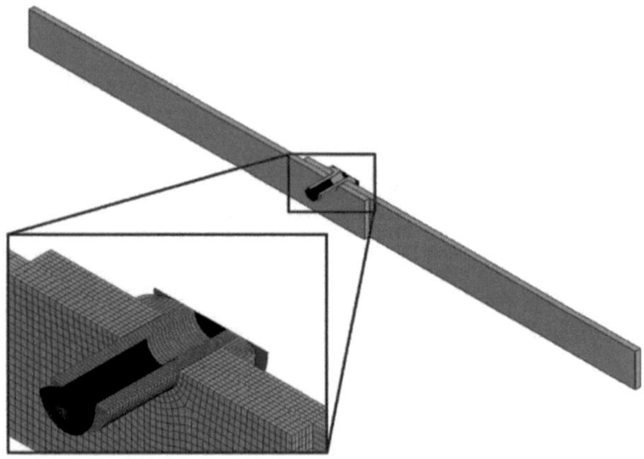

Fig. 33 Finite element model of the riveted joint

independent behavior was assumed for the adhesive (i.e. the mode II parameters were assumed to be equal to the mode I parameters).

Concerning the simple riveted joints, finite element simulations were carried out for the tuning of the ductile damage parameters. The finite element model is shown in Fig. 33. The plates and the rivet were modeled with 3D deformable elements while analytical rigid surfaces were used to represent the rivet mandrel and the reference surface: in order to take into account the residual stresses that occur during the rivet set up, the forming of the rivet was also simulated. More details on this choice are given in [15] while here only the principal results are discussed.

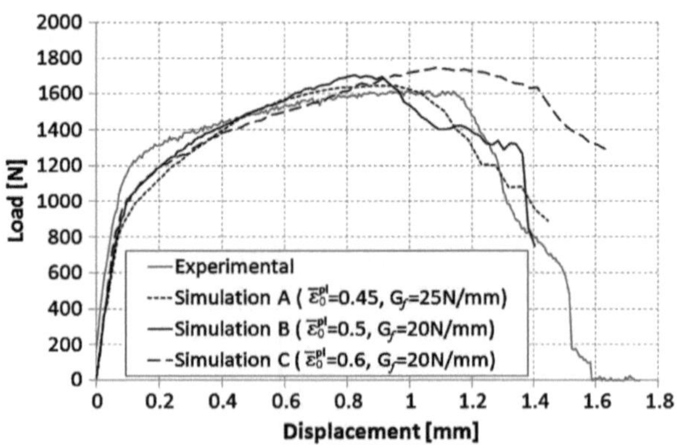

Fig. 34 Comparison between experimental and finite element analysis load–displacement curves of the riveted joint obtained with the DD model

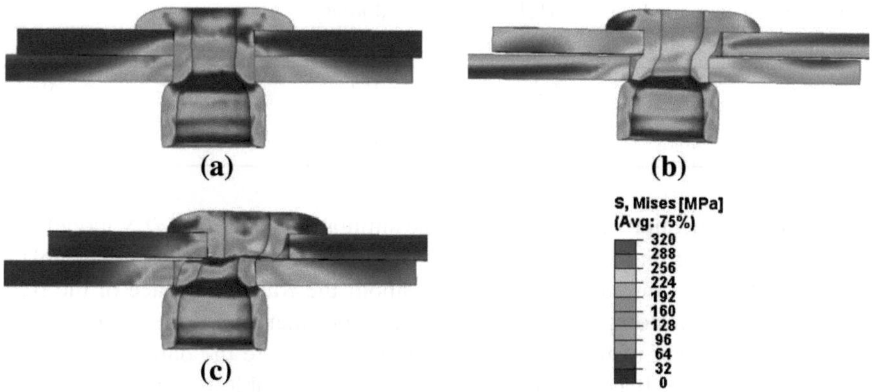

Fig. 35 Simulation of the shear test of the simple riveted joint: **a** residual stress after forming of the rivet; **b** onset of damage; **c** failure

Both the parameters of the DD and the GTN models were identified using an inverse method based on the fitting of the experimental load-elongation of simple riveted joints. Concerning the DD model, different combinations of parameters were tested, and the best results are shown in Fig. 34. The numerical load versus displacement curves shown in Fig. 34 are all in good agreement with the experimental ones. Figure 35 shows some stress contour maps on the deformed shape of simple PR joints. Figure 35a is before any displacement is applied, therefore it shows the residual stresses due to rivet forming. The rivet is then loaded until the onset of damage (Fig. 35b). The damage is greater where the maximum deformation occurs and, therefore, in the green-coloured zone in the middle of the rivet. Figure 35b corresponds to a value of displacement of 0.6–0.7 mm in the graph of Fig. 34. In Fig. 35c, the damage propagates to the entire section and final fracture occurs (blue-coloured elements mean completely damaged elements).

Fig. 36 Comparison between experimental and numerical load–displacement curves of a riveted joint obtained with the GTN model

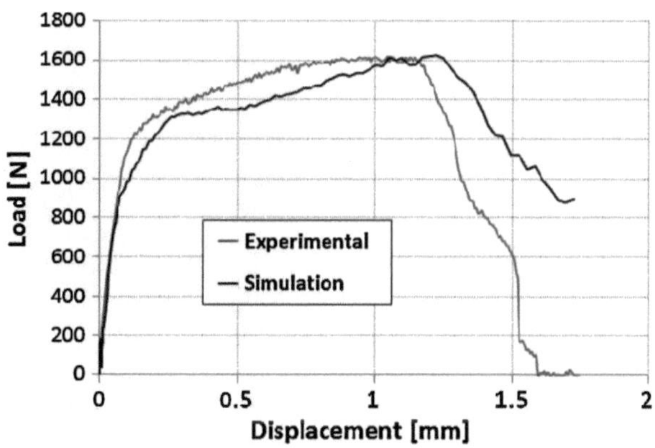

Table 6 Parameters used for the GTN model (riveted joint)

Parameter	Value	Parameter	Value	Parameter	Value
q_1	2	ε_N	0.4	f_F	0.35
q_2	1.1	s_N	0.08	f_c	0.0006
q_3	4	f_N	0.1		

The same procedure was followed for the simulation using the GTN model. The results are shown in Fig. 36 and the set of parameters obtained is given in Table 6.

An interesting comparison can be made about the fracture surface of the rivet. Figure 37 shows a comparison between the experimental fracture surface of a riveted joint and that resulting from the simulation, where the damaged elements have been removed. It is evident that the simulation is able to reproduce closely the fracture surface obtained experimentally. In both experiment and simulation, two patterns can be noticed: in the load direction (position A in Fig. 37), the surface is flat because it is sheared by the plates, while perpendicularly to the load direction (position B in Fig. 37) the surface is rougher since it undergoes a significant deformation before failure.

Failure of the hybrid joints can be therefore simulated by assigning the parameters of the damage models calibrated previously on simple joints, to the different constituents of the hybrid joints.

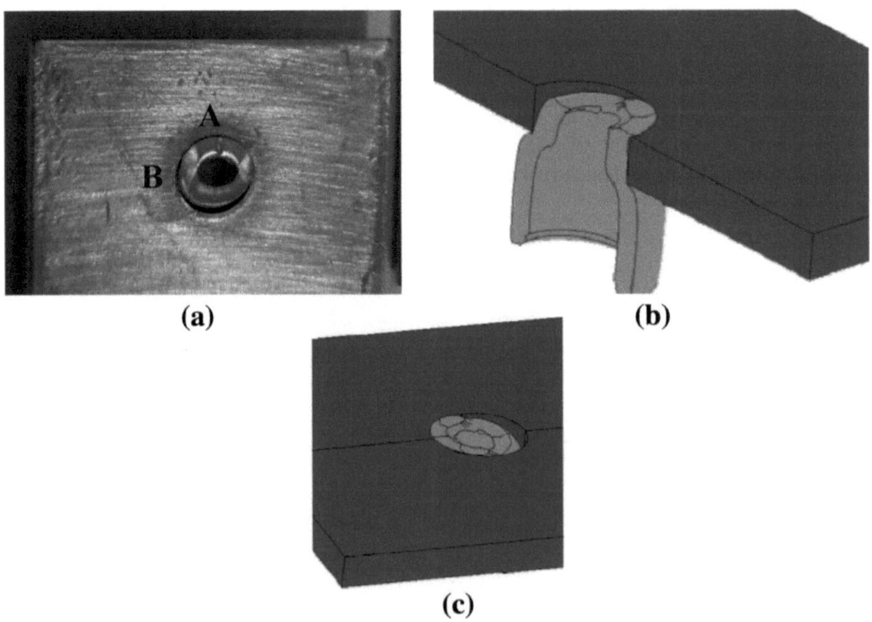

Fig. 37 Comparison between fracture surfaces obtaines experimentally (**a**) and by the FE simulation (**b**, **c**)

Fig. 38 Numerical load versus displacement curves from simulations of the hybrid PR-bonded joint using the DD model and the GTN model

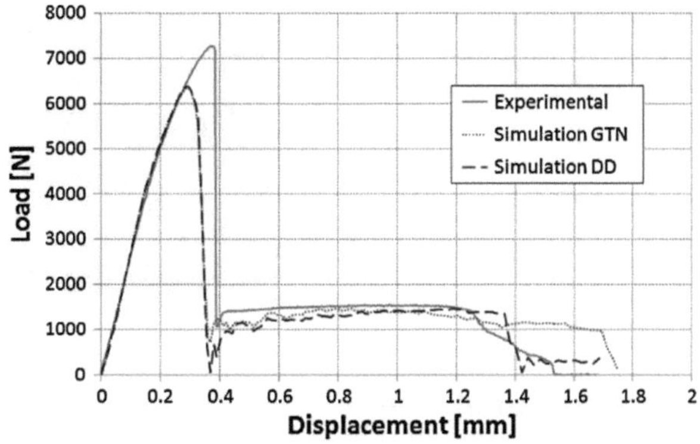

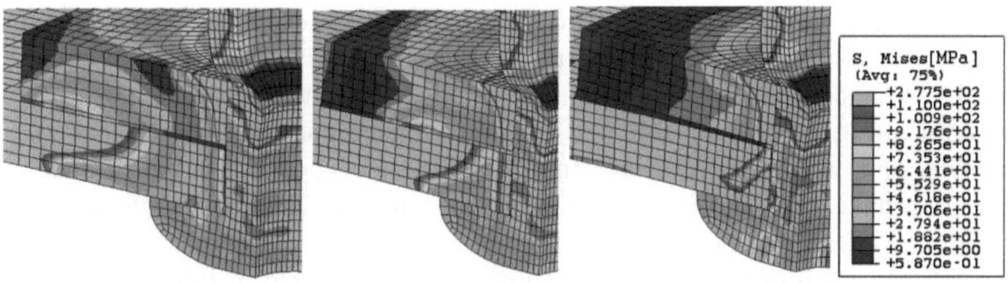

Fig. 39 Progressive failure of the adhesive layer for the PR bonded joint simulated using the DD model

Figure 38 compares the load–displacement simulations of the PR bonded joint using the DD model, the GTN model and the results of the experimental test. In both simulations, the CZ model was used to simulate the adhesive layer. A good agreement is found between the two simulations and the experimental test, especially regarding initial stiffness, displacement at failure and behaviour after the failure of the adhesive layer, which occurs close to a displacement of 0.4 mm. The peak load in both simulation is only about 12% lower than the experimental one.

The difference between the experimental and the simulated behaviour can be related to the use of the same cohesive zone parameters for mode I and mode II, while adhesives likely show a mode II fracture toughness higher than mode I fracture toughness. The load oscillations after the failure of the adhesive are due to the sudden release of strain energy when the adhesive fails. In Fig. 39, the progressive growth of the adhesive failure in the rivet bonded joint is shown. The fracture initiates at both the ends of the overlap and propagates toward the center of the joint. When the adhesive is completely failed, the load is sustained by the rivet, that behaves as a rivet in a simple riveted joint.

It appears therefore that either DD or GTN damage models can be used for the simulation of the PR bonded joint and, moreover, that the deformation and failure behaviour of the hybrid joint can be properly modeled using the damage models identified previously on simple riveted and adhesive joints.

Fig. 40 US school bus: an example of application of the rivet bonding technology

6 Examples of Use

The hybrid joining technique is used when some requirements cannot be obtained by a single joining technology: this could involve the strength but also some other factor as cost or the feasibility in general. Therefore, a rivet bonded joint could be used as replacement or repair of simple bonded joints with an extremely simple manufacturing procedure, although it does not give the same performance of a welded, or a weld bonded joint. Rivet bonding is a good choice when different materials have to be joined (aluminum, magnesium, steel and composite material), or in case of aesthetics requirements. Some other relevant applications concern aircraft and automotive panel fabrication and the assemble of truck cabs, and in general in other automotive aluminum frames due to its capacity to withstand different kind of dynamic and impact loads. Rivet bonding is also used for the internal structures of aircrafts. Several examples of application of the rivet bonding technology can be found: two of them concern the automotive and the bus construction industry. Rivet bonding is indeed largely used for the construction of the aluminum monocoque of the XJ Jaguar series car: here the use of hybrid techniques led to a weight reduction in the order of 40% and to a stiffness enhancement of 60% with respect to the previous models. The rivet bonding is also used for the assembly of school buses (Fig. 40), as replacement of traditional simple riveting. In particular, a strong reduction of the number of rivets can be obtained producing a noticeable reduction of assembly cost. Moreover, the adhesive also works as a sealant, leading to a significant reduction of the presence of leaks, to a better vibration damping and to an aesthetic improvement.

Moving to shipbuilding industry the rivet bonding is widely used for sealing purposes, especially when the strength of the sealant is not sufficient. In this case the rivet is used to sustain the load and silicone or polyurethane are the best choice for the sealing purpose.

7 Conclusions

In this chapter the technological features of rivet bonding technology are discussed: initially their advantages are examined in comparison with the industrial requirements with the aim to define the most worthwhile fields of application. Later on, the manufacturing processes are shown, with particular emphasis on the comparison with other traditional joining techniques like welding. The strength of hybrid rivet-bonded joints in comparison with that of simple bonded joints is discussed for different geometrical and environmental conditions. For both the pop rivet (PR) bonded and self-piercing rivet (SPR) bonded joints, it appears that at room temperature the performance is predominately given by the adhesive (a strong and stiff epoxy adhesive in this case), while the mechanical fastening significantly takes part to carry the load when the joint is subjected to high temperatures. Since the SPR is quite strong, its contribution to the hybrid joint strength is correspondingly high, even though the SPR set up process leads to a significant bond area reduction. In order to better understand the mechanical behaviour of hybrid joints, their failure mechanism are discussed, and in particular the failure of pop rivet bonded joints is simulated using appropriate damage models for the adhesive and the metal parts. Finally, some example of use of rivet bonded joints are shown, giving emphasis to the expected advantages.

The rivet bonding technology appears as a valid alternative to the traditional joining techniques, especially when different materials such as aluminum alloy and steel have to be joined giving the strength of the adhesive bonding but with an easier manufacturing process.

References

1. Akira, K.: A study for vehicle body weight reduction. Rivet-bonding between an aluminum alloy sheet and a steel sheet. In: Proceedings of JSAE Annual Congress (1999)
2. Bishopp J.: In: Cognard, Phillipe (ed.), Handbook of Adhesives and Sealants, Chap. 4, vol. 1, pp. 163–214. Elsevier, Amsterdam (2005)
3. Camanho, P.P., Matthews, F.L.: Stress analysis and strength prediction of mechanically fastened joints in FRP: a review. Compos. Part A Appl. Sci. Manuf. **28A**(6), 529 (1997)
4. Chan, W.S., Vedhagiri, S.: Analysis of composite bolted/bonded joints used in repairing. J. Compos. Mater. **35**(12), 1045–61 (2001)
5. Chu, C., Needleman, A.: Void nucleation effects in biaxially stretched sheets. J. Eng. Mater. Technol. **102**, 249 (1980)
6. Darwish, S.M.H., Ghanya, A.: Critical assessment of weld-bonded technologies. J. Mater. Process. Technol. **105**, 221-229 (2000)
7. Gomez, S., Onoro, J., Pecharromàn, J.: A simple mechanical model of a structural hybrid adhesive/riveted single lap joint. Int. J. Adhes. Adhes. **27**, 263–267 (2007)
8. Gurson, A.: Continuum theory of ductile rupture by void nucleation and growth, 1. Yield criteria and flow rules for porous ductile media. J. Eng. Mater. Technol. **99**, 2 (1977)
9. Haraga, K., Taguchi, K., Yoda, K., Nakashim, Y.: Assembly technique for control panel enclosures with the combined use of adhesive and rivets and the reduction of energy consumption. Int. J. Adhes. Adhes. **23**, 371–376 (2003)
10. Hillerborg, A., Modeer, M., Petersson, P.E.: Analysis of crack formation and crack growth in concrete by means of fracture mechanics and finite elements. Cem. Concr. Res. **6**, 773 (1976)
11. Hooputra, H., Gese, H., Dell, H., Werner, H.: A comprehensive failure model for crashworthiness simulation of aluminium extrusions. Int. J. Crashworthiness **9**, 449 (2004)

12. Kweon, J.H., Jung, J.W., Kim, T.H., Choi, J.H., Kim, D.H.: Failure of carbon composite to aluminum joints with combined mechanical fastening and adhesive bonding. Compos. Struct. **75**, 192–198 (2006)
13. Moroni, F., Pirondi, A., Kleiner, F.: Experimental analysis and comparison of the strength of simple and hybrid structural joints. Int. J. Adhes. Adhes. **30**, 367–379 (2010)
14. Moroni, F.: Testing and modeling of the strength of structural hybrid joints. PhD Thesis, University of Parma (2010)
15. Pirondi, A., Moroni, F.: Clinch-bonded and rivet-bonded hybrid joints: application of damage models for simulation of forming and failure. J. Adhes. Sci. Technol. **23**, 1547–1574 (2009)
16. Rice, J., Tracey, D.: On the ductile enlargement of voids in triaxial stress fields. J. Mech. Phys. Solids **17**, 201 (1969)
17. Tvergaard, V.: Influence of voids on shear band instabilities under plane strain conditions. Int. J. Fracture **17**, 389 (1981)
18. Tvergaard, V., Needleman, A.: Analysis of the cup-cone fracture in a round tensile bar. Acta Metall. **32**, 157 (1984)
19. Wingfield, J.R.J.: Treatment of composite surfaces for adhesive bonding. Int. J. Adhes. Adhes. **13-3**, 151–156 (1993)
20. Volkersen, O.: Die nietkraftverteilung in zugbeanspruchten Nietverbindungen mit konstanten laschenquerschnitten. Luftfahrtforschung 15, 41 (1938)
21. Wegman, R.F.: Surface Preparation Techniques for Adhesive Bonding. William Andrew Publishing, Norwich (2001)

Science of Clinch–Adhesive Joints

Alessandro Pirondi and Fabrizio Moroni

Abstract The idea of "hybrid joints" arises from the need of joining together materials, producing a tough, stiff and lightweight joint. The core concept is: use two different joining techniques, with the aim to increase the strength and gather the advantages of each technique, reducing the drawbacks. This chapter is devoted to clinch–adhesive (also called clinchbonded) joints, which is of particular interest for the chassis manufacturing in vehicles (automotive, trucks, buses, railways) but also of the nautical and ropeways applications. First, an overview of clinching, adhesive bonding and clinch–adhesive joint characteristics is given. The experimental details and the design of experiments plan are then described and the result are shown and discussed comparing the performances in terms of stiffness, strength and energy absorption to failure of simple and hybrid joints. Since one of the most important needs of the industry is the possibility to simulate mechanical behavior of the joint up to failure, damage models have been applied to the finite element analysis of the simple and hybrid joints and the results are shown in the last paragraph.

1 Introduction

In this work hybrid joints mean joints produced with two different kinds of joining techniques, one of them being adhesive bonding. The first application of this idea

A. Pirondi (✉) and F. Moroni
Dipartimento di Ingegneria Industriale, Università degli Studi di Parma, viale G.P. Usberti, 181/A, 43124 Parma, Italy
e-mail: alessandro.pirondi@unipr.it

F. Moroni
e-mail: moroni@ied.unipr.it

Adv Struct Mater (2011) 6: 109–147
DOI: 10.1007/8611_2010_36

Published Online: 31 August 2010

was in a Russian airplane in 1960. In particular in the Antonov AN-74 the resistance spot welding was combined with the adhesive bonding. Ever since, this kind of joints have had a relevant development and nowadays they are used in applications where high strength, stiffness, energy absorption at failure and contemporarily a lightweight design and low costs are required.

In this chapter, the focus is made on a specific type of hybrid joint, coming from the contemporary use of clinching and adhesive bonding (clinch–adhesive, also called here clinchbonded joints). Clinching and adhesive bonding are introduced and described first, then literature works about clinch-bonding are presented and also some technological details concerning the manufacturing are given.

1.1 Clinching

Clinching, or press joining, is a high-speed mechanical fastening technique for point joining of sheet metal using no consumables. It is used for joining metal sheets of 0.5 to about 3 mm in thickness, up to a total joint thickness, depending on the materials, of about 6 mm. With clinching, the sheets are clamped between a die and a blankholder (Fig. 1, Step 1). The punch locally pushes the sheets down into the die (Step 2). The sheets are squeezed between the punch and die, expelling material sideways (Step 3) and forming an interlocking button (Step 4). This interlocking button holds the sheets together.

The joining process involves the plastic deformation of the sheets, therefore, the materials should have sufficient ductility to avoid cracking. The process may produce a circular dot (in this case the dot is obtained by simple plastic deformation) or a rectangular dot (sheets are partially cut by shear deformation), as shown in Fig. 2.

Fig. 1 Clinching process [2]

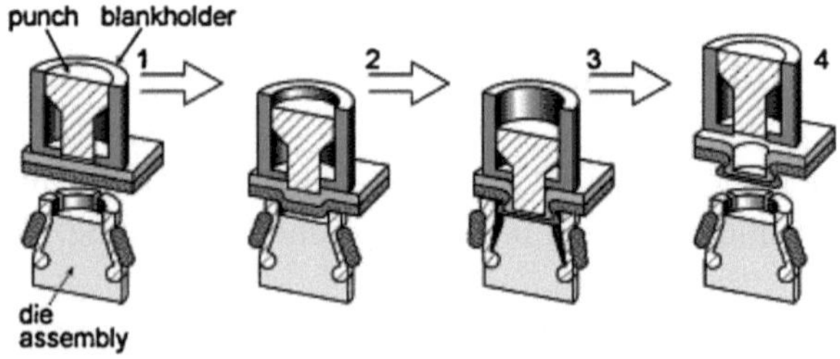

Fig. 2 Section of circular and rectangular (http://www.bohlloff.it)

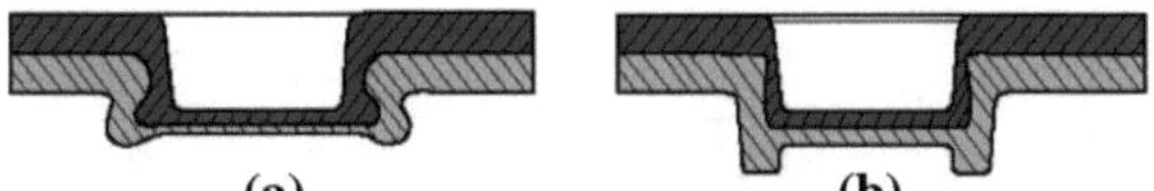

Fig. 3 Section of a "segmented matrix" (**a**) and of a "one piece matrix" (**b**) clinched dot (http://www.bohlloff.it)

Moreover, concerning the circular dots, two kinds of dots can be obtained depending on the die geometry. In the so called segmented clinching, the die is made with three (sometimes four) parts which can expand during the movement of the punch. The opening of the die allows the metal to flow and the typical shape of a clinched joint obtained in this case is shown in Fig. 3a. In the fixed clinching case the die is made by a single part, therefore, no relative movement and metal flow are allowed. This produces the shape shown in Fig. 3b.

It can be observed that in the first geometry, the shape of the sheet ensures a better interlocking in respect to the second one: this increases the strength of the joint and therefore, in industrial applications, the segmented matrix clinching is the most widely used. The equipment normally comprises a C-frame with a hydraulic force application to the punch. The forces involved [19] in the punch are in the range of 10–100 kN for a steel sheet of thickness in the range of 0.5–4 mm.

Some advantages of the clinching technique concern the rapidity of the process, the absence of consumables and of pre-drilled holes. In the case of automotive parts assembly, clinching replaces techniques such as resistance spot welding as well as the use of other mechanical fasteners, such as riveting. As many industries (e.g. automotive) are driven by weight-reduction, manufacturers are moving towards thinner and stronger materials. This may affect the ductility of the sheet materials, impairing their suitability for clinching. This issue can be overcome by improving the optimization of the process (i.e. modifying the tool strength and geometry) or by using hybrid joining techniques.

Typically the strength of clinched joints is evaluated by shear (or pull) and peel test. Shear is almost always stronger than peel as shown in Fig. 4 and therefore

Fig. 4 Comparison of shear and peel strength clinched joints for different sheet thickness (http://www.norlok.com/main.html)

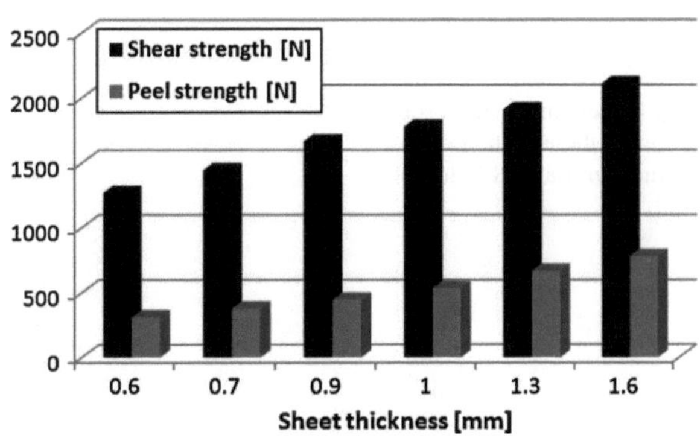

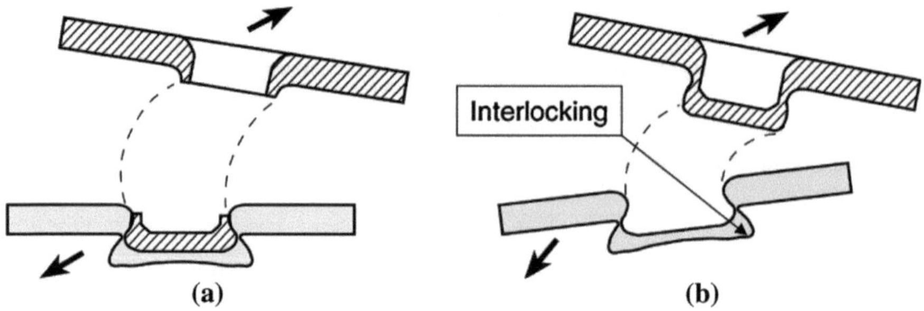

Fig. 5 Failure of the neck (**a**) and opening failure mode (**b**) of a clinched dot [21]

during the design phase a shear-loaded geometry should always be preferred to a peel-loaded one.

In general, three modes of failure can be observed for a clinched joint. The first mode is the failure of the neck as shown in Fig. 5a. Clinched joints fail in this manner when the thickness of the neck is too small. The second mode involves an opening like a press-stud (see Fig. 5b). This kind of failure happens when the interlocking is not strong enough.

The third mode is a combination of the first and second ones: in this mode, initially, one edge of the joint fails, whilst the other is deformed. Subsequently, the upper sheet is partially pulled out and at the same time the fracture moves toward the deformed edge. This mode, involving both previous modes, has been evaluated to be the best one [20].

The shear strength of the joint is essentially based on the material thickness in the neck of the joint. A sufficient "S" shape (Fig. 6) is needed in order to keep the sheets together while the joint is subjected to the shear force.

The quality of the joint is industrially evaluated by measuring the material thickness in the middle of the base of the joint (dimension X in Fig. 6): in particular a joint is accepted if this measure is equal to one third the sum of the thickness of the sheets (with a tolerance of $\pm 15\%$). This method guarantees a first approximated evaluation of the joint manufacturing, even if for a better analysis the complete geometry of the dot must be taken into account.

Fig. 6 Schematic representation of the clinched dot section with an "S" slope [21]

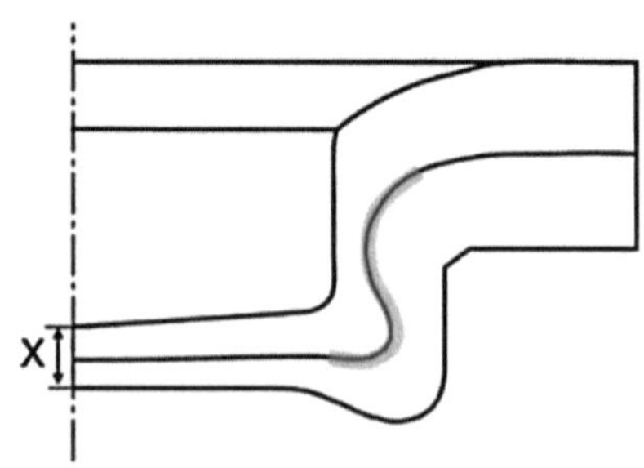

1.2 Adhesive Bonding

Adhesive bonding is a modern assembly technique where two similar or non-similar materials (metals, plastics, composites, etc.) are joined using an adhesive.

In many applications, adhesive bonding is being evaluated to replace traditional assembly techniques. The main advantages provided by adhesive bonding are:

- more uniform stress distribution in respect to welding or mechanical fastening
- weight reduction
- less components (no rivets or bolts)
- different materials with dissimilar thicknesses can be assembled
- lower aesthetic impact in respect to welding or mechanical fastening
- vibration damping
- energetically efficient
- easy to automate

But there are also some limitations:

- difficult to assess the durability of a bonded structure
- surface pre-treatment is usually required
- limited heat resistance
- difficult disassembly
- handling, curing time and temperature

The strength of attachment, or adhesion, between an adhesive and its substrate depends on many factors, including the means by which this occurs. Adhesion may occur either mechanically, that is the adhesive penetrates into small pores of the substrate, or works by chemical mechanisms. In some cases, an actual chemical bond occurs between adhesive and substrate. In other cases electrostatic forces, as in static electricity, hold the substances together. Another method involves van der Waal's forces which develop between each molecules. Bonding can also occur by moisture-aided diffusion of the glue into the substrate, followed by hardening.

When designing a bonded joint, one must take into account how the adhesive withstands the stress type. In general, it is stated that adhesives resist well to compression, pure tensile stress, and shear, while they are rather sensitive to cleavage and peeling.

This assertion is only partially true. In fact, rather than "which kind of stress the adhesive has to withstand" is better to know "what is the stress distribution in the adhesive layer" and then design to reduce the peel stress. Joint geometry and deformation compatibility between adherends yield local stresses that may be many times greater than the average stress. It is immediate to understand that in case of peeling the stresses are extremely concentrated. A lower but significant stress concentration can be met in the case of cleavage. The stress distribution in single lap joints, which represent the most common bonded joint geometry, is extensively analyzed in the literature. The stresses (both peel and shear) are concentrated at the ends of the overlap, while in the middle of the joint the stresses are very low: this leads to the conclusion that an increase of the overlap length, certainly produces an improvement of the joint strength, but the gain in strength is not proportional to the overlap, because the stress is predominantly carried by the end zones.

1.2.1 Influence of the Temperature

Adhesives are predominantly of polymeric nature and the mechanical behaviour of polymers is strictly dependent on the temperature. In order to understand if an adhesive can withstand a certain level of temperature, the knowledge of the glass transition temperature T_g is fundamental. This is the temperature at which there is significantly more molecular mobility than at lower temperatures (i.e., the molecules have sufficient thermal energy to be considered mobile). T_g is a property of polymers that depends on the chemical composition and the degree of crosslinking or molecular interaction. For a good bond strength and creep resistance, the T_g of the adhesive should be well above the maximum temperature that it will see in service. However, peel strength will be low if the T_g is high. It is very difficult to provide an adhesive that has high peel strength with good cohesive strength and temperature resistance.

It is also relevant to notice that the higher is the cure temperature, the higher will be the T_g. Therefore, apart from the chemical composition, a hot cured adhesive (typically one component) has better temperature performances in comparison with a room temperature cured adhesive (typically two components). Anyway, the chemical composition of epoxies can be modified in order to produce high temperature resistant adhesives.

1.2.2 Ageing

In the case of adhesives the ageing can be produced by exposure to:

- High temperature
- Moisture
- Chemically aggressive environment

In the first case, prolonged exposure to elevated temperatures may cause several reactions in the adhesive. These mechanisms can weaken the bond, both in the bulk and at the interface. The reactions that affect the bulk material are oxidation and pyrolysis. But the thermal aging can also affect the interface by causing chemical and/or physical changes of the adhesion phenomena. If heating brings a non-crosslinked adhesive far above its glass transition temperature, the molecules will become so flexible that their cohesive strength will drastically decrease. In this flexible, mobile condition, the adhesive is susceptible to creep and increase chemical or moisture penetration. Generally, prolonged heating at an excessive temperature will have the following effects on a crosslinked adhesive:

- Split polymer molecules (chain scission) causing lower molecular weight and degraded cohesive strength
- Continued crosslinking resulting in bond embrittlement and shrinkage
- Evaporation of plasticizer resulting in bond embrittlement
- Oxidation (if oxygen or a metal oxide interface is present) resulting in lower cohesive strength and weak boundary layers

In the second case (moisture exposure), it has to be stated that the water is one of the substances which gives the greatest problems in terms of environmental stability for many adhesive joints. Water is a problem because it is very polar and permeates most polymers. Other common fluids, such as lubricants and fuels, are of low or zero polarity and do not permeate and weaken thermosetting adhesive or sealant joints significantly. Moisture can affect an adhesive bond in two distinctive ways:

- Moisture can degrade the properties of the bulk adhesive. Internal degradation within the bulk adhesive occurs primarily by absorption of water molecules into the polymer structure. Water permeation in polymers generally lowers the glass transition temperature of the polymer by reducing the forces between molecules.
- Moisture can degrade the adhesion properties at the interface. Water can also permeate the adhesive and preferentially migrate to the interface region displacing the adhesive at the bond interface. It is the most common cause of adhesive strength reduction in moist environments

For most bonded metal joints which see outdoor service, corrosive environments are a more serious problem than the presence of moisture. The degradation mechanism is corrosion of the metal interface resulting in a weak boundary layer. Surface preparation methods and primers that make the adherend less corrosive are commonly employed to retard the degradation of adhesive joints in these environments. In this case, the most dangerous and common aggressive chemical agent is salt spray.

1.3 Clinch–Adhesive Joints

Differently from weld-, rivet- and bolt-adhesive joints for which a number of literature studies are available, clinchbonding has received little attention so far. The effectiveness of this kind of hybrid joint has been tested by Del Gamba [5]. In this work, the tensile shear strength and the peel strength of simple clinched and clinchbonded joints (see Fig. 7) was evaluated.

Figure 8 shows the comparison between the load versus displacement plots of single lap tests carried out on simple clinched joints and clinchbonded joints. In the first case, the joints behave linearly until point A. Here the material of the clinched button begins to yield. At B, the maximum load is reached and the clinched button is completely yielded. At C, a failure nucleates in the most stressed zone, and

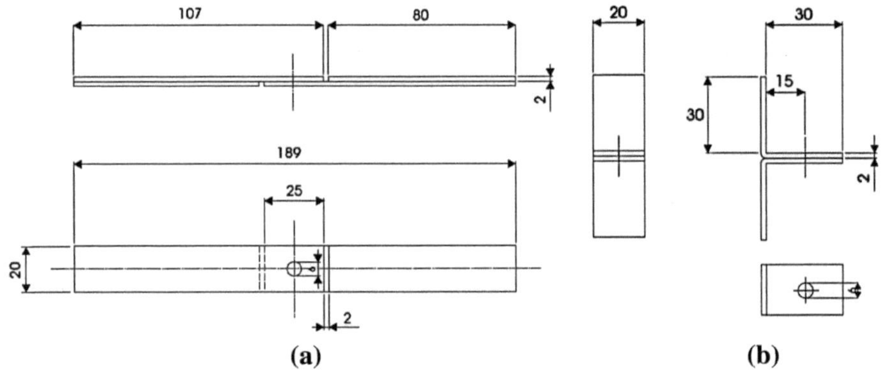

Fig. 7 Geometry of tensile shear (**a**) and peel (**b**) joints. Dimensions in mm [5]

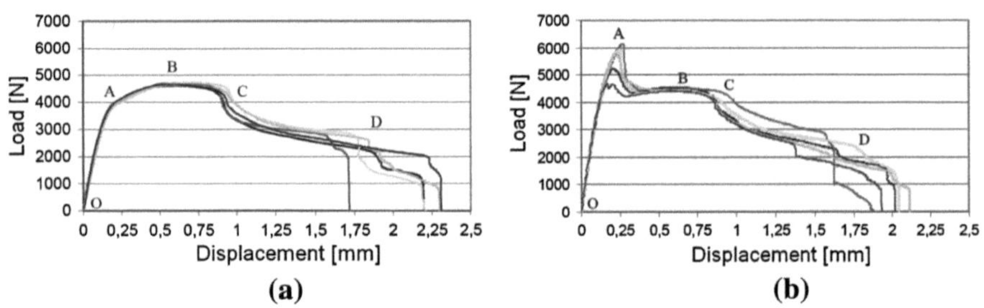

Fig. 8 Load versus displacement curves of simple clinched (**a**) and clinchbonded single lap joints (**b**) [5]

increasing the applied displacement, it propagates until D, when the final rupture happens. The behaviour of clinchbonded joints is rather similar to that of simple clinched joints, except for the initial linear elastic behaviour. Indeed in this case the joint is stiffer, and a higher peak load is reached in comparison to simple cliched joints. This difference is due to the presence of the adhesive, which contributes to carry the load. In this case the strength is predominantly due to the adhesive strength; at A the adhesive fails, and the load drops down to the curve typical of a simple clinched joint.

The comparison between peel tests carried out on simple clinched and clinch bonded joints is given in Fig. 9. Considering simple clinched joints, a linear elastic behaviour followed by the yielding of the material forming the interlock button is seen. At A the failure nucleates and propagates involving the complete section of the clinched dot until the final rupture at C. Considering the hybrid joint, it can be noticed that, except for small differences at a very low values of displacement, the joints behave like simple clinched joints. In fact, with this geometry, the stresses are concentrated in a small zone close to the end of the overlap, therefore, when the load is initially applied only the adhesive is stressed, while the clinched dot is

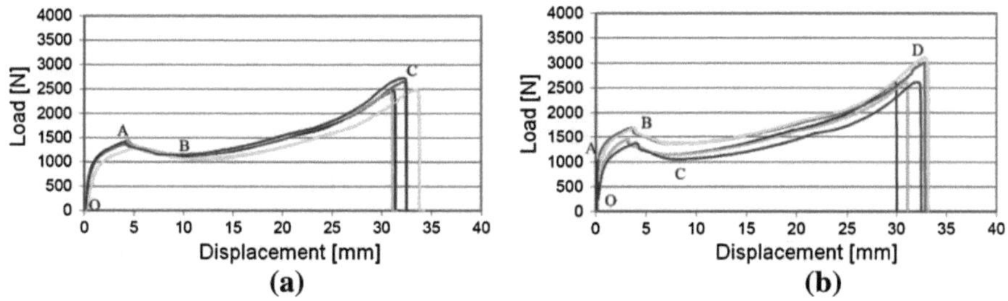

Fig. 9 Load versus displacement curves of simple clinched (**a**) and clinchbonded peel joints (**b**) [5]

completely unloaded. This produces a rapid failure of the adhesive layer and after that the joint behaves like a simple clinched joint.

Some important remarks appear from this experience: in order to have a hybrid joint with a better performance than simple joints, the two joining methods must work together, and carry load at the same time. In order to obtain this condition, the geometry and also the mechanical properties of substrates and adhesive must be correctly taken into account.

2 Experimental Details

2.1 Definition of Specimens: Materials and Dimensions

Since shear is the most frequent loading condition, the single lap joint geometry was adopted. As high strength steels cannot be easily clinched when the sheet thickness is >1.5 mm, a S275 galvanized construction steel and a 5052 aluminum alloy typically used in folding were selected. More details on the material properties are given in Table 1

Concerning the adhesive, the one-component epoxy adhesive Terokal 5077 from Henkel was selected. This is a heat-curing adhesive, widely used in the automotive field due to the good impact peel resistance. One of the most relevant features of this adhesive is the possibility to bond with good adhesion strength oily metal surfaces: it allows to reduce the cleaning treatment which sometimes could be difficult to integrate in an industrial process. Moreover, this adhesive

Table 1 Mechanical properties of selected materials (http://www.matweb.com)

Material	Standard denomination	Modulus of elasticity (GPa)	Yield strength (MPa)	Ultimate strength (MPa)	Elongation at break (%)	Surface treatment
Steel	S275	206	275	410	12	Galvanized
Aluminum	5052	70.3	193	228	22	None

(a) (b)

Fig. 10 Specimens with one (**a**) or two (**b**) fastened spot representing a pitch of 60 and 30 mm, respectively

can be used in combination also with resistance spot welding. The technical data sheet shows a value of shear strength that is higher than 20 MPa. From a survey of nautical, shipbuilding, automotive and public transport vehicles, it was decided to select two different substrate thicknesses, equal to 1.5 and 2.5 mm. Considering the adhesive shear strength and the tensile strength of the weakest plate, the overlap length was defined equal to 15 mm in order to force the failure to occur in the overlap and not in the substrates, even though these latter may eventually deform plastically. This overlap length is also enough to place in the clinched dot. Also the pitch of clinching was considered. In order to simplify the cutting procedure of the plates, the different pitches were simulated with a different number of dots in joints of the same width, as shown in Fig. 10. In order to represent pitches of 30 and 60 mm, a joint width of 60 mm was selected and a single dot simulated a pitch of 60 mm, while two dots simulate a pitch of 30 mm.

While resistance spot welding is the most used technique for joining substrates made of the same material, riveting or clinching are much more suited in general to join different materials. Following this remark, clinchbonding was studied only in the case of *heterogeneous joints*, that is with steel on one side and aluminium on the other side.

2.2 Environmental Conditions

Regarding environmental conditions, a temperature range from −30°C until 90°C was considered and it was decided to evaluate also the influence of ageing. The ageing was carried out following the German standard test VDA 621-415. This test consists in a sequence of cycles (the number of cycles is user defined) at different environmental conditions. In detail, each cycle consisted in:

24 h exposure to salt spray according DIN 50021-SS

- temperature equal to 35°C
- concentration of spray equal to 5 parts by mass of sodium chloride in 95 parts of water
- pH of the salt solution at 25°C between 6.5 and 7.2

96 h exposure to a changing climate according to DIN 5007-KFW. 96 h split in four cycles, each cycle consisted in:

- 8 hours to a heated, saturated mixture of air and water vapor, at a temperature of 40°C and a relative humidity of 100%
- 16 h exposure at a temperature of 23°C and a relative humidity of 50% (according to DIN 50014)

48 h exposure to ambient climate according to DIN 50014, at a temperature of 23°C and a relative humidity of 50%.

Following suggestions of Henkel and referring to literature works, a number of four cycles was set. Concerning temperature, in order to obtain a better description of its influence, the levels −30°C, 23°C (also called room temperature, or T_{amb}), 60°C and 90°C were tested.

2.3 DoE Plan

The experiments have been planned according to the DoE methodology in order to evaluate the influence of the factors considered on the strength of clinched, adhesive and clinch–adhesive joints. In particular, the analysis is addressed to evaluate the influence of factors over the strength (F_{max}), the stiffness (K) and the energy absorption (En).

Referring to Fig. 11, the strength is the maximum load reached during the test, the stiffness is the slope of the best straight line fit for the initial linear behaviour, while the energy absorption is the area under the curve until final failure.

The factors and the respective values considered here are given in Table 2. The temperature levels are represented using two other factors D and E, as shown in Table 3

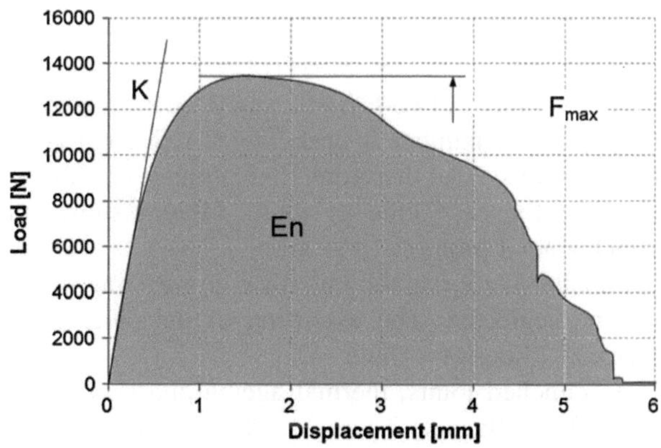

Fig. 11 Outputs considered in DoE

Table 2 Definition of factors and their levels

Symbol	Factors	Level (−)	Level (+)
A	Thickness (mm)	1.5	2.5
B	Pitch (mm)	30	60
C	Ageing	No	Yes
T	Temperature (°C)	−30/23/60/90	

Table 3 Definition of temperature levels joints

T	Temp 1 (*D*)	Temp 2 (*E*)
−30	−	−
23	−	+
60	+	−
90	+	+

Table 4 Design table of the experimental plan for clinchbonded joints

	Sheet thickness *A*	Pitch *B*	Ageing *C*	Temp 1 *D*	Temp 2 *E*	Temperature *T*	Treatment combination
1	−1	−1	−1	−1	1	23	e
2	−1	−1	−1	1	−1	60	d
3	−1	−1	1	−1	−1	−30	c
4	−1	−1	1	1	1	90	cde
5	−1	1	−1	−1	−1	−30	b
6	−1	1	−1	1	1	90	bde
7	−1	1	1	−1	1	23	bce
8	−1	1	1	1	−1	60	bcd
9	1	−1	−1	−1	−1	−30	a
10	1	−1	−1	1	1	90	ade
11	1	−1	1	−1	1	23	ace
12	1	−1	1	1	−1	60	acd
13	1	1	−1	−1	1	23	abe
14	1	1	−1	1	−1	60	abd
15	1	1	1	−1	−1	−30	abc
16	1	1	1	1	1	90	abcde

The number of two-level factors to be evaluated is therefore equal to 5. In order to reduce the number of tests, a 1/2 reduction is applied. The design results in a 2^{5-1} analysis and therefore 16 treatments (combinations of factors levels) should be done for each kind of joint. Table 4 shows the resulting design plan for clinchbonded joints.

In adhesive joints the pitch has no meaning, therefore the column of the pitch can be neglected. The exclusion of the corresponding factor (*B*) yields a full factorial 2^4 design (Table 5).

In clinched joints, thermal ageing and service temperature have a reasonably negligible influence. Therefore, the corresponding factors (*C*, *D*, *E*) are dropped

Table 5 Design table of the experimental plan for adhesive joints

	Sheet thickness *A*	Pitch *B*	Ageing *C*	Temp 1 *D*	Temp 2 *E*	Temperature *T*	Treatment combination
1	−1	*−1*	−1	−1	1	23	e
2	−1	*−1*	−1	1	−1	60	d
3	−1	*−1*	1	−1	−1	−30	c
4	−1	*−1*	1	1	1	90	cde
5	−1	*1*	−1	−1	−1	−30	0
6	−1	*1*	−1	1	1	90	de
7	−1	*1*	1	−1	1	23	ce
8	−1	*1*	1	1	−1	60	cd
9	1	*−1*	−1	−1	−1	−30	a
10	1	*−1*	−1	1	1	90	ade
11	1	*−1*	1	−1	1	23	ace
12	1	*−1*	1	1	−1	60	acd
13	1	*1*	−1	−1	1	23	ae
14	1	*1*	−1	1	−1	60	ad
15	1	*1*	1	−1	−1	−30	ac
16	1	*1*	1	1	1	90	acde

The column highlighted in italics represents that pitch is not relevant for this kind of joint

Table 6 Design table of the experimental plan for clinched joints

	Sheet thickness *A*	Pitch *B*	Ageing *C*	Temp 1 *D*	Temp 2 *E*	Temperature *T*	Treatment combination
1	**−1**	**−1**	***−1***	***−1***	***1***	***23***	**0**
2	**−1**	**−1**	***−1***	***1***	***−1***	***60***	**0**
3	**−1**	**−1**	***1***	***−1***	***−1***	***−30***	**0**
4	−1	−1	*1*	*1*	*1*	*90*	0
5	**−1**	**1**	***−1***	***−1***	***−1***	***−30***	**b**
6	**−1**	**1**	***−1***	***1***	***1***	***90***	**b**
7	**−1**	**1**	***1***	***−1***	***1***	***23***	**b**
8	−1	1	*1*	*1*	*−1*	*60*	b
9	**1**	**−1**	***−1***	***−1***	***−1***	***−30***	**a**
10	**1**	**−1**	***−1***	***1***	***1***	***90***	**a**
11	**1**	**−1**	***1***	***−1***	***1***	***23***	**a**
12	1	−1	*1*	*1*	*−1*	*60*	a
13	**1**	**1**	***−1***	***−1***	***1***	***23***	**ab**
14	**1**	**1**	***−1***	***1***	***−1***	***60***	**ab**
15	**1**	**1**	***1***	***−1***	***−1***	***−30***	**ab**
16	1	1	*1*	*1*	*1*	*90*	ab

The combinations not meaningful for clinched joints are bold and in italics are thermal ageing and service temperature which have a reasonably negligible influence

from the plan and consequently the table changes to Table 6, which is a simple 2^2 design where the two factors are the thickness and the pitch.

Finally, the values of F_{max}, K and En coming from the experimental plan were analysed in terms of intercept (average value of all the experiments) and estimated coefficients (effects) of the relevant factors and factor interactions, using the ANOVA (analysis of variance) technique. The value of the coefficient corresponds to the variation with respect to the intercept when a factor or interaction changes from a low (−) to a high (+) level.

2.4 Joint Manufacturing

All of the joints were manufactured without any surface preparation, as it is desirable in industrial mass production. A device to ensure the correct position was used. The sequence of steps followed was

- Heating up of adhesive and adherends
- Holding of the first sheet in the positioning device
- Laying and spreading of adhesive
- Positioning and holding of the second sheet (Fig. 12)
- Clinching (only for clinched and clinchbonded joints)
- Curing in oven (only for bonded and clinchbonded joints)
- Cooling down to room temperature

The curing cycle consisted in heating up specimens from ambient temperature to 160°C for 20 min followed by the curing of 30 min at 160°C. After curing, specimens were left to cool down in laboratory ambient.

Concerning the clinched joints, different tool dimensions and different sizes of the clinching machine were used following the suggestions of the supplier. These are shown in detail in Table 7. J_m is the depth of the matrix, while the maximum force is the maximum load produced by the pneumatic piston.

Fig. 12 Specimen positioned and held in the positioning device

Table 7 Clinch joining parameters as a function of plate thickness

	1.5 mm thickness	2.5 mm thickness
Machine Size	Rivclinch P35 Pass	Rivclinch P75
Matrix	SR 503	SR 804
Jm (mm)	1.2	1.7
Punch	SR 42	SR 70
Max force (kN)	35	105

Machine, matrix and punch characteristics can be found on http://www.bollhoff.it

2.5 Experimental Apparatus

The whole amount of specimens were tested using an electromechanical uniaxial Instron testing machine. The tests were carried out with a 2 mm/min crosshead speed. Results are given in terms of displacement (with a maximum range of 100 mm) and in terms of applied load, measured using a load cell with a load range of 100 kN. Temperatures of 60 and 90°C were obtained using an electrical heater, while the −30°C temperature was obtained using liquid nitrogen vapor. In order to ensure an uniform temperature distribution in the specimens, but at the same time in order to avoid significant material degradation due to the thermal ageing, the specimens were left for half an hour at the test temperature before testing.

3 Experimental Results

3.1 Clinched Joints

In the subsequent pages, the mechanical response of simple clinched joints is analyzed. First of all, Fig. 13 shows an example of the load versus displacement plot obtained in the tests.

Fig. 13 Load versus displacement curve of a clinched joint in the case of 1.5 mm plate thickness and 30 mm pitch

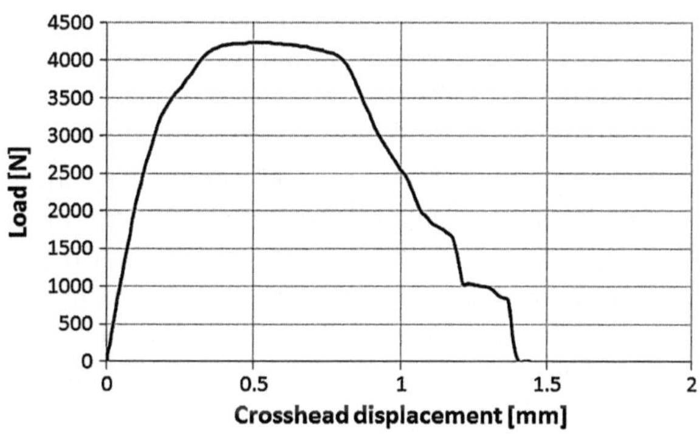

Fig. 14 Fracture surface of a simple clinched joint

Factor	Coefficient estimate Stiffness [kN/mm]	Strength [kN]	Energy [J]
Intercept	25.31	6.75	10.10
A - Thickness	4.02	3.43	6.89
B - Pitch	-1.91	-2.17	-3.92
AB	-	-1.07	-2.93

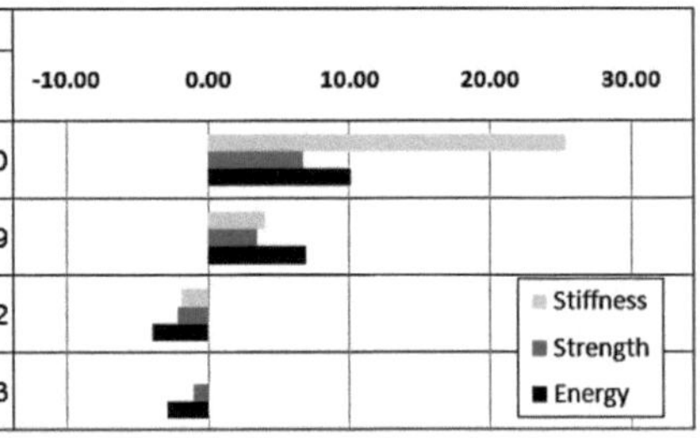

Fig. 15 Estimated coefficients for the stiffness of simple clinched joints

An example of the fracture surface of a simple clinched joint is given in Fig. 14. It appears that, as the button is plastically deformed, the interlock of the button is partially released and the final fracture happens due to shear failure of the internal part of the button.

The values of the intercept and of the estimated coefficients for the stiffness, strength and energy absorption of clinched joints are shown in Fig. 15. It appears that the most relevant factor is the thickness, followed by the pitch. Concerning F_{max} and En, the interaction of pitch and stiffness is also relevant.

Figure 16 shows instead the prediction, based on the ANOVA results, of the stiffness as a function of the factors effect listed in Fig. 15.

3.2 Adhesive Joints

Concerning the stiffness, Fig. 17 shows the estimated effects of the significant factors. It can be immediately noticed that the intercept has a value much higher than the other factors. This means that the stiffness of bonded joints is not much influenced by the geometry and the environmental factors. The most relevant factor is the thickness, followed by the factors representing the temperature. It should also be noticed that the ageing slightly affects the stiffness of bonded joints, since it appears only in the second or third order interactions.

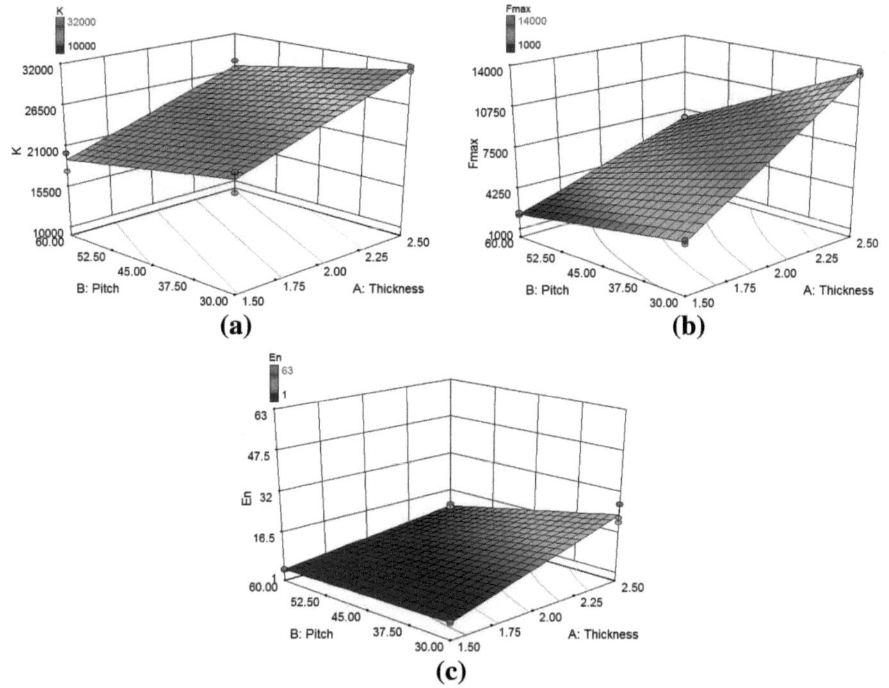

Fig. 16 Prediction of **a** stiffness, **b** strength and **c** energy absorption as a function of pitch and plate thickness for clinched joints

Factor	Coefficient estimate [N/mm]
Intercept	28802
A - Thickness	2096
D - Temp 1	-1174
E - Temp 2	-762
CD	-516
DE	-438
CDE	-592

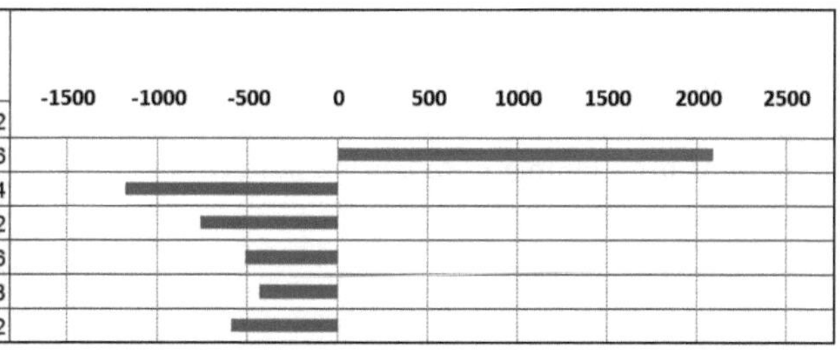

Fig. 17 Estimated coefficients for the stiffness of adhesive joints

Figure 18 shows the prediction of the stiffness for simple bonded joints as a function of the temperature factors, for different plates thicknesses. It appears that the temperature moderately affects the stiffness, and moreover, that a reduction of the thickness leads to a reduction of the stiffness in the order of 3–4000 N/mm, which represents only 10% of the joint stiffness.

The influence of the relevant factors for the strength in the case of simple bonded joints is shown in Fig. 19. Apart from the substrate (not evaluated in this case), the trend of the influences are similar. The most influent factors are again those involving the temperature, followed by the ageing and the thickness.

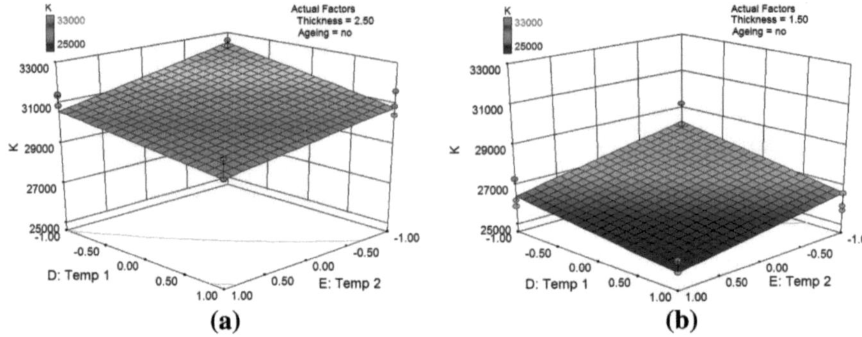

Fig. 18 Prediction of the stiffness of adhesive joints as a function of temperature for **a** 2.5 mm plate thickness and **b** 1.5 mm plate thickness

Factor	Coefficient estimate [N]
Intercept	12334
A - Thickness	504
C - Ageing	-987
D - Temp 1	-2189
E - Temp 2	-698
AC	-423
BD	-447
DE	-685

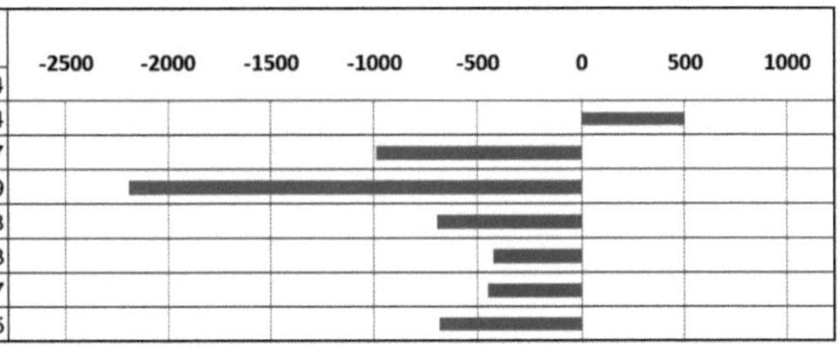

Fig. 19 Estimated coefficients for the strength of adhesive joints

The prediction of the strength of bonded joints as a function of the temperature is illustrated in Fig. 20 for non-aged and aged joints. It can be seen that the ageing significantly affects the performances and, at the same time, also the reduction of strength at high temperature can be noticed.

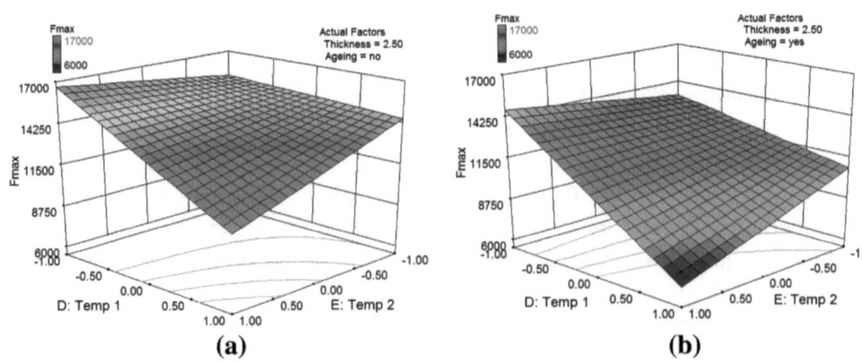

Fig. 20 Prediction of the strength of adhesive joints as a function of temperature for **a** non aged and **b** aged joints

Factor	Coefficient estimate [J]
Intercept	14.85
A - Thickness	-7.63
D - Temp 1	-6.81
AD	4.99

Fig. 21 Estimated coefficients for the energy absorption of adhesive joints

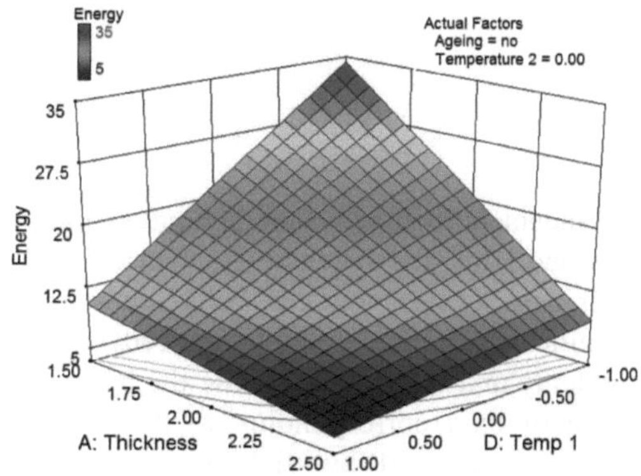

Fig. 22 Prediction of the energy absorption of adhesive joints as a function of thickness and Temp 1 temperature factor

Concerning energy absorption, it appears that the most relevant factors are, essentially, the thickness, the level of the Temp 1 factor and their interaction (Fig. 21). The influence of the temperature can be easily understood: the higher is the temperature, the lower will be the strength, and therefore, the lower will be the energy absorption. The thickness negatively affects the energy absorption: in fact, as already shown for homogenous aluminum joints, the lower is the thickness, the higher is the plastic deformation of the plates, and therefore, the higher is the energy absorption.

Using the estimated coefficient of the factors shown in Fig. 21, the prediction of the energy absorption of simple bonded joints is given in Fig. 22. It can be noticed that the influence of thickness is relatively low at high temperature, while it rapidly increases with the reduction of temperature (effect linked to the interaction AD): this is because at low temperature the joints strength increases and therefore the plastic deformation increases too.

3.3 Clinch–Adhesive Joints

In the case of hybrid joints, similarly to two springs in parallel, where the stiffness and strength of the two springs is predominantly given by the stiffest and strongest spring, in clinchbonded joints the stiffness and the strength are predominantly given by the adhesive. This is confirmed by Fig. 23, where the load versus

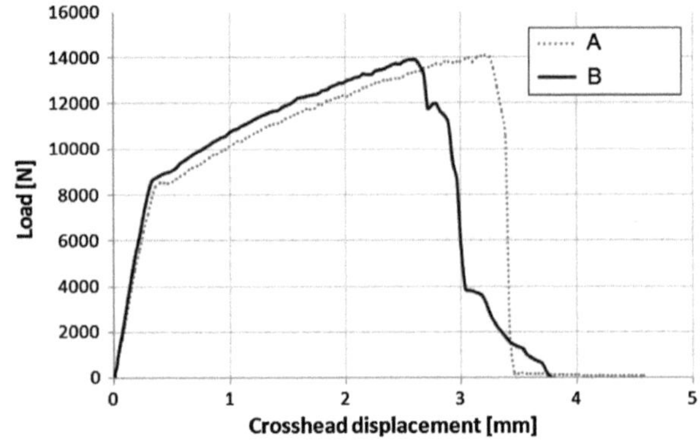

Fig. 23 Load versus displacement curve of an adhesive joint (*A*) and of a hybrid clinchbonded (*B*) joint, in the case of 1.5 mm plate thickness and 30 mm pitch, at room temperature

displacement behaviour of a bonded joint is compared with that of a hybrid fastened joint, for the same geometrical and environmental conditions.

It can be seen that the behaviour until the maximum load is very similar. Significant differences can be noticed after the peak of load: the bonded joint fails suddenly and completely, while the hybrid joint has a sudden drop until a certain value of load corresponding to the clinch behavior. The initial purely elastic behavior followed by yielding and subsequent hardening indicates that the aluminum plates (which is weaker than steel) accumulate plastic deformation.

Therefore, it can be concluded that in clinchbonded joints, the aluminum plate is plastically deformed until the adhesive fails, with a small deformation of the overlap zone. When the adhesive fails, the overlapped adherends start to slide with respect to each other, leading to the failure of the clinch.

As previously discussed, the stiffness and the strength of hybrid heterogeneous joints are predominately due to the presence of the adhesive. Analyzing Fig. 24, it is evident that the intercept value shows only a little improvement with respect to adhesive joints ($K = 28800$ N/mm, $F_{max} = 12334$ N).

	Coefficient estimate		
Factor	Stiffness [kN/mm]	Strength [kN]	Energy [J]
Intercept	30.30	11.57	20.1
A - Thickness	2.34	1.02	
B - Pitch	-0.72		-3.7
C - Ageing		-0.98	-2.8
D - Temp 1	-0.95	-1.77	-7.9
E - Temp 2		-0.79	-2.6
AB			-4.7
AD			3.6
CD			3.0
DE		-0.74	

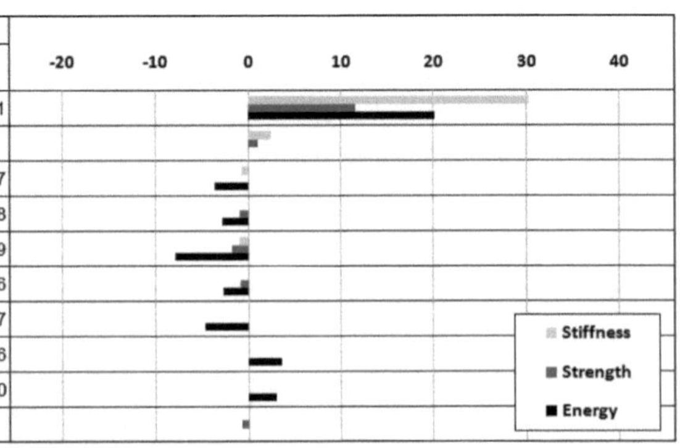

Fig. 24 Estimated coefficients for the stiffness of clinchbonded joints

Fig. 25 Fracture surface of a clinchbonded joint

It can be noticed that clinching reduces significantly the adhesive bonding area, as shown in Fig. 25, and this justifies the tendentially weaker response at low temperatures with respect to the adhesive joint.

Concerning the energy absorption, more difference can be found than when comparing adhesive and hybrid joints in terms of stiffness and strength. In fact, the energy absorption accounts also for the contribution of clinching after the adhesive failure. Therefore, the energy absorption of clinchbonded joints is approximately the sum of the energy absorption of simple adhesive and simple clinched joints.

Similarly to adhesive joints, clinchbonded joints are influenced by the thickness and by the temperature, although the presence of the clinch dot reduces the influence of the temperature with respect to adhesive joints. The pitch is also a significant factor and it is characterized by a negative coefficient, meaning that the clinch dot contributes to increase the joints stiffness. Using the coefficients shown in Fig. 24 the behaviour of clinchbonded joints is depicted in Fig. 26.

3.4 Comparison Between Simple and Hybrid Joints

The intercept (average value of all of the experiments) is shown in Fig. 27 and it can be noticed that the clinchbonded joint yields a small reduction in strength in comparison with the adhesive joint, which is the best one.

In other words, the reduction of bonded area caused by clinching in clinchbonded joints is not compensated since the clinch dot is much weaker and also less stiff than the adhesive. This is not the case regarding stiffness, where the hybrid joint yields slightly better performances. Finally, regarding the energy absorption, the clinchbonded joint gives a noticeable improvement with respect to both clinched and adhesive joints. Indeed, in the hybrid joint the load-elongation behaviour exhibits two well-differentiated regions: the first one characterized by high stiffness and strength (main contribution coming from the adhesive) and the second one after adhesive failure, with stiffness and strength corresponding to those of the mechanical fastening. For this reason, the energy absorption is very

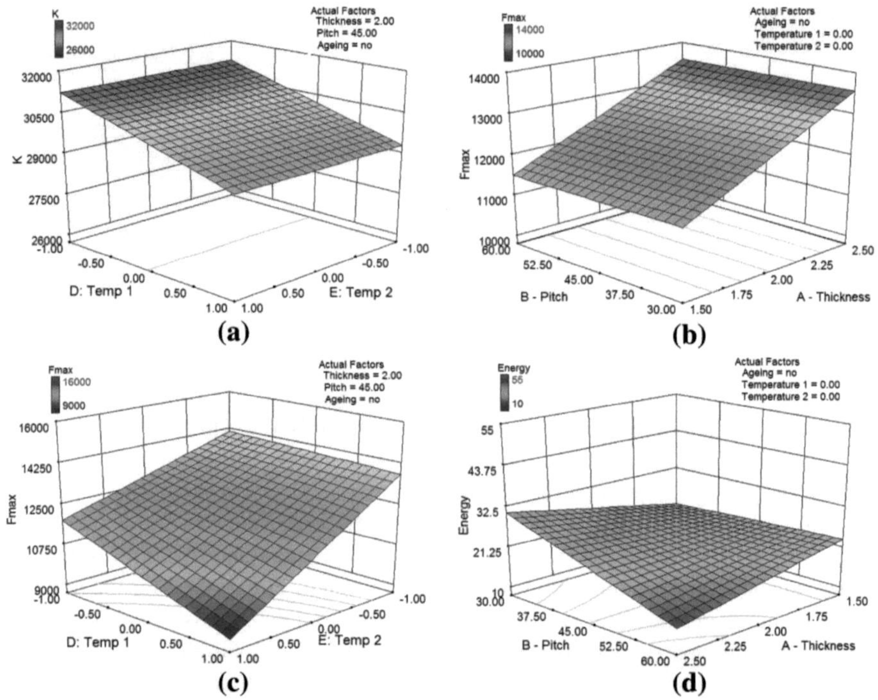

Fig. 26 Prediction of the stiffness as a function of temperature (**a**), strength as a function of geometrical parameters (**b**), temperature (**c**), energy absorption as a function of temperature (**d**) for clinchbonded joints

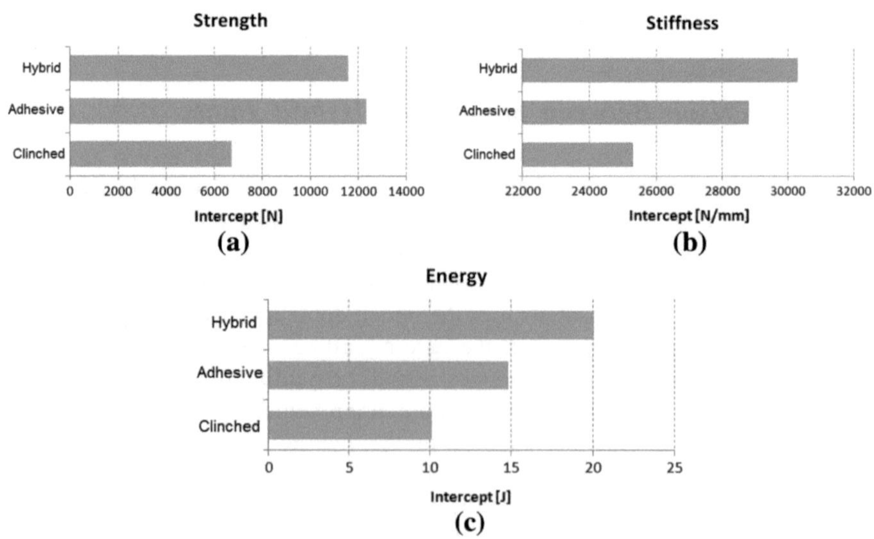

Fig. 27 Intercepts of: **a** strength, **b** stiffness, **c** energy absorption of the joints

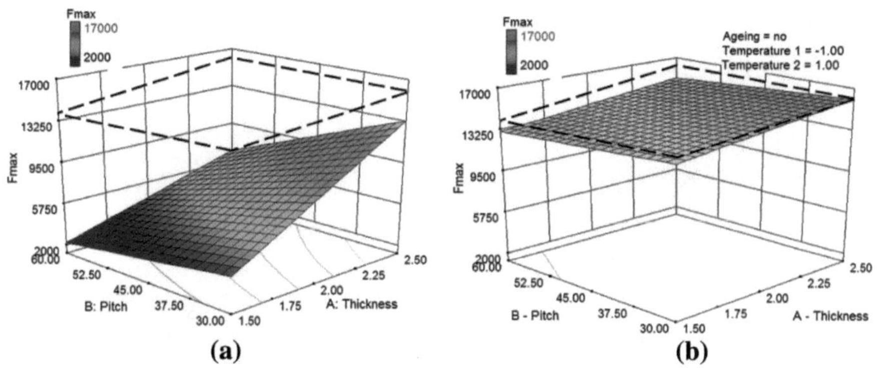

Fig. 28 Strength at room temperature as a function of thickness and pitch of 1.5 mm-thick adherends: **a** clinched, **b** clinchbonded joints. *Dotted lines* refer to bonded joints

close to the sum of the simple bonded and fastened joints, respectively. This fact can be explained also in terms of relative stiffness of adhesive and fastener. Since the bondline is large and thin, it is stiffer than the fastening, therefore, the majority of the load is carried by the adhesive layer until it breaks, then fastening takes over. The strength of adhesive, clinched and hybrid joints is compared in Fig. 28 as a function of geometrical parameters.

As expected, the strength of clinched joints is affected by the pitch and the plate thickness, while the strength of clinchbonded joints is, alike adhesive joints, only slightly dependent on geometrical factors. It can be noticed also that the response surface of adhesive and clinchbonded joints are close to each other. This probably means that for the temperature considered in Fig. 28 (23°C), the strength of the clinchbonded joints is predominantly related to the contribution of the adhesive.

In Fig. 29 the strength is shown as a function of the temperature. The plate thickness in Fig. 29 is 2.5 mm in order to highlight the contribution of the clinch, which increases for thick plates. The advantage of the clinchbonded joint is

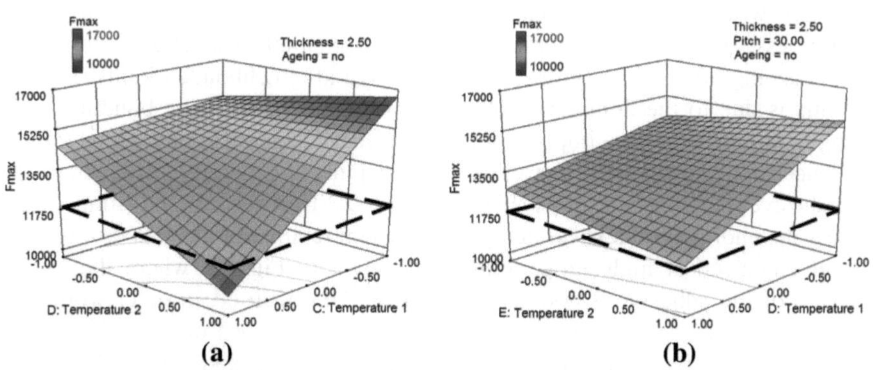

Fig. 29 Strength as a function of temperature of 2.5 mm plate thickness: **a** adhesive, **b** 30 mm pitch clinchbonded joints. *The dotted lines* refer to clinched joints

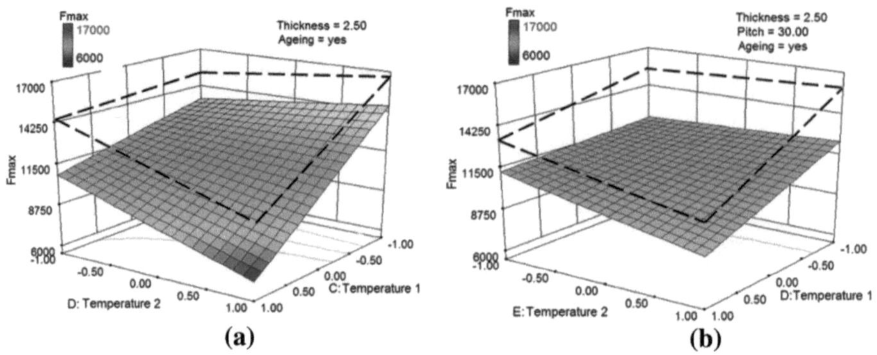

Fig. 30 Strength as a function of temperature of aged, 2.5 mm plate thickness: **a** adhesive, **b** 30 mm pitch clinchbonded joints. *The dotted lines* refer to non-aged condition

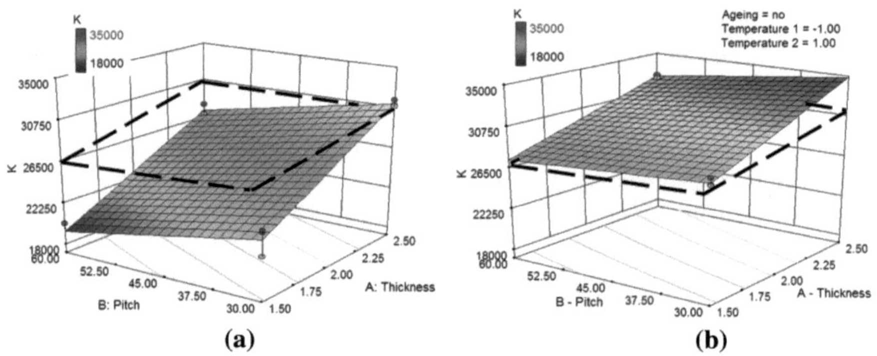

Fig. 31 Stiffness at room temperature as a function of thickness and pitch of: **a** clinched, **b** clinchbonded joints. *Dotted lines* refer to adhesive joints

evident with respect to the clinched one, while the situation is different when compared with the adhesive joint: at low temperatures the clinchbonded joint is slightly less strong, while at high temperatures it guarantees a better performance.

The ageing affects in a similar way adhesive and clinchbonded joints (Fig. 30) and this is due to the dominance of the contribution of adhesive bonding to the strength. Hybrid bonding simply reduces the effect of ageing at high temperatures.

Concerning the stiffness, the clinchbonded joint is generally stiffer than both the adhesive and clinched joints (Fig. 31), with a difference ranging from about 3 to 10% depending on thickness and pitch (at room temperature). This difference is further enhanced at high temperatures, as shown in Fig. 32, where the loss of stiffness of the adhesive is counterbalanced by the clinch or, in other words, the rigidity of adhesive layer and of the clinch get close to each other.

The stiffness of aged and non-aged adhesive and clinchbonded joints is affected in a different way depending on the temperature (Fig. 33). In general, the hybrid joint is less sensitive to the temperature than the adhesive joint under both aged

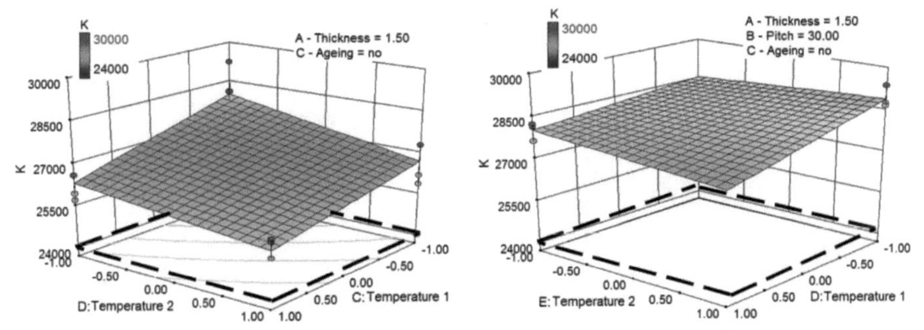

(a) **(b)**

Fig. 32 Stiffness as a function of temperature of 1.5 mm plate thickness: **a** adhesive, **b** 30 mm pitch clinchbonded joints. *Dotted lines* refer to clinched joints

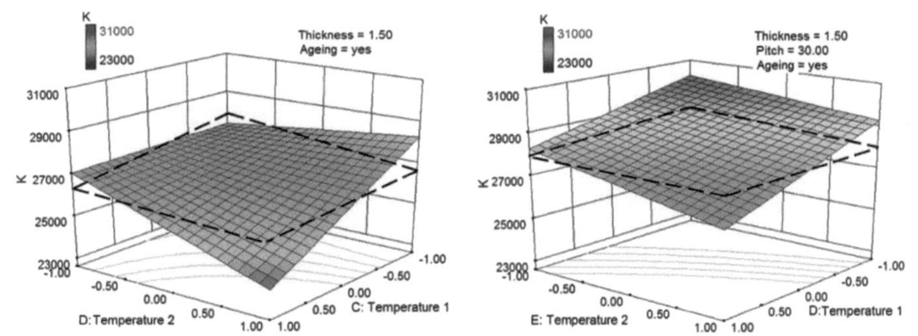

(a) **(b)**

Fig. 33 Stiffness as a function of temperature of 1.5 mm plate thickness: **a** adhesive, **b** 30 mm pitch clinchbonded joints. *Dotted lines* refer to the non-aged condition

and non-aged conditions and, interestingly, ageing causes an increment of strength at temperatures lower than room temperature.

Finally, the clinchbonded joint shows always a higher energy absorption than both adhesive and clinched joints, Fig. 34. It is evident also that, for the motivations described at the beginning of this section, the clinchbonded joint energy absorption tends to the sum of the values of simple joints.

Only at very high temperatures, the value of energy absorption of clinchbonded joints becomes close to that of clinched joints (as shown in Fig. 35), because of the decrease of adhesive strength (see Fig. 29).

The ageing affects strongly the energy absorption of adhesive joints (Fig. 36). After the peak load, they do not carry any further deformation, in other words no more energy can be absorbed. Therefore, a reduction of strength is accompanied by a reduction of energy absorption. In the case of the clinchbonded joint, when the adhesive fails the clinch continues to carry the load allowing large elongations,

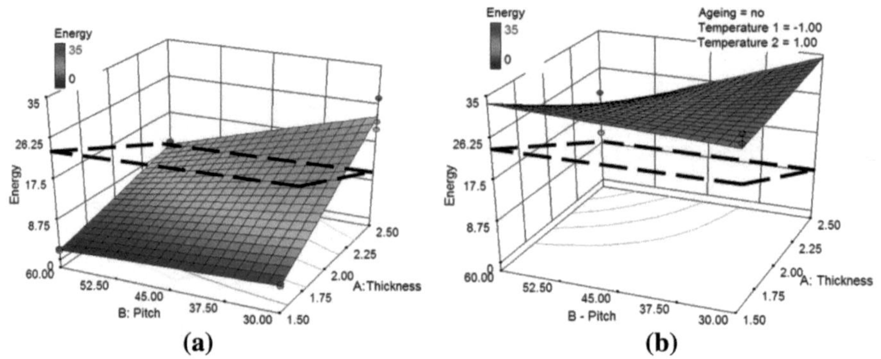

Fig. 34 Energy absorption at room temperature as a function of thickness and pitch of: **a** clinched, **b** clinchbonded joints. *Dotted lines* refer to adhesive joints

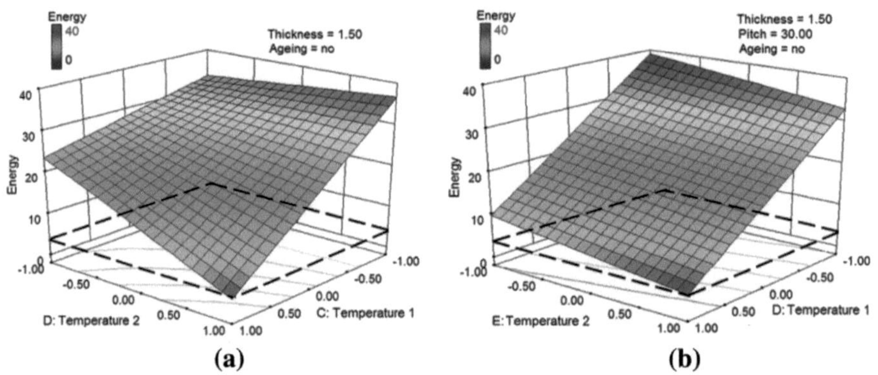

Fig. 35 Energy absorption as a function of temperature of 1.5 mm plate thickness: **a** adhesive, **b** 60 mm pitch clinchbonded joints. *Dotted lines* refer to clinched joints

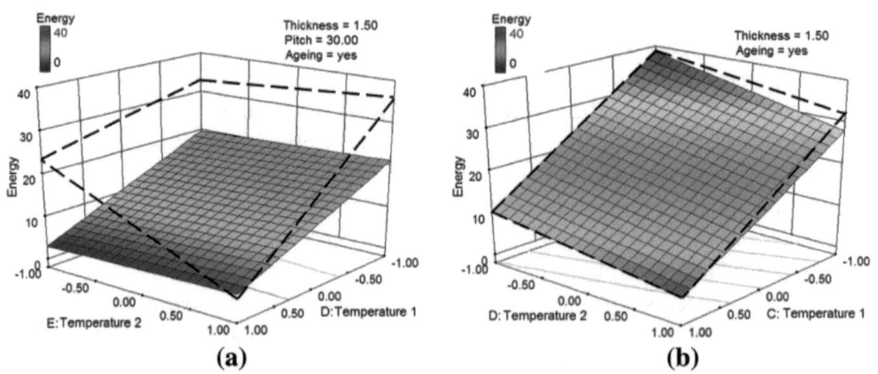

Fig. 36 Energy absorption as a function of temperature of 1.5 mm plate thickness: **a** aged adhesive, **b** 60 mm pitch aged clinchbonded joints. *Dotted lines* refer to non-aged conditions

hence energy absorption. Since ageing does not affect the amount of energy required to break the clinched dot, it affects moderately the energy absorption of the clinchbonded joint.

4 Joint Modelling

During the design of a product, the knowledge of its mechanical behaviour is fundamental. In case of simple components like a shaft or a beam, analytical solutions are available to predict the stresses and evaluate the failure load. When the component becomes more complicated, the use of the finite element method (FEM) or other numerical techniques is necessary. Following this trend, the possibility to predict accurately the behaviour of joints appears from the needs expressed by the industry.

The main objective of the work done in this section was to simulate the failure process of a clinchbonded joint, including also the damage of the material for both the adhesive and the metallic plates. In particular, the damage development in the metallic parts was accounted for by means of the Gurson–Tveergard–Needleman (GTN) and, concerning the adhesive layer, using the cohesive zone (CZ) model. The damage models parameters were obtained by inverse analysis, fitting experimental load-elongation data of simple joints with the numerical analyses. The calibrated parameters were then used to predict deformation and failure of the clinchbonded joints.

4.1 Literature Analysis

Failure of adhesive joints is a subject dealt with since long in the literature which will not be discussed here for the sake of brevity. However, no model has been developed, to the knowledge of the authors, neither on the simulation of failure of clinched joints nor of clinchbonded joints. Only a few works deal with the simulation of forming of clinched joints which are described next.

The forming process of a clinch joint was studied by Hamel et al. [6]. This work proposes a fast and efficient FEM methodology to simulate clinch forming, in respect to process parameters, by developing remeshing and field mapping tools, in order to accurately account for the material non-linearities, contact interactions and large deformations involved in the forming process. The authors studied the influence of friction, materials and geometry of tools in order to verify their influence over the joint section shape.

A similar modeling work was done by de Paula et al. [4] with the aim to understand the influence of punch and die geometries on the joint shape. They concluded that the best geometries is characterized by a reduced die depth and a slightly conical punch (configuration d). The authors also verified the strength of the joints by detaching plates and good results were found in comparison with experimental tests. The simple elasto-plastic behaviour of the material was used, and damage was not considered. Therefore, the failure is simply the failure of the interlocking, while the intrinsic failure of the sheet cannot be simulated.

4.2 Damage Models

4.2.1 GTN Model

The model is based on a homogenized representation of a voided material. This corresponds to the definition of a void volume or porosity fraction, *f*, defined as the ratio of the volume of all cavities in a material element to its total volume. In other words, a porous material can be thought as a material without cavities, where the mechanical response is modified by *f*.

Using this representation, the yield condition is given by the equation:

$$\Phi = \frac{\sigma_{eq}^2}{\sigma_0^2} + 2q_1 f^* \cosh\left(\frac{3}{2} q_2 \frac{\sigma_m}{\sigma_0}\right) - 1 - (q_3 f^*)^2 = 0 \tag{1}$$

where Φ is the plastic potential, σ_0 is the yield stress, σ_{eq} is the von Mises equivalent stress, σ_m is the hydrostatic component of the stress state, q_1, q_2 and q_3 are three coefficients that can be tuned by comparison with a finite element analysis of a unit cell containing a void. The coalescence of voids is simulated by accelerating artificially the voids growth predicted by the model. In other words, an effective porosity f^* is introduced:

$$f^* = \begin{cases} f & \text{if } f < f_c \\ f_c + \delta(f - f_c) & \text{otherwise} \end{cases} \tag{2}$$

where f_c is the critical porosity ratio at the onset of void coalescence. The nucleation of voids is taken into account by defining the void-volume fraction rate $\dot{f}$ as the sum of a void growth rate $\dot{f}_{\text{growth}}$ and a void nucleation rate $\dot{f}_{\text{nucl}}$. Therefore,

$$\dot{f} = \dot{f}_{\text{growth}} + \dot{f}_{\text{nucl}} \quad \text{with} \quad f(t_0) = f_0 \tag{3}$$

where f_0 is the initial void volume fraction. The growth term is obtained from the conservation of mass by means of the plastic flow rate $\dot{\varepsilon}_{kk}^{pl}$,

$$\dot{f}_{\text{growth}} = (1 - f)\dot{\varepsilon}_{kk}^{\text{pl}}, \tag{4}$$

and the nucleation term $\dot{f}_{\text{nucl}}$ is commonly adopted from an empirical approach by assuming a normal distribution of void nucleating particles as a function of plastic flow $\dot{\bar{\varepsilon}}^{pl}$ [3].

$$\dot{f}_{\text{nucl}} = A\dot{\bar{\varepsilon}}^{pl} \tag{5}$$

In this equation A is defined as

$$A = \frac{f_N}{s_N\sqrt{2\pi}}\exp\left[-\frac{1}{2}\left(\frac{\bar{\varepsilon}^{pl} - \varepsilon_N}{s_N}\right)^2\right] \tag{6}$$

where ε_N is the mean value of the normal distribution, S_N represents the average value of the normal distribution and f_N is the volume fraction of the nucleated voids. Voids nucleate up to a volume void fraction equal to f_F where final failure of material is reached.

4.2.2 CZ Model

The underlying idea of the CZ model is to combine an energetic criterion to simulate crack propagation with a stress criterion for damage initiation. The crack growth is described by equations that relate the opening and shear displacement of the damaged material ahead of the crack tip to the stress applied to the crack faces. More or less complicated shapes can be assigned to this relationship, for example a trapezoidal or a three-parameter law [7]. Based on previous experience of Pirondi [15], in this work a trapezoidal shape was assumed, as shown in Figure 37, where

$$\lambda = u_n/\delta_c. \tag{7}$$

where u_n is the opening of crack faces and δ_c the value of u_n at failure. The dissipated fracture energy Γ (area under the curve) is equal to

$$\Gamma = \frac{1}{2}\sigma_m\delta_c[1 - \lambda_1 + \lambda_2] \tag{8}$$

where the values of λ_1 and λ_2 define the shape of the trapezium.

Considering also the tangential displacement u_t, a more complex model was developed by Needleman [11] and later extended by Tvergaard and Hutchinson [18]. A damage variable d was used to modify the initial elastic behaviour:

$$d = \max\sqrt{\langle\upsilon_n\rangle^2 + \upsilon_t^2} \tag{9}$$

Fig. 37 Trapezoidal stress–displacement relation of the cohesive law used for the adhesive

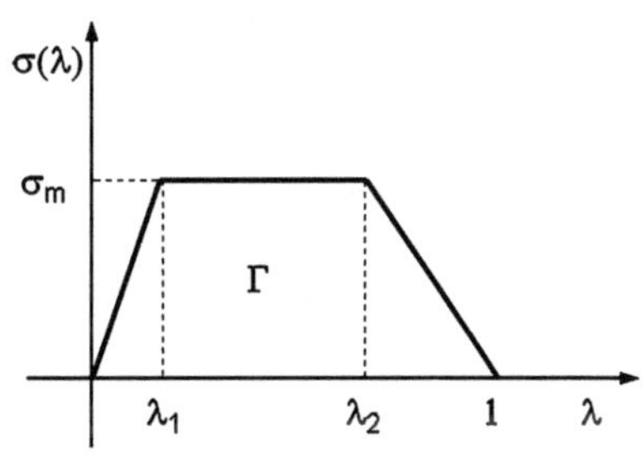

Fig. 38 Fracture surfaces of a clinched specimen: the joint fails do to the failure of the material around the button [5]

where

$$\upsilon_n = u_n/\delta_n \quad \text{and} \quad \upsilon_t = u_t/\delta_t \tag{10}$$

and δ_n (or δ_t) is the critical opening for a pure mode I (or mode II). After the onset of damage, the normal (S_n) and tangential (S_t) stress components can be written as

$$\left\{ \begin{array}{c} S_n \\ S_t \end{array} \right\} = F(d) \left\{ \begin{array}{c} E\upsilon_n \\ G\upsilon_t \end{array} \right\} \tag{11}$$

where *F*(*d*) is a decreasing function of the damage variable. *E* and *G* are the initial Young's and shear elastic moduli respectively. Concerning the mode I, a negative displacement (i.e. a compressive stress) does not increase the damage.

The parameters to be calibrated are the stress peak (σ_m), the fracture energy (Γ) and the critical value of the displacement (δ_c). These parameters must be calibrated for opening and shear modes. Complex relationships for the mixed-mode behaviour can be found in the literature [7, 8]). In this work, as an approximation, the same behaviour was assumed for both mode I and mode II, and therefore, using a traction-separation law, the values of parameters are equal for both the opening and the shear modes.

4.3 Reference Experiments

At the time the simulation were performed, the experimental phase previously discussed was not yet carried out, therefore, tests on simple adhesive joints were performed, while the clinched and clinch bonded joints data were taken from literature.

4.3.1 Clinched Joint

The numerical simulations were compared qualitatively with literature results [5]. The joint geometry is shown in Fig. 7a.

The clinched joints were fabricated following the ASTM 3165-91. The construction steel used has nominally a yield stress of 295 MPa and a tensile strength of 350 MPa. The test results obtained are shown in Fig. 8a while the failure surfaces are illustrated in Fig. 38: it can be seen that the button is deformed and the failure happens due to the shear failure of the button.

Fig. 39 Fracture surfaces of the CB joint (Del Gamba 2006)

4.3.2 Adhesive Joint

To obtain the cohesive law of a thin layer of adhesive in pure mode I, fracture toughness tests were carried out on double cantilever beam (DCB) specimens according to ASTM D 3433. The adhesive used was Henkel Hysol 9466, a two-component epoxy adhesive. The specimens were bonded and cured at room temperature. Although the adhesive needs only 24 h to develop most of the strength, all of the tests were carried out at least one week after the joint production, to ensure a complete curing. Before the application of the adhesive, the surfaces of the substrate were grit blasted and then cleaned with acetone.

The initial crack length (a_0) was controlled with a Teflon tape. In order to develop a natural sharp crack, the specimen was fatigue pre-cracked. The crack length was evaluated by means of the compliance calibration method (CCM). The crack opening was measured by means of a clip gauge. The critical strain energy release rate G_c is given by

$$G_c = \frac{P^2}{2b}\left(\frac{\partial C}{\partial a}\right) = \frac{P^2 a^2}{bEI}\left(1 + \frac{1}{\lambda_\sigma a}\right)^2 \tag{12}$$

where P is the maximum load sustained by the specimen, a the crack length, I the moment of inertia of each substrate and λ_σ a coefficient depending on the geometry and materials [9]. The joints always showed a mixed interfacial-cohesive failure.

4.3.3 Hybrid Clinchbonded Joint

The hybrid clinchbonded joint experimental results were taken from the work of Del Gamba [5]. The geometry of the joint was the same as that shown in Fig. 7a for a simple clinched joint, including also a thin layer of adhesive between the plates. The adhesive used was Henkel Hysol 9461. This is a thixotropic, two-component epoxy adhesive quite similar to Henkel Hysol 9466. The adhesive increases the maximum load (up to 30%), and after the failure of the adhesive, the joint behaves like a simple clinched one (Fig. 8b).

Figure 39 shows the failure surfaces of the clinchbonded joint. The rupture of the clinched dot is similar to that of a simple clinched joint. Moreover, an interfacial failure was identified after testing.

4.4 Finite Element Models

3D numerical models were built for clinched and clinchbonded joints, while two-dimensional models were used for the adhesive joints. Common features of 3D models are the use of hexahedral, eight-noded iso-parametric elements, with reduced integration and hybrid formulation. Where possible, symmetry conditions were used in order to reduce the computational effort. In the case of the adhesive layer in hybrid joints, hexahedral cohesive elements with traction-separation nonlinear response were used. Clinched and clinchbonded joints were analyzed using the explicit version of the ABAQUS® ver. 6.7 software. With the aim to reduce the drawback coming from the high level of distortion typically present in case of large plastic deformation, the arbitrary Lagrangian–Eulerian adaptive meshing was used: this meshing technique combines the features of pure Lagrangian analysis and pure Eulerian analysis. Initially the simulation works as simple Lagrangian model where the material points are coincident with the mesh; when the mesh distortion reaches a certain value (defined by the user), the solver moves the mesh over the material point in order to reduce the mesh distortion.

The use of explicit solver was required, due to the non-linearities shown by the simulations. The explicit procedure integrates through time by using many small time increments: the stability time increment limit for the solution (with no damping) is given in terms of the highest frequency of the system ($\omega_{\max}$) as

$$\Delta t \leq \frac{2}{\omega_{\max}} \tag{13}$$

An approximation to the stability limit is often written as the smallest transit time of a dilatational wave across any of the elements in the mesh

$$\Delta t \approx \frac{L_{\min}}{c_d} \tag{14}$$

where $L_{\min}$ is the smallest element dimension in the mesh and c_{d} is the dilatational wave speed that depends on the density (ρ) of the material and on the elastic modulus (E).

$$c_{\mathrm{d}} = \sqrt{\frac{E}{\rho}} \tag{15}$$

The time increment used in an analysis must be smaller than the stability limit. Failure to use a small enough time increment will result in an unstable solution. When the solution becomes unstable, the time history response of solution variables such as displacements will usually oscillate with increasing amplitudes. In case of small elements, the time increment of the analysis is reduced proportionally, therefore the number of increments is increased leading to a high computational time. In other words, the smaller is the minimum element dimension, the higher is the computational effort. In order to reduce this drawback, the mass scaling procedure can increase locally the material density, thus the dilatational wave speed. In order to tune mass scaling and avoid excessive mass increase, that could bring to overestimation of the inertias and dynamic effect, the analysis was run and then the deformation energy was compared with the kinetic energy. For these analyses, the observed time increment was accepted only if the deformation energy was at least three orders of magnitude higher than the kinetic energy, otherwise the time increment was reduced. All of the analyses were conducted considering the geometrical non-linearities, and in order to reproduce the experimental boundary conditions, the clinched and clinchbonded joints were pinned at one end and a displacement was applied to the other end.

4.4.1 Clinched Joint

The simulation was divided in steps representing the phases of the forming process and the subsequent shear test (see Fig. 40).

In the first step, the holder pushes the plates against the matrix. In the second step, the punch moves down and squeezes the plates into the matrix cavity. In the third step, the punch continues to move down and the segments of the matrix begin to open, allowing the metal to flow. In the fourth step, the contacts involving rigid surfaces are disabled in order to simulate the removal of the joint from the clinching machine. The last step simulates the shear test and therefore an appropriate relative displacement is applied to the plates.

The plates were modeled with 8 deformable elements in the plate thickness direction, while the matrix, the holder and the punch were modeled as analytical rigid surfaces.

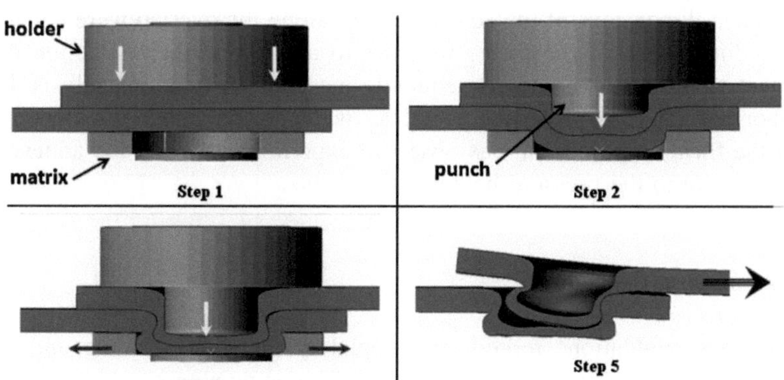

Fig. 40 Steps of analysis in the finite element model for the clinched joint

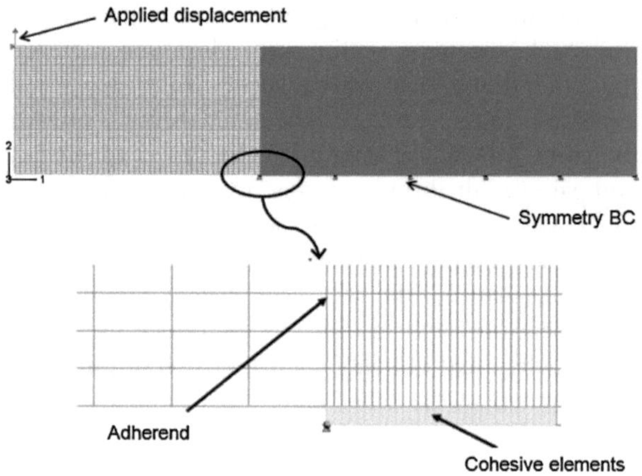

Fig. 41 Geometry, mesh and boundary conditions of the finite element model for the fracture test DCB simulation

4.4.2 Adhesive Joint

In order to obtain the parameters of the CZ model for the adhesive layer, the DCB tests were simulated. The results of this simulation were compared with the fracture tests descried previously, in order to find the values of fracture energy Γ and maximum tension $\sigma_{\max}$ by an inverse method. The geometry mesh and the boundary conditions of the model are shown in Fig.41.

The steel adherends were modeled by means of 2D quadrilateral, four nodes and plane-strain elements with reduced integration. The adhesive was simulated with a layer of 2D cohesive elements, with a trapezoidal traction-separation law.

4.4.3 Clinchbonded Joint

The clinchbonded joint geometry was the same as those of the simple clinched joint, including also a layer of cohesive elements to simulate the adhesive layer. The element dimensions of the adhesive layer along the overlap were equal to the element dimensions of the plates in order to avoid a further reduction of time increment of the solution and consequently increase the time of analysis. For the clinchbonded joint, no cohesive elements were placed under the punch, since during the forming process it was observed experimentally that the adhesive was spread out due to the pressure between the sheets.

4.4.4 Annealing Procedure

The previous simulations reproduce completely the joint forming and testing. In terms of computational effort, the possibility to simulate only the testing, using as the initial clinch geometry the shape of the clinch at the end of the forming, represents an attractive solution. Although this produces a strong reduction of the time of analysis, this procedure leads to the complete neglection of the residual stresses and damage distribution occurring during the forming process. The understanding of the influence of the forming process over the mechanical behaviour of the joints, is therefore fundamental. In order to understand the relevance of the forming process on the behaviour of clinched joints, an annealing step was added in selected cases. The annealing step was introduced between the forming process and the test steps. Annealing is a feature of the ABAQUS® solver that works like a full annealing heat-treatment. Without changing the deformed shape of the model after forming, all of the stress components, hardening and damage variables are reset to zero. With this procedure, the importance of simulating the forming process can be assessed. The damage sustained during the forming process is also better understood. Unfortunately, the procedure cannot be applied to cohesive elements, therefore this kind of analysis was limited to clinched joints. However, since the adhesive was cured after the clinch was formed, processing effects should affect hybrid joints in the same way as simple joints.

4.5 Calibration of Damage and Failure Models

4.5.1 GTN Model of the Clinched Joint

The clinched joint simulation was directly carried out using parameters taken from literature [1, 11–13, 17] where only f_N, f_F and f_c were slightly modified to avoid excessive distortion of the elements and analysis convergence difficulties (Table 8).

Figure 42 shows the load vs. displacement curve for the clinched joint simulation. The simulation reproduces qualitatively well the different test stages identified experimentally (Fig. 8a). However, due to the unknown material properties of metal plates, differences were identified in the maximum load and corresponding displacement.

4.5.2 Influence of the Forming Process

Using the damage parameters previously defined, the simulation including the annealing step were carried out and the results compared with those without the

Table 8 Parameters of the GTN damage model used for the clinched joint

Parameter	Value	Parameter	Value	Parameter	Value
q_1	1.5	ε_N	0.1	f_F	0.98
q_2	1	s_N	0.075	f_c	0.015
q_3	2.25	f_N	0.09		

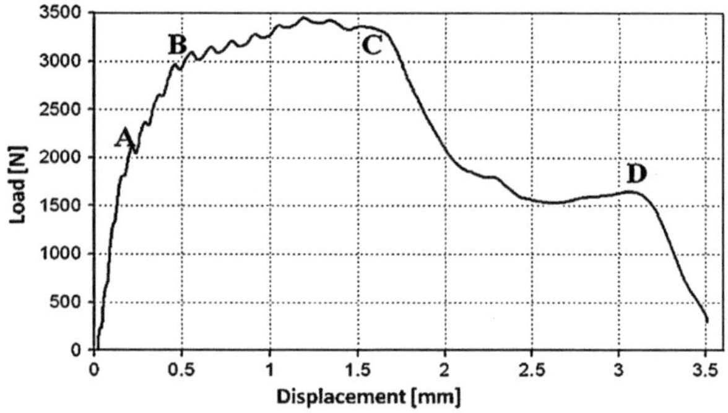

Fig. 42 Numerical load vs. displacement curve of the clinched joint with the GTN model

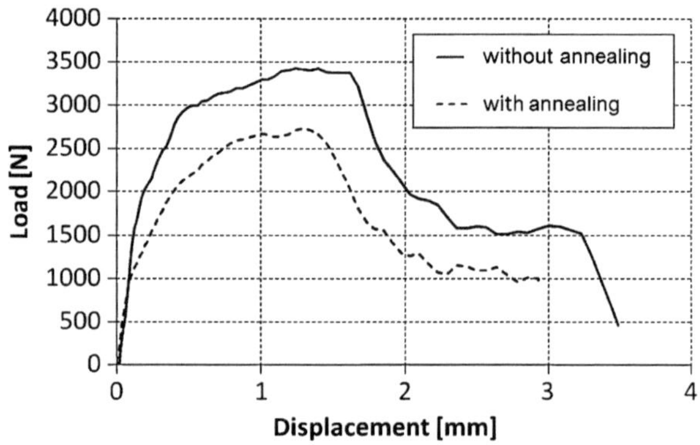

Fig. 43 Numerical load–displacement curve with and without annealing for the clinched joint with the GTN model

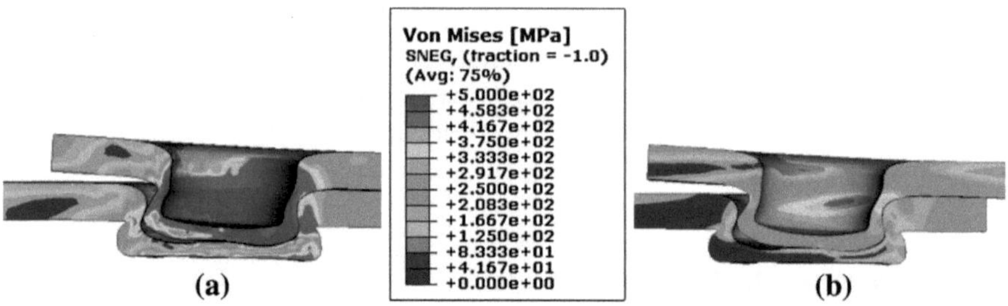

Fig. 44 Influence of the annealing step on the clinched joint: deformed shape and von Mises stress distribution for simulation without (**a**) and with the annealing step (**b**)

annealing step. Figure 43 shows the numerical load versus displacement curve for the simulation with and without the annealing step.

A different behaviour of the joint was identified in this case. In fact, although the damage variable was relieved, the maximum load is smaller for the model with the annealing step, where the contact stresses were also eliminated. Since friction in the clinched joint plays an important role concerning the plates sliding, its suppression reduces the joint strength. This is confirmed by comparing Figure 44a and b.

Table 9 CZ model parameters for the adhesive Hysol 9466

CZM parameter	
Γ_0 (J/m^2)	725
σ_m (MPa)	60
λ_1	0.2
λ_2	0.5

The two figures show the von Mises equivalent stress distributions for the same displacement with (b) and without (a) annealing. In the second case, the interfacial stress are relieved and there is a geometrical interlock. When the load is applied, the contact on the leftmost part of the button disappear and the strength of the joint is simply given by the mechanical strength of the rightmost part of the button: in other words there is no contribution of the interlocking and this produces a reduction of the joint strength.

4.5.3 CZ Model of the Adhesive Joint

The set of parameters which gave the best matching to the experimental results are summarized in Table 9.

The comparison between the results of the simulation and the experimental data is shown in Fig. 45. As previously mentioned, this cohesive law was also considered to simulate the adhesive layer behaviour in shear, as a first approximation.

4.6 Simulation of Hybrid Joints

Failure of the hybrid joints was simulated by assigning the parameters of the damage models calibrated previously on clinched and adhesive joints, respectively, to the different materials of the hybrid joint. Figure 46 shows a comparison between the simulations of a simple clinched joint and a hybrid clinchbonded joint.

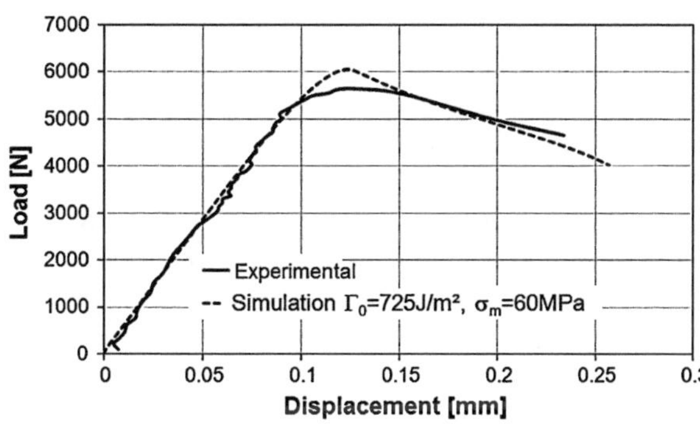

Fig. 45 Experimental and numerical simulation load–displacement curves comparison for the DCB test after the numerical fitting procedure

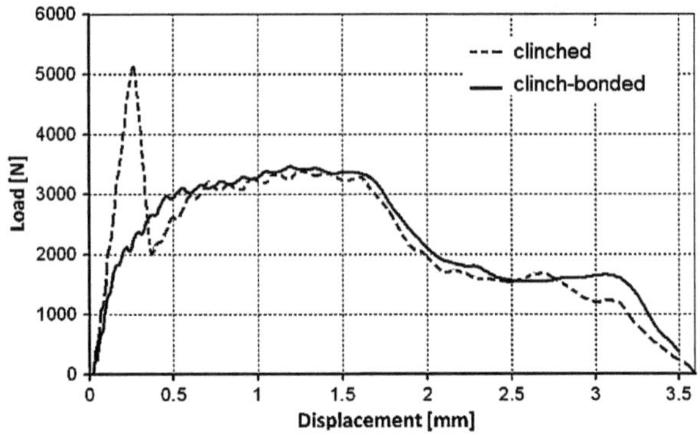

Fig. 46 Numerical load–displacement curves for the clinched and clinchbonded joints simulated with the GTN model

In the case of the clinchbonded joint, the initial peak load is due to the adhesive and after that, the joint behaves like a simple clinched joint. This result is in good agreement with the results found by Del Gamba [5] shown in Fig. 8b.

5 Conclusions

In this work the mechanical behaviour of clinchbonded joints is deeply analyzed: in particular the main effort is addressed to understand under which conditions the hybrid joint is more efficient than simple clinched and adhesive joints. The analysis has been focused on joints where plates of different materials (aluminum and steel in this case) are joined to each other, where the commonly used resistance welding cannot be applied. To perform the investigation, a test plan was defined according to the design of experiment (DoE) methodology. From the test results, it appears that in the clinchbonded joint the strength corresponds to the strength of the simple bonded joint since adhesive bonding is much stronger and stiffer than the mechanical fastening. However, when the adhesive performance is reduced by environmental effects (high temperature, ageing) the presence of a clinch dot mitigates the influence of these effects over the strength in comparison with simple bonded joints. Concerning stiffness and energy absorption, an improvement due to the contemporary presence of both the joining techniques is always found. This is a relevant remark since the improvement of stiffness and of the energy absorption in case of impact are two of the most important goals expected by the industries.

The last part of the work deals with the finite element simulation of the forming and failure process of clinchbonded joints using damage models. In particular, the parameters of damage models are identified by comparison of finite element simulations and experiments conducted on simple clinched and adhesive joints. These parameters are then used for the simulation of the hybrid joint. The results of the simulation are in good agreement with the experimental results of the hybrid joints tests, therefore it is possible to simulate the mechanical behaviour and failure of clinchbonded joints starting from the knowledge of the mechanical behaviour and failure of clinched and adhesive joints.

References

1. Aravas, N.: On the Numerical integration of a class of pressure-dependent plasticity models. Int. J. Numer. Methods Eng. **24**, 1395–1416 (1987)
2. Carboni, M., Beretta, S., Monno, M.: Fatigue behaviour of tensile-shear loaded clinched joints. Eng. Fract. Mech. **73**, 178–190 (2006)
3. Chu, C., Needleman, A.: Void nucleation effects in biaxially stretched sheets. J. Eng. Mater. Tech. **102**, 249–256 (1980)
4. de Paula, A.A., Aguilar, M.T.P., Pertence, A.E.M., Cetlin, P.R.: Finite element simulations of the clinch joining of metallic sheets. J. Mater. Process. Technol. **182**, 352–357 (2007)
5. Del Gamba, L.: Studio teorico e sperimentale del comportamento di unioni ottenute per clinciatura con o senza incollaggio (Experimental and numerical study of the behaviour of clinched and clinchbonded joints). Graduation Thesis, University of Pisa, Pisa, Italy (2005)
6. Hamel, V., Roelandt, J.M., Gacel, J.N., Schmit, F.: Finite element modeling of clinch forming with automatic remeshing. Comput. Struct. **77**, 185–200 (2000)
7. Li, S., Thouless, M.D., Waas, A.M., Schroeder, J.A., Zavattieri, P.D. Use of a cohesive-zone model to analyze the fracture of a fiber-reinforced polymer-matrix composite. Compos. Sci. Tech. **65**, 281–293 (2005)
8. Li, S., Thouless, M.D., Waas, A.M., Schroeder, J.A., Zavattieri, P.D.: Mixed-mode cohesive-zone models for fracture of an adhesively-bonded polymer-matrix composite. Eng. Fract. Mech. **73**, 64–78 (2006)
9. Krenk, S. (1992) Energy release rate of symmetric adhesive joints. Eng. Fract. Mech. **43**, 549–559
10. Needleman, A.: A numerical study of necking in circular cylindrical bars. J. Mech. Phys. Solids **20**, 111–127 (1972)
11. Needleman, A.: A continuum model for void nucleation by inclusion debonding. J. Appl. Mech. **54**, 525–531 (1987)
12. Needleman, A., Tvergaard, V.: Analysis of the cup-cone fracture in a round tensile bar. Acta Metall. **32**, 157–169 (1984)
13. Needleman, A., Tvergaard, V.: Material Strain-Rate Sensitivity in the Round Tensile Bar. Brown University Report, Division of Engineering, Providence, USA (1985)
14. Petrie, E.M.: Handbook of Adhesives and Sealants. McGraw-Hill, UK (2000)
15. Pirondi, A.: Influence of cohesive law shape and adhesive modelling on the failure prediction of T-peel joints. In: Proceedings of XVI European Congress of Fracture, Alexandroupolis, Greece (2006)
16. SIMULIA: ABAQUS® User's Manual, Ver. 6.7 (2007)
17. Tvergaard, V.: Influence of voids on shear band instabilities under plane strain conditions. Int. J. Fract. **17**, 389–407 (1981)
18. Tvergaard, V., Hutchinson, J.: The relation between crack growth resistance and fracture process parameters in elastic-plastic solids. J. Mech. Phys. Solids **40**, 1377–1397 (1992)
19. Varis, J.: Ohutlevyjen Puristusliittäminen (Clinching of Sheet Metals). Federation of Finnish Metal, Engineering and Electrotechnical Industries, MET, Helsinki, Finland (1997)
20. Varis, J.: Kuumasinkityn lujan rakenneteräksen puristusliittaminen (Clinching of zinc-coated high strength structural steel). Research Report No. 27, Lappeenranta University of Technology, Department of Mechanical Engineering, Lappeenranta, Finland (1998)
21. Varis J., Lepisto J.: A simple testing-based procedure and simulation of the clinching process using finite element analysis for establishing clinching parameters. Thin-Walled Struct. **41,** 691–709 (2003)

Technology of Clinch–Adhesive Joints

Tomasz Sadowski and Tadeusz Balawender

Abstract The technology of clinching is rather old. The first patent for clinching was granted in Germany in 1897 (He in Int. J. Adv. Manuf. Technol. 48:607–612, 2010; Varis in J. Mater. Proc. Technol. 132:242–249, 2003). However, pure clinching was not used on an industrial scale until the 80s of the twentieth century. Shortage of this cold-formed mechanical fastening can be removed by hybrid joining involving clinching and adhesion techniques. It is a modern and innovative technology allowing connection of different types of materials to create durable and reliable light constructions. However, its practical implementation is still very limited. Aspects associated with the automotive, aeronautical and aerospace industries which could benefit from this technique are described in this chapter. The application of clinching together with adhesive joining leads to an improvement (in comparison to a simple joint): of the quality, rigidity and the load capacity, dumping of noise and vibration, pressure tightness and corrosion protection. This entails a significant increase of: long-term static strength; the amplitude of force under fatigue test; the energy required to the rupture of the hybrid joint under static, dynamic and impact loading. This chapter aims to describe the basic technological aspect of the creation of clinch–adhesive joints and different types of joining. Basic parameters that need to be taken into account in the designing process are also presented. A comparison of experimental testing of the hybrid joint with simple clinching for a combination of different joining

T. Sadowski (✉)
Department of Solid Mechanics, Faculty of Civil Engineering and Architecture, Lublin University of Technology, Nadbystrzycka 40, 20-618 Lublin, Poland
e-mail: t.sadowski@pollub.pl

T. Balawender
Department of Materials Forming and Processing, Faculty of Mechanical Engineering and Aeronautics, Rzeszow University of Technology, A. Powstańców 8, 35-959 Rzeszów, Poland
e-mail: tbalaw@prz.edu.pl

Adv Struct Mater (2011) 6: 149–176
DOI: 10.1007/8611_2010_45

Published Online: 7 January 2011

materials underlines the advantages of the application of hybrid joints. The formulated conclusions can be useful for the application of this new fastening technology in practice.

1 Introduction

Clinching technology (mechanical interlock) is a method of joining different metal parts (mainly sheets) by a process of local deformation without the use of any additional elements (e.g. [1–4]) with the application of a punch and a die. The idea of clinching different materials is rather old—the first German patent was filed in 1897 (DRP-Nr 98517). The application of this technology on an industrial scale started in the 80s of the twentieth century. In particular, the automobile industry uses clinching for joining parts of the body of a vehicle. Prospective application of this technology in aerospace requires more attention.

The mechanical interlock with the application of the TOX clinching technique (TOX® PRESSOTECHNIK GmbH & Co. KG, Weingarten, Germany) is shown schematically in Fig. 1. The required tooling set consists of: a punch, a die and a blank holder. The mechanical joining of two or more blanks (e.g. metal sheets) is only based on the accurate movement of the punch into the die. The sheet metals are deformed locally without the use of any additional elements. The joint strength is due to: force locking, material locking and "S" shape locking, Fig. 2. The quality of the clinching process strongly depends on precisely selected tools. The proper description of this process should include: tool geometries, parameter optimisation, and FEA simulation of the process [5–10]. In this chapter, specific requirements concerning the clinching process in industry (aeronautical, aerospace and automotive) are discussed. The basic modes of joints failure of clinched joints are described. The disadvantage of this type of joint in case of fastening of two or more pieces of materials can be the initiation of a corrosion process at interfaces of joined parts due to environmental factors: physical and chemical.

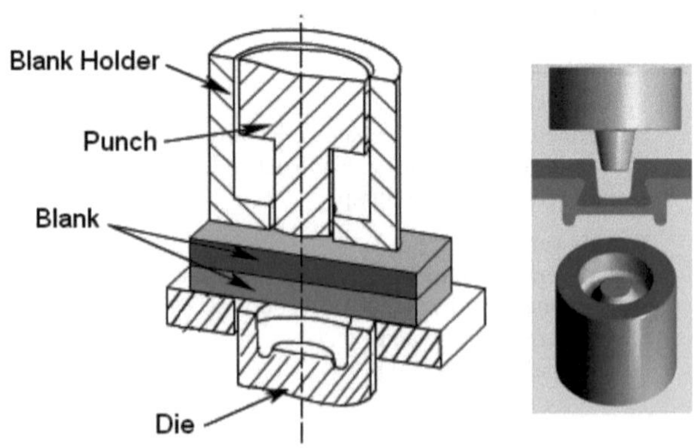

Fig. 1 The TOX clinching model

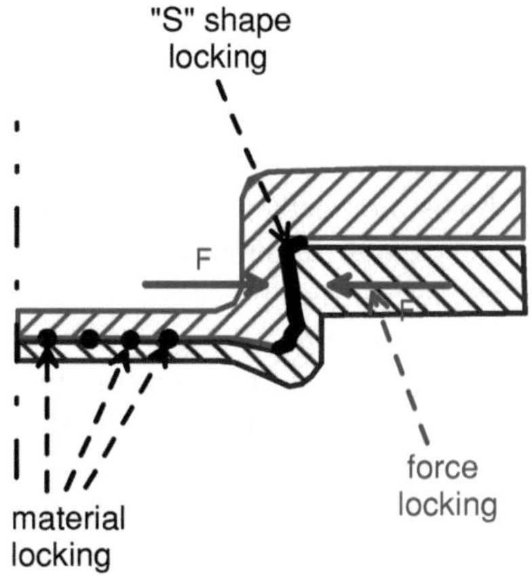

Fig. 2 The mechanisms of locking of joining parts during the clinching process

The main subject of this chapter is a description of the new hybrid joining technology, i.e. clinch–adhesive joint. The idea of combining these two simple techniques, clinching and adhesive bonding, leads to numerous advantages in comparison to both simple methods. The hybrid joining:

- compensates for the disadvantages of two single techniques,
- allows a fixation of the joining materials (blanks) until the adhesive is cured,
- increases the joint strength of both, e.g. shear strength or peeling resistance,
- improves the pressure tightness and corrosion resistance.

The authors' experimental results of joint strength and durability of different clinch–adhesive joints made of different materials are described to illustrate the potentiality of this technique. The failure mechanisms associated with the clinch–adhesive technique are visualized experimentally. Finally, some examples of applications are discussed and conclusions are formulated.

2 Specific Requirements in Industry

The clinch–adhesive technique may be useful mainly in the automotive industry. However, wider application in aerospace and aeronautical industries should be investigated after an elaboration of the proper technology and an experimental verification of numerical models. Each application of the clinch–adhesive technique leads to a formulation of the specific requirements to guarantee a proper design of the joint, and further a manufacture of safe and durable fastening. The basic technological problem is the optimisation of the clinching process—cold forming operations—by the application of the suitable shape of the tools. The proper shape of the manufacturing process leads to improvements of joint strength under different types of loading: mechanical, temperature and environmental or aggressive effects.

The application of pure clinching technology has many advantages for the automotive (mass production) and aeronautics (small series production) industry.

This is important in order to satisfy the production criteria and requirements in industry (e.g. [1]):

- clinching is the green assembly method—no fumes, emissions or high current
- ergonomic for operation and easy for automation—no needs for, e.g. pre-drilling holes, no needs for pre-treatment
- very good lifetime of the tooling set (the punch, the die and the blank holder)
- manufacturing process consumes little energy in relation to spot-welding
- simple set-up
- a similar speed of operation in comparison to spot-welding technique
- allows the joining of different materials (including a multi-layer, an insulation layer) with large differences in thickness
- allows the joining of pre-coated materials, rubber gasket materials and sandwich plates
- easy for non-destructive testing—quality of joint can be checked visually or with the application of simple measurement techniques
- economically attractive—low capital and operation costs
- the clinch type of joint causes more favourable stress distributions and stress concentrations associated with design.

However, clinching has several limitations in practice, like:

- the technique requires access to both sides of the joint
- restricted access to joint areas due to size of the gun
- weak gas and fluid tightness
- weak prevention of corrosion due to surface irregularities.

The benefits of adhesive bonding technique in automotive and aeronautics industries can be specified in the following:

- a force transmission through the joint over large areas (not in a localised contact point)
- no thermal influence on the joined materials microstructure
- suitable for different types of materials and for non-metals
- a good absorption of energy, good noise-dumping properties
- high shear strength
- seals the joint against moisture and aggressive environment
- low cost of production.

The main disadvantages of this simple joint are:

- the technology requires heat curing (when thermosetting adhesive is used)
- relatively weak peeling forces
- sensitiveness to ageing process
- limited strength under thermal loading.

A combination of clinch and adhesive bonding techniques allows us both to take advantage of the positive features and lessen as much as possible the negative factors. It is, however, important to recognise that hybrid joining techniques need to be adapted to each individual case if maximum reliability is to be achieved. The clinch–adhesive bonding provides a more robust solution for joining structural parts. For example, the ability of adhesives to seal joints against moisture and ability to dump noise are highly desirable. The addition of clinching increases the weak peel strength of the adhesive joint, thus resulting in a joint which offers the best of both techniques.

However, from the point of view of the production, the new technological process involves a second processing operation (adhesive bonding) with its associated implications for the control of the process and additional costs.

The considerable benefits that are ensured due to the combination of clinching and adhesive bonding techniques lead to a marked increase of the viability of both techniques in the production of parts of cars and airplanes.

The new clinch–adhesive technique should meet requirements concerning:

- application of more ductile adhesives (low temperature adhesive—LTA and high temperature adhesive—HTA)
- resistance to low impact damage during hail storm
- resistance to ageing under fatigue and stress corrosion (chemical or environmental, e.g. the presence of slat water)
- withstanding low and high temperatures (from −55°C to 200°C) [11, 12].

3 Adhesive Layer Characteristics Required

The types of adhesive layers in clinch–adhesive joints are presented in Fig. 3. One can distinguish two types of interfaces at the joined materials—created by different types of glues: two component adhesive for metal joining (Fig. 3a) and acrylic pressure sensitive tape (Fig. 3b).

In the first case (Fig. 3a), the strength of the hybrid joints highly depends on the type of applied adhesives. One can distinguish brittle (strong and stiff) or ductile (flexible) adhesives. However, the application of an adhesive exhibiting visible

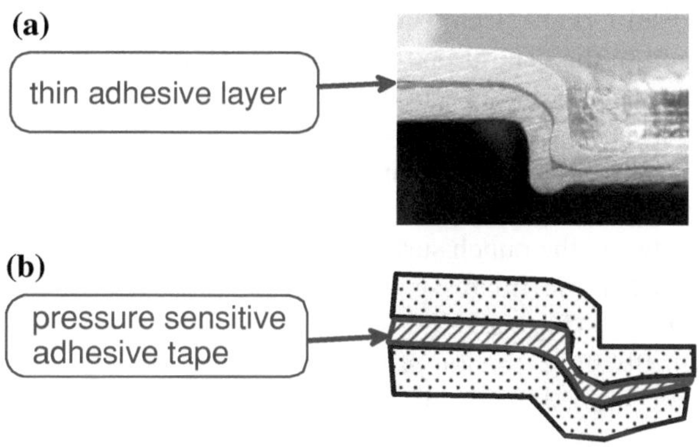

Fig. 3 Types of adhesive layers in hybrid joints: **a** two component adhesive [17], **b** pressure sensitive adhesive type [15]

ductility leads to a decrease in the joint strength. The new concept of mixed adhesive joints —a combination of two different types of adhesives: brittle and ductile—seems therefore to be a good solution to optimise the joint strength, particularly when the joint is subjected to a wide range of temperatures from −50°C to 200°C [13, 14]. The mechanical properties of adhesives strongly depend on the temperature level. At the temperature glass transition point (T_g) a rapid reduction in the Young's modulus and strength can be observed as the temperature increases. The adhesive can no longer carry a substantial load, what can cause a failure of the joint. This temperature fixes the limit of practical hybrid joint application. In relation to the operational temperature influencing the behaviour of the structural joint, one can therefore use LTA or HTA materials for adhesive layers.

The second type of hybrid joint (Fig. 3b)—the application of adhesive tapes—is discussed in Geiss et al. [15] and Brockmann et al. [16]. The authors propose using pressure-sensitive adhesive tape or structural bonding tape. The advantages of this approach include: quick-fix properties of joined components, good viscoelastic properties under impact loading and vibration damping properties. The proposed type of hybrid joints exhibit pressure-sensitive properties from ambient temperatures to 140°C and can be cured to develop good structural properties. However, until now the used pressure-sensitive adhesive joints are limited by their creep under static mechanical load and relatively low strength at elevated temperatures.

4 Manufacture of the Joint

The clinching process proceeds in several stages, as illustrated in Fig. 4 for a circle shape tools geometry. The thickness of the joined sheets ranges from 0.4 to 8 mm for mild steel. Typically the sheet thickness varies from 0.2 to 4 mm, however, there is no requirement of equal thickness of joined sheets. The necessary level of force to create clinching varies between 10 and 100 kN.

The joining process during clinching is caused by metal flow of metallic sheets. The first stage begins when the joined sheets are subjected to blank holder force (BHF), (Figs. 1 and 4a). The joined sheets adhere closely to the surface of the die and to the central bottom point of the punch. It is very important that the area of the punch (cross section) should be approximately equal to 65–70% of the area of the die to create clinching without local incision. In the next step (Fig. 4b), the punch moves downwards and the process of plastic deformation starts in the metallic blank sheets. The deformation is influenced by the friction coefficient between the punch surface and the metallic blanks and is continued until the metal reaches the impression of the die (Fig. 4b). In the next stage of the clinching process (Fig. 4c), the joined sheet blanks gradually fill the die impression and finally the extrusion of the metal sheets takes place. The metal begins to flow and the straight walls of the joint are subjected to a thickness reduction (due to compression) and creation of a specific "S" shape (form locking of the blanks).

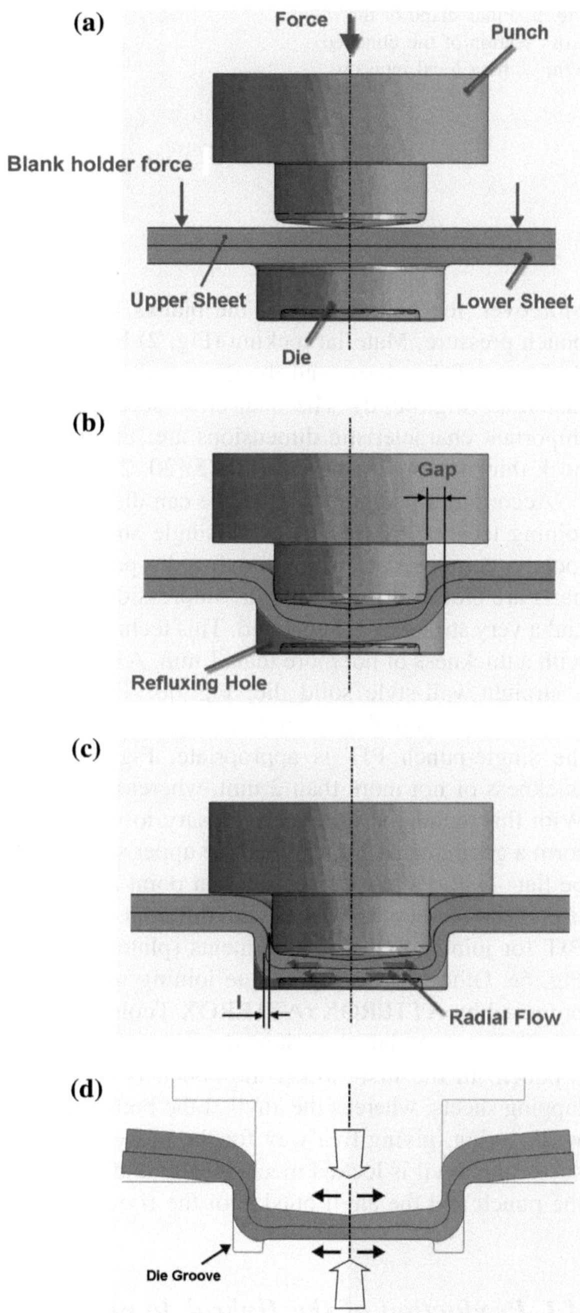

Fig. 4 Stages of the clinch joining process in the straight-wall-style die press joining technology

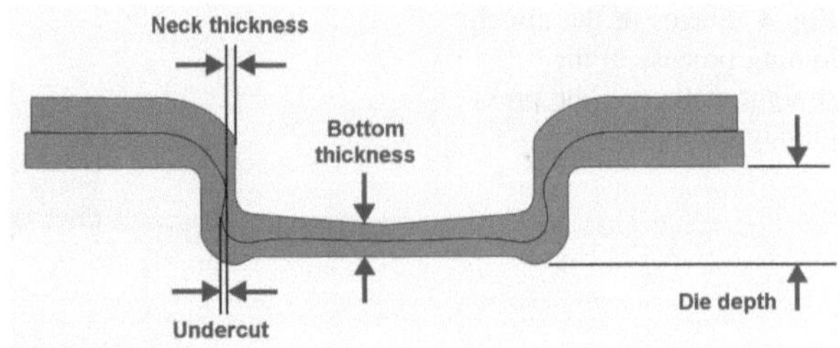

Fig. 5 Final shape of the cross section of the clinched joint without local incision

Moreover, the bottom part of the blanks flow in the radial direction due to the punch pressure. Material locking (Fig. 2) begins with continuous reduction of the thickness of the sheets and the die grooves are filled by extruding metallic sheet materials (Fig. 4c, d). The final cross section of the clinched joint and the most important characteristic dimensions are: the bottom thickness, the die depth, the neck thickness and undercut (Fig. 5)[20, 22].

According to Nong et al. [8] one can distinguish several specific types of press joining technology (PJT) for the single stroke technique, Fig. 6. The first set of tools is composed of a few moving die parts, Fig. 6a. During clinching, the die parts are closed, forming the die impression. After processing, the die parts open and a very strong joint is created. This technology can be applied to metallic sheets with a thickness of not more than 3 mm. A simpler technology is the application of a straight-wall-style solid die, Fig. 6b. The advantage of this technique is its simplicity and durability. In case of joining metal sheets of different thicknesses, the single-punch PJT is appropriate, Fig. 6c. The upper sheet should have a thickness of not more than 2 mm, whereas the bottom one—not less than 6 mm. With this technique it is not necessary to use any die. Single-punch is enough to form a good quality joint. When the upper surface of the clinched materials need to be flat—without any bulge—the flat point PJT is applied, Fig. 6d. This method is useful for a joint of materials with different thicknesses. The last technique is plank PJT for joining very thick elements (plates with a thickness higher than 4 mm), Fig. 6e. Other interesting unique joining method—double stroke clinching—was proposed by ATTEROX (ATTEROX Tools S.A., Renens/Lausanne, Switzerland), Fig. 7. The materials to be joined are pushed into a rigid die with the movement of a punch. In the first stroke, the punch is active and creates a preform of overlapping sheets, whereas the anvil at the bottom of the die is held in place only by a weak spring, giving free way for the movement of the punch. During the second stroke the anvil is locked mechanically and the preform is then squeezed between the punch and the anvil outside of the rigid die, creating a rivet-like joint.

4.1 Production of the Hybrid Joints

The clinching tools described above can be applied in the case of hybrid joints: clinch–adhesive, which are created by putting together simple joints. The hybrid

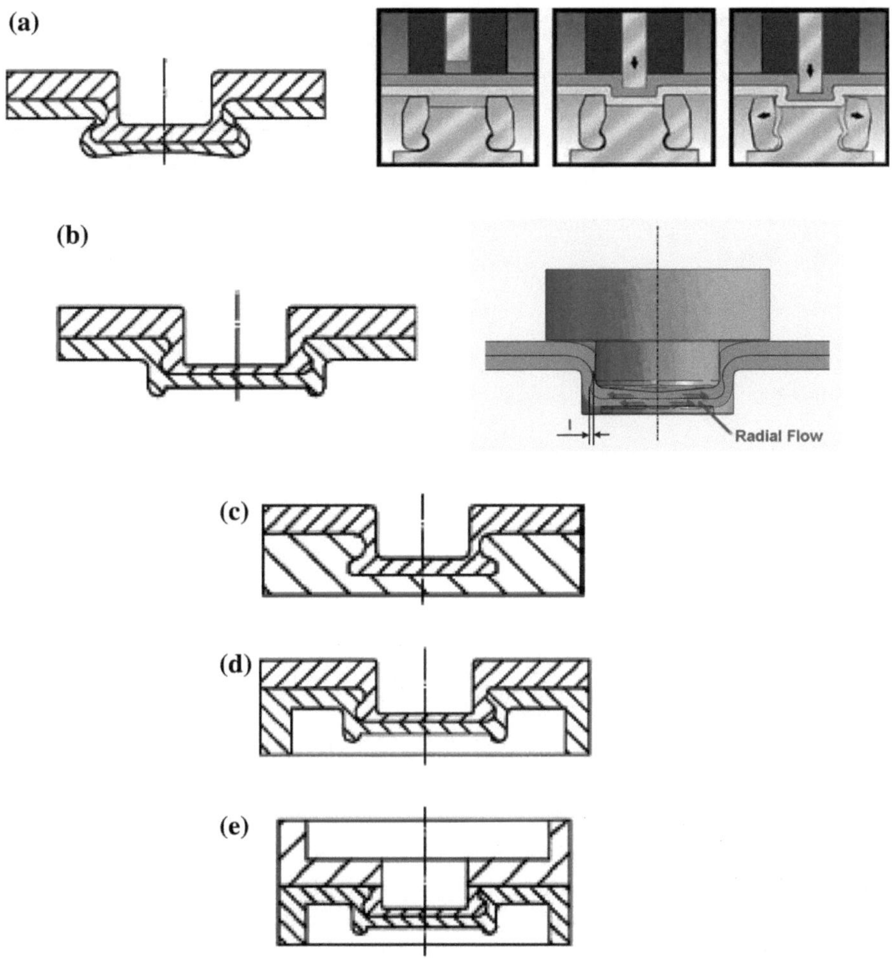

Fig. 6 Types of clinching techniques in relation to the form of the die shape by the single-stroke technique: **a** divisible-style die press joining, **b** straight-wall-style solid die PJT, **c** single punch PJT, **d** flat point PJT, **e** plank PJT

joints can be fabricated in one process or by multiple processes. In the last case, the multiple processes can be performed by producing the simple joints either simultaneously or consecutively—one after the other. In the second case, the chronological order strongly influences the mechanical and thermal properties of the hybrid joint. One can distinguish three variants of production:

- in the variant 1 the sequence of manufacturing operations is the following:
  1. the adhesive is applied to one of the components to be joined
  2. the components are placed together
  3. the joined materials are subjected immediately to the clinching process, which causes the outflow of the adhesive

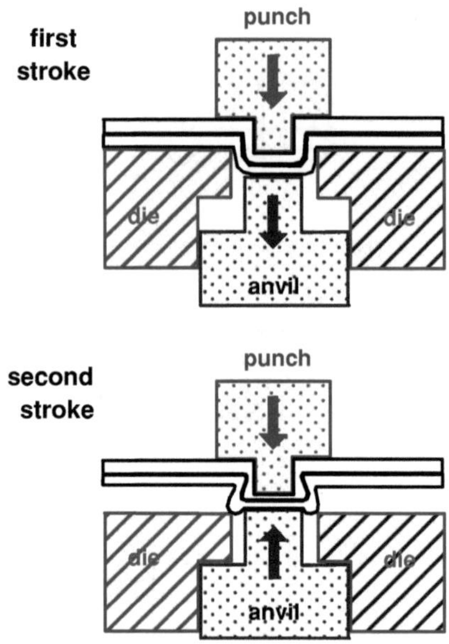

Fig. 7 Double-stroke technique by ATTEROX

4. the joint is subjected to the curing process.

- the variant 2 is called "injection method" and consists of the following operations:

1. place joined parts together
2. clinch the components together
3. apply the low viscosity adhesive in the clinch place—the adhesive penetrates the joined parts (including crevices) due to capillary action
4. the joint is subjected to the curing process

- in the variant 3 the operations in fabrication are the following:

1. the adhesive is applied to one of the components to be joined
2. the components are placed together
3. the joint is subjected to the curing process
4. the joint is subjected to clinching.

The advantage of the clinch–adhesive joining is that it allows the use of not only fluid and pasty adhesives, but also of adhesive foils and bands, Fig. 8. Clinching without local incision can be used with pasty adhesives, whereas the application of an adhesive foil or band is more appropriate when small incisions are created during the manufacturing process.

In comparison to pure clinching, Fig. 2, one can notice that in hybrid fastening an additional strengthening mechanism of the joint is created by the adhesive in the clinching place and also in the overlapping area of the hybrid joint, Fig. 9.

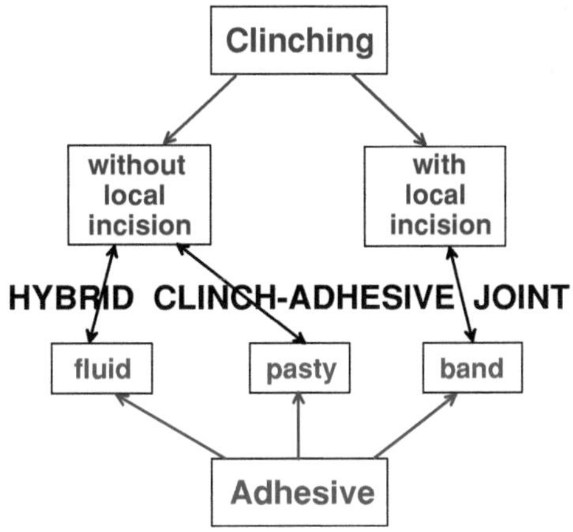

Fig. 8 Fabrication of clinch–adhesive joints

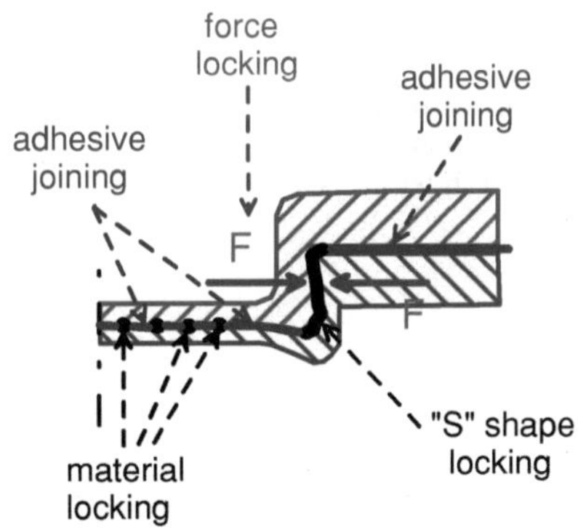

Fig. 9 Mechanisms of clinch–adhesive joining

It significantly increases the load capacity of the joint and its energy absorption during failure.

When the area of the punch (cross section) is much higher than 70% of the area of the die, a certain part of the side surface of the clinch contains local incisions, Fig. 10. This causes the global weakening of the connection of the fastened materials due to material discontinuities, which create local stress concentrations. This technology cannot therefore be recommended for automotive and aeronautical structural applications.

The basic question from the manufacturing point of view is how to optimise the production process in order to get a good quality of the clinching joint, which can be estimated by test separation methods: peel–tension test and tensile–shear test (e.g. [3–10]), Fig. 11. It has been established that the quality of the joint depends on many geometrical fastening process parameters. Figure 12 [5] presents the whole set of parameters to take into account in the clinching optimisation process. The quality of a joint should be examined taking into account: the bottom thickness of the joint—X, the thickness of the undercut—u and the thicknesses of two necks—N_1 and N_2 (also Fig. 5). Jayasekara et al. [5] performed a detailed analysis

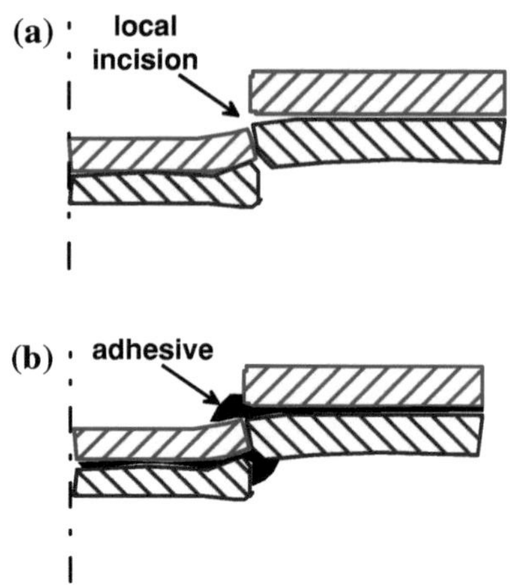

Fig. 10 Clinch joints with local incision: **a** pure clinching, **b** clinch–adhesive joint

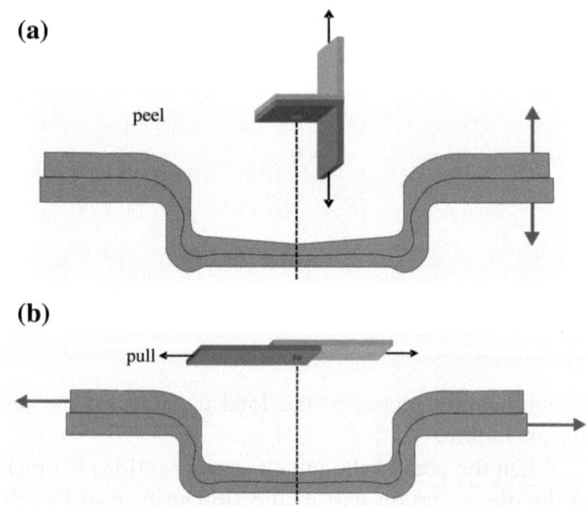

Fig. 11 Test methods for estimation of quality of clinched joints: **a** peel–tension, **b** tensile–shear

of the optimum joining conditions for two sheets made of AA5754 aluminum alloy, of thicknesses equal 0.5 mm. The main conclusions are the following:

- the ratio D_d/H (die diameter/die depth) is the most decisive parameter influencing the quality of a joint
- the bottom thickness "X" should be in the range of 40% of the initial thickness of both joined metal sheets
- the bottom thickness "X" strongly depends on the variation of the ratio D_d/H
- in the process simulation, two models of friction can be applied: Coulomb type or the constant shear friction.

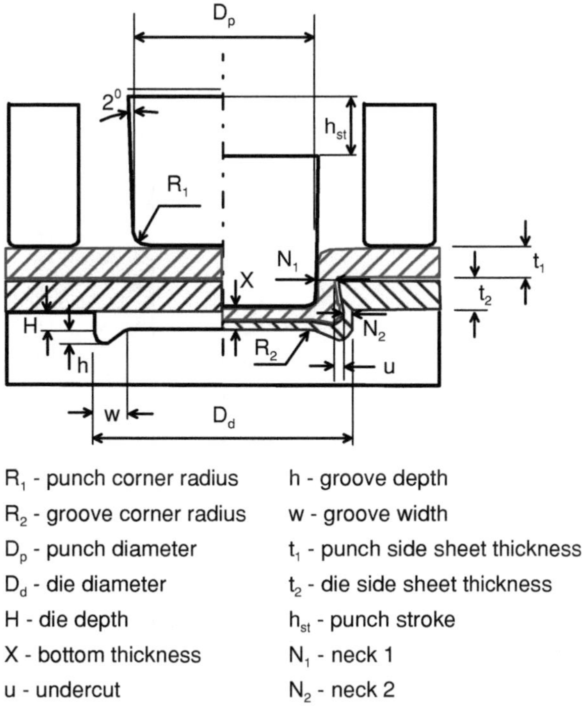

Fig. 12 Processing parameters of clinched joint [5]

The methods described above should be used for the estimation of the quality of hybrid joints including both clinching and adhesives. The question is if the adhesive film in the clinching point significantly influences the strength of the combined joining.

5 Strength and Durability of Clinch–Adhesive Joints

In the experimental verification of simple or hybrid joints one can distinguish two methods of measuring the strength of a clinch joint: "pull" (tensile–shear) and "peel" (peel–tension) (Fig. 11). For a clinch joint, pull is almost always stronger than peel. Both methods are checked in a tensile test. In the pull method the joint shearing strength is determined, whereas in the peel mode—the axial strength. The specimens' geometry for experiments done in Balawender et al. [17] is presented in Fig. 13.

The strength of a clinch joint depends essentially on four major factors:

- Type of material—the type of joined materials is very important for the clinch joint strength. The geometry of forming tools should be matched to the mechanical properties of joined materials. The shearing strength of clinch joints

Fig. 13 Geometry of the overlap specimen

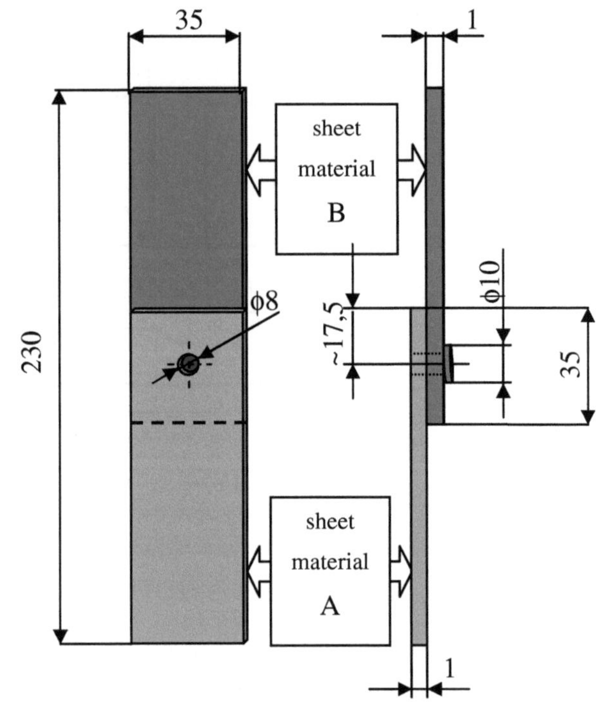

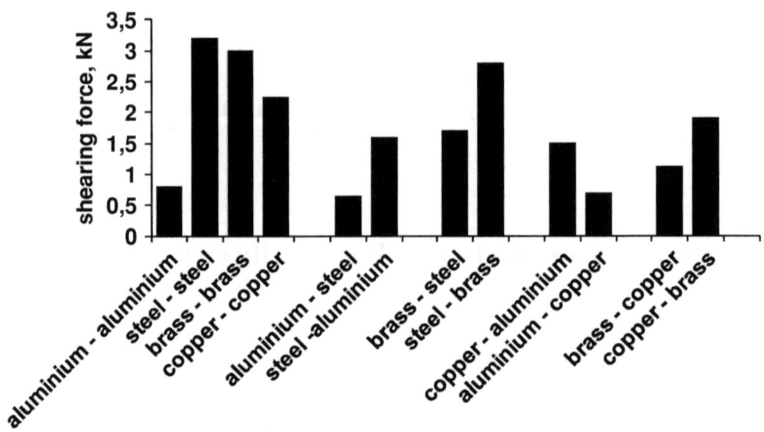

Fig. 14 Typical clinch joint shearing strengths obtained for different materials

obtained for pairs of different materials, made with the same clinching tool arrangement, are shown in Fig. 14 [17]. As it can be seen, for the same pair of materials two different values of shearing strength were obtained. The strength of joint is higher when the "stronger" material is on the punch side and the "weaker" on the die side, e.g. compare strength of aluminum–steel and steel–aluminum joints. This coincides with the result obtained by Abe et al. [18],

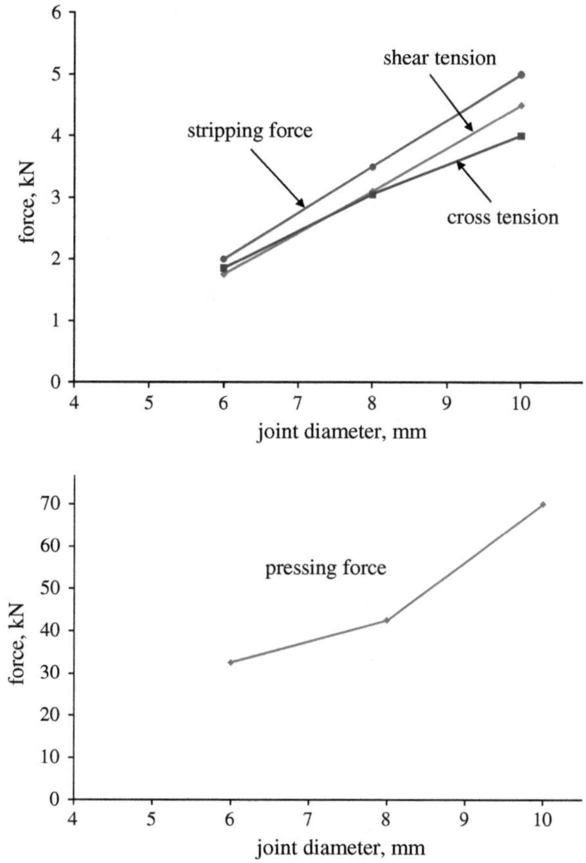

Fig. 15 Medium values of clinch joint strength (steel St12/St14)—based on TOX® catalog date

determined experimentally: when the strength of the lower sheet increases, the amount of interlock decreases due to large flow stress of the lower sheet.

- Material thickness—when the total sheet thickness increases, the strength of the clinch joint increases too, for the same tool combination.
- Clinch point size—as in other mechanical joining methods, a larger diameter of a clinch joint involves a greater joint strength. This relation determined by the TOX® PRESSOTECHNIK for the TOX® Round Joint is shown in Fig. 15 as a common graph and table.
- Material surface condition—the material surface condition has influence on the clinch joint strength; a dry surface should give a stronger joint than an oiled or greased one, however, in steel these effects are relatively minor while they have considerable influence in aluminum.

The clinch joint strength is determined by the neck-thickness N_1 and the undercut u, as illustrated in Fig. 5. Small undercuts lead to small joint strength because of the weak interlocking of the two sheets resulting in the separation of the upper sheet and the lower sheet. A thin neck causes the upper sheet in the neck to fracture. To increase the strength of the mechanical clinched joint, it is important to increase the undercut as well as the neck-thickness of the upper sheet.

In the case of clinch–adhesive joints the strength of the joint additionally depends on the adhesive interface structure (Fig. 9). Due to the squeezing of the joined materials just under the punch, the sheets flow in a radial direction and one can notice the varying thickness of the adhesive interface in the cross section, Fig. 3a. A local adhesive concentration takes place in the gorge of the joint, due to insufficient fulfillment of the die groove. This poses an additional difficulty in the formulation of the numerical model.

Experiments concerning the estimation of clinched and clinch–adhesive joints were described by Moroni et al. [19] and Balawender et al. [17].

In the first paper, the most advanced investigations concerning hybrid joints were carried out for specimens with substrates made of aluminium alloy (AA5052) and galvanized steel (S 275). The adhesive was Terokal 5077, a one component heat curing toughened epoxy adhesive. The tests were performed with an INSTRON 100 kN electro-mechanical testing machine with a crosshead speed of 2 mm/min. Additionally the effects of:

- temperature influence from −30°C to 90°C and
- ageing treatment by the salt spray

were investigated. The experimental analysis was conducted using the Design of Experiment (DoE) methodology in order to assess the influence of the material, geometrical factors and environment on static strength stiffness and energy absorption.

General conclusions from these investigations can be summarised as follows:

- The strength of the hybrid joints is determined essentially by the adhesive.
- The stiffness of hybrid joints is generally higher than that of simple joints by about 3–10% at room temperature. The loss of stiffness of the adhesive under high temperatures (90°C) is counterbalanced by clinching in the hybrid joint.
- The ageing effects on the strength of bonded joints are caused by the reduction of the adhesive strength. The hybrid joints including the mechanical interlock by clinching significantly reduce the effect of ageing at high temperatures, i.e. are less sensitive to a temperature rise.
- Hybrid joints have higher energy absorption than both simple bonded and mechanically fastened joints. Ageing substantially decreases energy absorption due to the brittle behaviour of the adhesive below its glass transition temperature. In the clinch-bonded joints, when the adhesive fails (after the peak load) the mechanical fastening continues to carry load, allowing a significant energy absorption increase.

Other types of materials for hybrid clinch–adhesive specimens were analysed by Balawender et al. [17]. Three sheet materials were used in the study: ETP-cooper, CuZn37 brass and low-carbon steel. Thickness of all sheets was 1 mm. The mechanical properties of the sheet materials were determined in uniaxial tensile tests and are collected in Table 1. The adhesive used in the tests was Dragon®

Table 1 Mechanical properties of materials used in the tests

Material	Sheet thickness (mm)	Yield stress, $R_{0.2}$ (MPa)	Ultimate tensile stress, R_m (MPa)	Elongation to failure, A (%)	Lankford parameter, r
CuZn37 brass	1.0	251	389	39	0.81
Copper	1.0	242	273	20	0.98
Deep-drawing steel	1.0	203	325	29	1.93

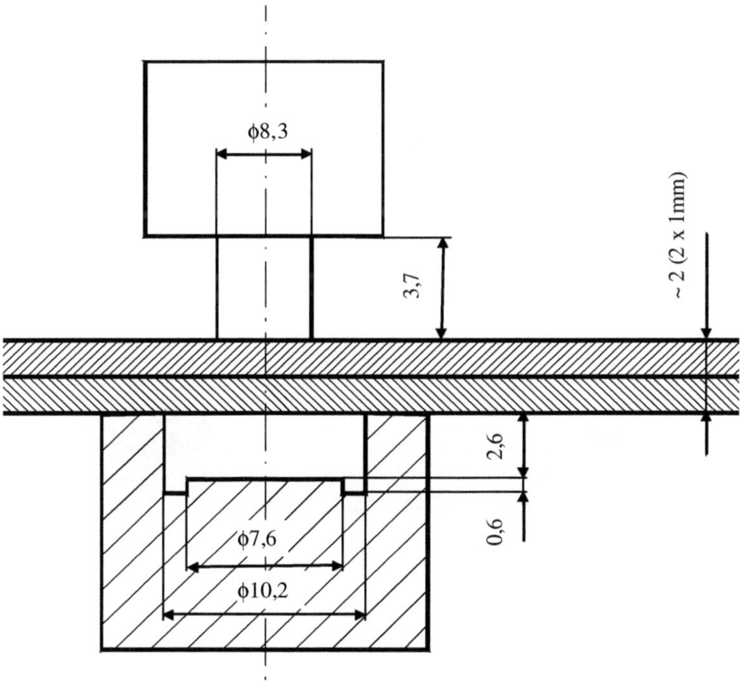

Fig. 16 Clinching tools geometry

(Chemical Factory DRAGON, Cracow, Poland), a two-component epoxy used for metal joining. The curing cycle of adhesive is 4 h at 20°C and it can be accelerated by warming up to 120°C. The full strength of the joints was obtained after 24 h.

The clinched lap joints were manufactured by a die and a punch arranged in a stamping attachment set on a 120 kN C-frame hydraulic press. The maximal press load was reduced by the overflow valve of the hydraulic system to about 90 kN. The geometry of the clinch tools is shown in Fig. 16. The tools were designed for joining two 1 mm thickness sheets. The position of the round indentation was symmetrical with respect to the specimen axis. The nominal diameter of the clinch bulging measured on the side of the die was equal 10 mm.

The maximal resistance to a shearing force was obtained when steel–steel sheets were joined and minimal for brass–brass sheets, Fig. 17.

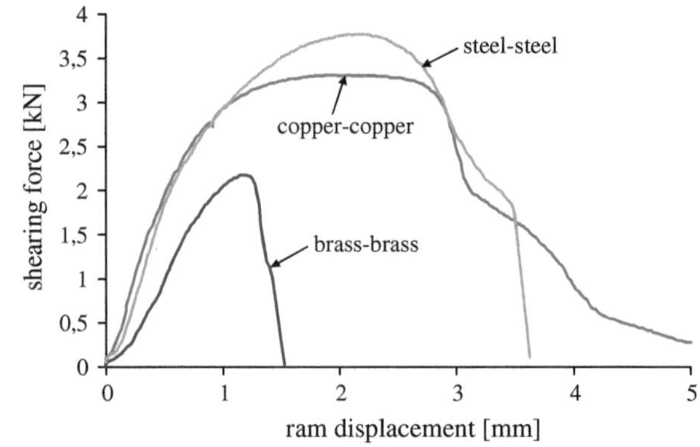

Fig. 17 Experimental load–displacement curves obtained for clinched joints

Table 2 Clinch joint parameters of hybrid lap joints

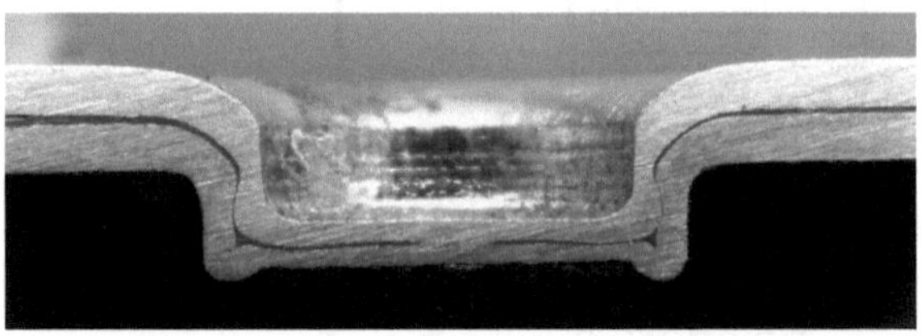

Joined materials	Thickness of the clinched bottom X (mm)	Min. thickness of the upper sheet th (mm)	Undercut u (mm)
Steel–brass	1.0	0.51	0.15
Brass–copper	0.96	0.45	0.18
Steel–copper	0.94	0.5	0.19

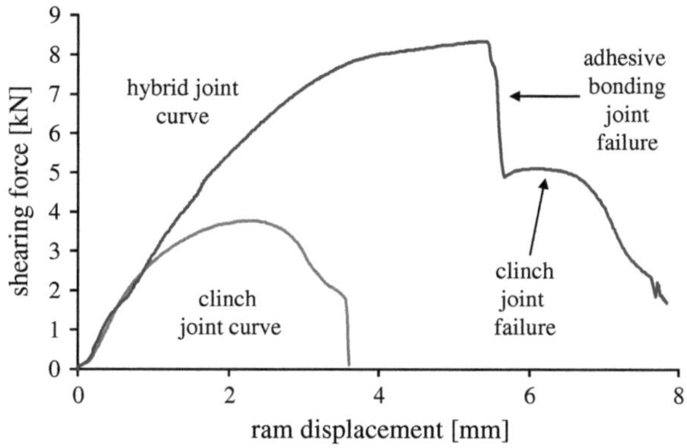

Fig. 18 Experimental load–displacement curves (clinched and hybrid) obtained for steel–steel joints

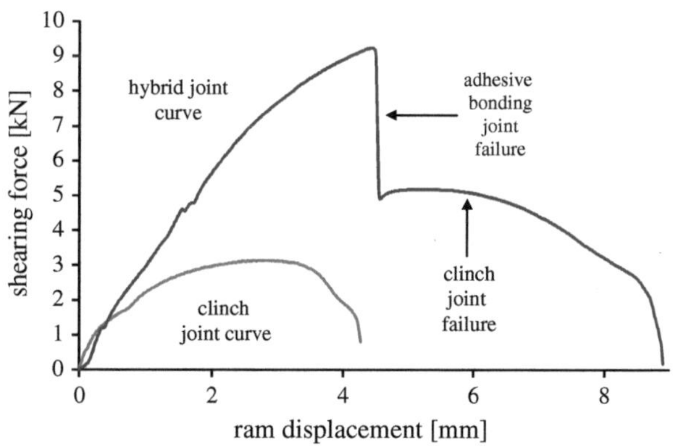

Fig. 19 Experimental load–displacement curves (clinched and hybrid) obtained for steel–copper joints

The hybrid joints were prepared with the application of the variant 3 of operations (Sect. 4) during manufacturing. The average value of the adhesive layer was equal to about 0.15 mm. The clinch joint parameters: X, th and u as well as a longitudinal section view of a hybrid single lap joint are shown in Table 2. As it can be seen, the parameters are comparable for steel–brass, brass–copper and steel–copper overlap. Differences between the clinch joint and the hybrid joint are visible in the bottom corner of the joint cavity as a fold; the adhesive caused a buckling of the lower material in the die groove. There is no difference in the shape of the upper material, which has been deformed by the punch; it is comparable with the one in the clinch joint without adhesive. The effect of this buckling on the shearing strength of hybrid lap joint can be negative.

The strength characteristics of hybrid joints compared with those of clinched joints are shown in Figs. 18, 19, 20, 21 and 22. Materials which create a strong clinched joint, i.e. steel–steel and steel–copper, do not create a hybrid joint with a high response to the shearing force. But strength characteristics of these materials have a special feature—when adhesive bonding joint fails, the clinching joint still keeps the materials connected, which results in high elongations (Figs. 18 and 19). When the shearing strength of clinched joint decreases (in relation to steel–steel

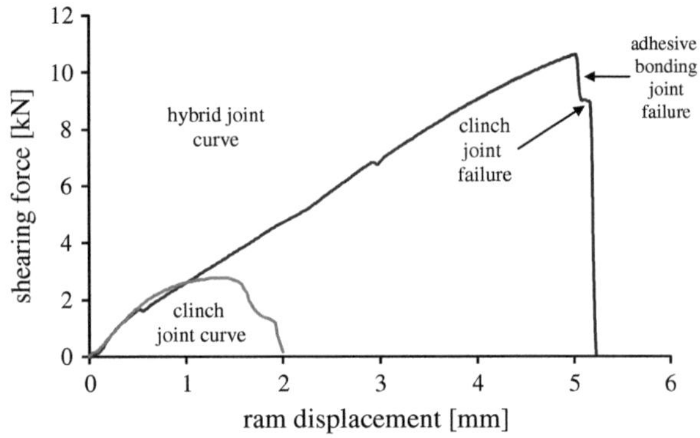

Fig. 20 Experimental load–displacement curves (clinched and hybrid) obtained for brass–copper joints

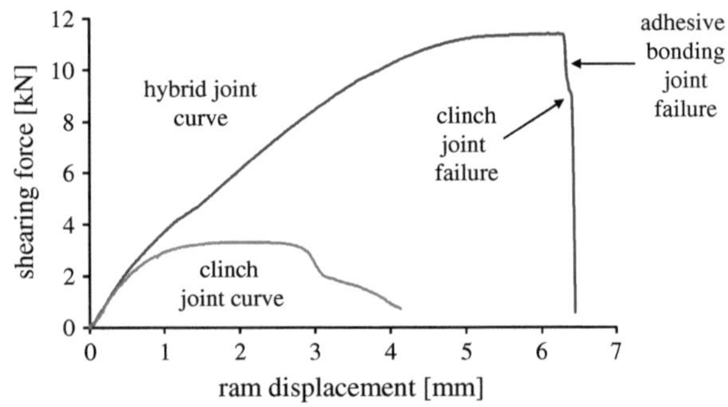

Fig. 21 Experimental load–displacement curves (clinched and hybrid) obtained for copper–copper joints

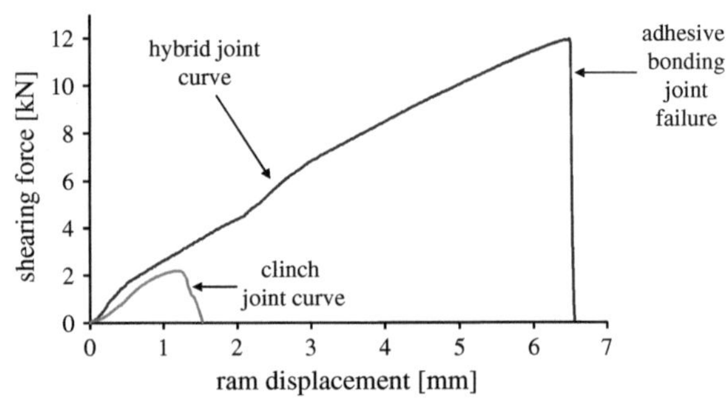

Fig. 22 Experimental load–displacement curves (clinched and hybrid) obtained for brass–brass joints

joint), the strength of hybrid joint increases (Figs. 20, 21 and 22) and the clinch joint failure, observed as a step in the strength characteristic, is not so distinct (in case of brass–brass joint, it is not visible at all).

Figure 23 shows a hybrid specimen made of steel and copper with an adhesive layer. The mode of failure is typical, i.e. the part of the softer material is subjected to bending. On the fracture surfaces, the adhesive layer remains adhered mainly on the steel, which means that the shear strength of the adhesive used in this experiment is much higher in comparison with the copper sheet yield stress.

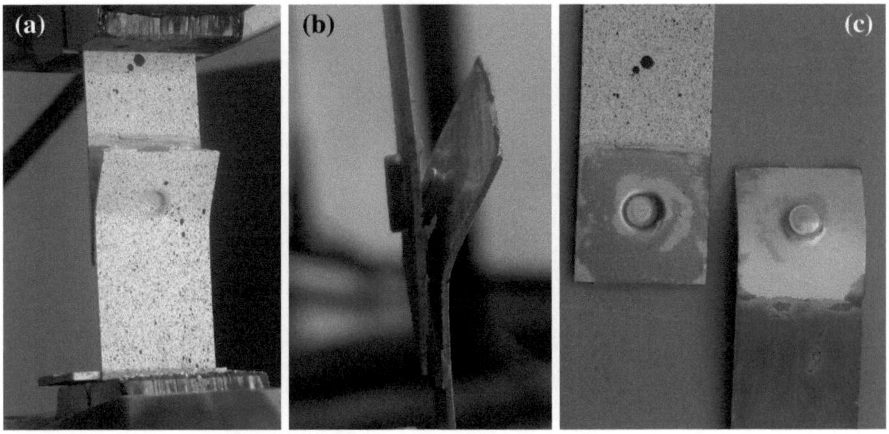

Fig. 23 Failure modes of clinch–adhesive steel–copper joints

Table 3 Energy absorption of the clinched and hybrid lap joints

Joined materials	Energy absorption (J)			
	Clinch joint	Hybrid joint		Total hybrid joint
		Before max. shearing force point	After max. shearing force point	
Steel–copper	10.4	24.9	19.3	44.2
Steel–steel	9.6	31.5	10.5	42.1
Copper–copper	9.9	46.9	3.7	50.6
Brass–copper	3.7	28.9	1.9	30.8
Brass–brass	1.9	44.0	0.4	44.5

Only in the region of the clinching and the corners of the overlap joint area fragments of the adhesive remain on the surface of the copper sheet.

Table 3 gives of the energy absorption (EA) of all considered types of joints. With the application of the adhesive, the EA of hybrid joints manufactured from steel or steel–copper increase EA about 4.5 times in relation to pure clinching. It is interesting to notice that this increase up to the maximum peak of the force–displacement diagram is equal to 2.5–3 times. In the case of the copper–copper, brass–copper and brass–brass joints, almost the whole energy accumulation takes place up to the force maximum. The addition of the adhesive to the copper–copper joint is the most effective way to increase EA. Pure clinch joints made of brass and brass–copper are very weak when it comes to EA accumulation. Introduction of the adhesive layer diametrically changes the mechanical performance of these joints. It is particularly visible in the case of the brass–brass joints, where the increase of EA is approximately 20 times. To conclude the analysis of the EA parameter for the joints efficiency, one can state that the most effective application of the adhesive is in the case of the copper–copper and brass–brass joints.

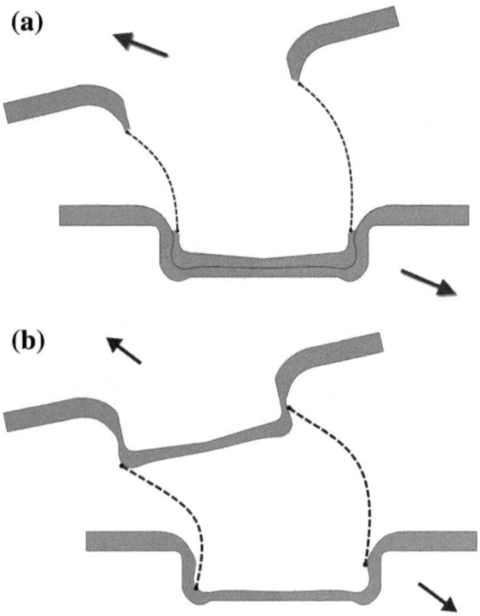

Fig. 24 Failure modes of clinch joints: **a** neck fracture mode, **b** press-stud fastener mode (button separation)

6 Types of Failures

Typical failure modes of clinch joints under the tensile–shear (Figs. 11b and 13) are presented in Fig. 24 (e.g. [3, 4]). In the first mode, the thickness of the neck N_1 (Fig. 12) is too small and the loading results in the failure of the neck. Such a situation takes place because of wrong production technology, i.e. the tools diameter is too small or the die is too deep. This leads to the formation of a long region of the joint neck and further to a crack nucleation. In the second mode, the loaded joint opens as a press-stud fastener. This is due to an excessive value of the parameter X, i.e. insufficient deformation during clinching produces a minor "S" shape interlocking (Fig. 9) and this leads to the second mode of failure. The third mode of failure is a combination of the two previous ones. In this mode, one edge of the joint fails and the other is deformed as in the first stage. This is because of local non-homogeneity of the clinch dimension and the internal material structure during the manufacturing process which can create local incisions or crevices (Fig. 10).

Our experimental investigations indicate that some more failure modes under tensile–shear (Figs. 11b and 13) can be observed, Fig. 25:

- a mode shown in Fig. 25a occurs when the clinch joint is deformed and proceeds with a small deformation of the sheets. It is the result of an excessive bottom thickness X which produces small interlocking of the sheets (the undercut u is small)

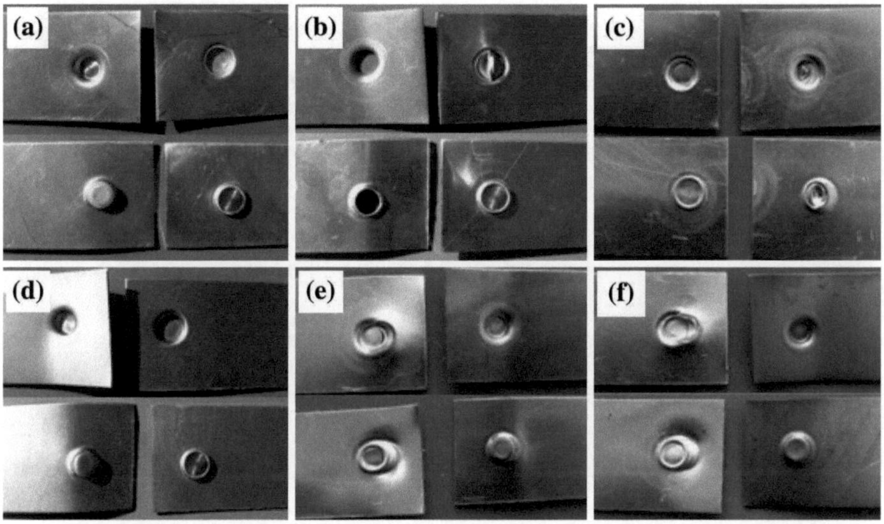

Fig. 25 Experimentally observed failure modes for different types of clinch joints: **a** joint of brass sheets, **b** joint of copper sheets, **c** joint of aluminum sheets, **d** joint of deep-drawing steel sheets, **e** joint of copper–aluminum sheets, **f** joint of steel–aluminum sheets

- in the failure mode illustrated in Fig. 25b, there is insufficient material in the neck of the joint, and loading results in cracks of this region. There are two reasons for this type of failure: the clearance of tool diameters is too small or the die is too deep. This leads to an excessive elongation in the region of the joint neck and causes a crack formation
- the failure mode shown in Fig. 25c occurs when the sheets are folded in the punch side bottom. It is the result of small thickness of the clinch joint bottom X and takes place when joining materials exhibit high ductility, e.g. aluminum sheets
- the mode presented in Fig. 25d corresponds to the case when bending of sheets is observed. This occurs when a good quality clinch joint made of deep-drawing steel sheets is subjected to shear
- the last two modes of clinch joint failure are shown in Fig. 25e, f. This failure occurs when joining materials have different plastic properties, i.e. when the "strong" material is on the punch side and the weak one on the die side.

Lee et al. [6, 7] considered only the axial strength and pointed out two typical joint failure modes, i.e. the neck fracture mode and the press-stud fastener separation mode, Fig. 24. The neck fracture mode occurs as fracture of the upper sheet is the result of the thin neck; this fracture mode is very similar to the one of a tensile test (Fig. 25b). The press-stud fastener separation mode occurs when insufficient geometrical interlocking is performed during the clinching process and it is observed as the separation of the upper sheet and the lower sheet without fracture. This failure is similar to the one shown in Fig. 25a.

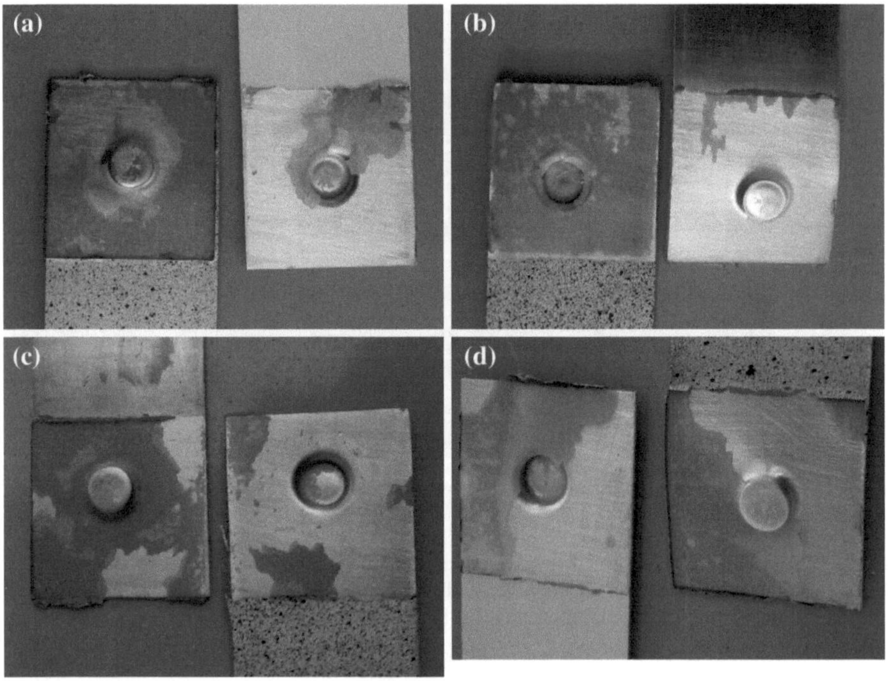

Fig. 26 Experimentally observed failure modes for different types of hybrid clinch–adhesive joints: **a** joint of deep-drawing steel–copper sheets (clinching from the steel side), **b** joint of deep-drawing steel–copper sheets (clinching from the copper side), **c** joint of deep-drawing steel sheets, **d** joint of copper sheets

Figure 26 presents failure modes under tensile–shear of hybrid joints with a brittle adhesive. They are very similar to the corresponding simple clinching joints. All joints were opened in the press-stud process. However, the most important part is the sheets sequence during clinching. In case of clinching from the steel side (Fig. 26a), both joined sheets remain almost straight after the failure. The other situation takes place when the clinching process is from the copper side. Then bending of the copper side is observed.

It is important to point out that during the clinching process the following initial defects can occur:

- large necking of the upper sheet on the side wall surface on the punch side (Fig. 27a)
- crack on the corner of the die projection (Fig. 27b)
- no interlocking due to no flaring of the upper sheet in the lower sheet (Fig. 27c).

All these initial local discontinuities strongly influence the failure processes of clinched joints. The optimisation of the technology by the analysis of the crucial process parameters would therefore lead (Fig. 12) to the production of ideal joints without initial defects.

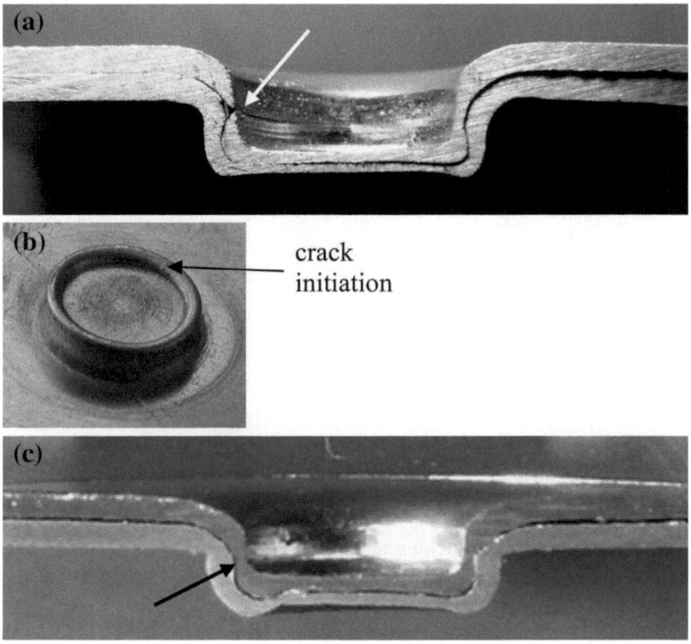

Fig. 27 The most often defects in the clinching process: **a** necking of upper sheet leading to crack formation, **b** cracks on the die side of clinch joint of aluminum–copper, **c** no interlocking of upper sheet (sheet materials: copper–steel)

7 Examples of Use

Applications of the clinch joining process in manufacturing industry date back to about 25 years ago. The world's leaders in this technology are TOX® PRESSOTECHNIK GMBH (TOX® PRESSOTECHNIK GmbH & Co. KG, Weingarten, Germany), ATTEXOR, Inc., Böllhoff Group (Wilhelm Böllhoff GmbH & Co. KG, Bielefeld, Germany) and others. The main industrial domains in which clinching is used, are:

- automotive industry: front and rear bonnets, doors, wings, air bags, seat frames, pedals, air conditioners, etc. (Fig. 28),
- electric and electronic equipment: computer housings, cable housings, contactors, etc.,
- building: door and window frames, floor and roof elements, elevators ventilation, etc.,
- household equipment: refrigerators, washing machines, micro-wave ovens, etc.,
- and others.

Hybrid clinch–adhesive joining is a very modern assembly technology and it is not used in practise nowadays. This technology is being intensively investigated

(a)

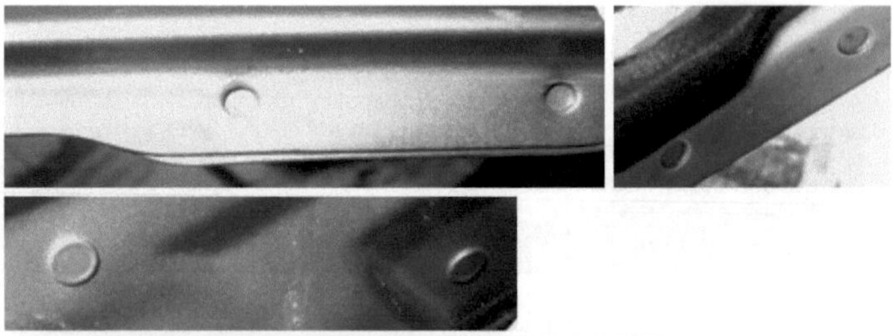

(b)

Fig. 28 Examples of use of the clinching technique—clinch joints in: **a** automobile engine lid, **b** ventilating duct panel

because of the expected advantages including low costs, flexibility and its environmental compatibility. The low mechanical strength of simple clinched joints (resistance to shear or tensile loading) compared to equivalent spot welds, forces to search for new joining solutions. Clinch–adhesive joining seems to be a very good combination which can enhance the advantages of both methods.

8 Conclusions and Future Trends

The clinch–adhesive technique may be useful for applications in the automotive, aeronautical and aerospace industry.

The mechanical strength of hybrid clinch–adhesive joints strongly depends on the adhesive and adherends' properties. When joined materials undergo finite plastic deformation, adhesive properties should match these conditions. The adhesive Dragon® used in the tests described in this chapter is a brittle and stiff adhesive. The adhesive layer did not therefore sustain plastic deformation during the clinching process and adhesive failure occurred around the clinching indentation. This additionally demonstrates the influence of the processing technology—in the variant 3 (described in page 11) when the clinching process is the last one—on the initial microdamage introduced to the adhesive interface by clinching. Local microdamage entails a decrease in the shear strength of the tested hybrid joints. The proper production of hybrid joints (Sect. 4) is therefore a crucial point for further engineering applications.

The introduction of adhesives into clinching joints essentially changes the mechanical strength. It was confirmed by experimental and numerical simulation results [17, 19]. But taking into account the wide range of commercial adhesives for professional usage (from stiff and strong to flexible and ductile) the adhesive choice should be carefully balanced. The choice criterion of adhesive for the hybrid clinch–adhesive joint seems to be similar to the one of a mixed adhesive joint. The clinch joint is like a strong adhesive in the middle of the overlap and a flexible adhesive should be used in the neighbourhood. Future investigations should focus on the determination of such joint compositions such as a clinch-flexible adhesive.

The clinch–adhesive joints with a single adhesive or a combination of several adhesives might be useful for the aeronautical industry. However, they should bear temperatures from −55 to +200°C and dynamic loading. More attention should be focused on fatigue testing, including environmental corrosion and ageing problems. Low impact and high impact tests are important for total characterisation of the presented hybrid joints. Future experimental and numerical investigations are necessary.

Acknowledgement The research leading to these results has received funding from: Financial support of Structural Funds in the Operational Programme—Innovative Economy (IE OP) financed from the European Regional Development Fund—Project "Modern material technologies in aerospace industry", No. POIG.0101.02-00-015/08 is gratefully acknowledged (RT-15: Unconventional technologies of joining elements of aeronautical constructions),The European Union Seventh Framework Programme (FP7/2007–2013), FP7-REGPOT-2009-1, under grant agreement No. 245479.

References

1. Barnes, T.A., Pashby, I.R.: Joining techniques for aluminium spaceframes used in automobiles. Part II—adhesive bonding and mechanical fasteners. J. Mater. Proc. Technol. **99**, 72–79 (2000)
2. Liebig, H.P., Bober, J., Beyer, R.: Connecting sheet metal by press joining. Bänder Bleche Rohre **25**(9), 7 (1984). Vogel-Verlag, Würzburg
3. Varis, J.: Ensuring the integrity in clinching process. J. Mater. Proc. Technol. **174**, 277–285 (2006)
4. Varis, J.P., Lepistö, J.: A simple testing-based procedure and simulation of the clinching process using finite element analysis for establishing clinching parameters. Thin Walled Struct. **41**, 691–709 (2003)
5. Jayasekara, V., Min, K.H., Noh, J.H., Kim, M.T., Seo, J.M., Lee, H.Y., Hwang, B.B.: Rigid-plastic and elastic–plastic finite element analysis on the clinching joint process of thin metal sheets. Met. Mater. Int. **16**, 339–347 (2010)
6. Lee, Ch.J., Kim, J.Y., Lee, S.K., Ko, D.Ch., Kim, B.M.: Design of mechanical clinching tools for joining of aluminium alloy sheets. Mat. Design **31**, 1854–1861 (2010)
7. Lee, Ch.J., Kim, J.Y., Lee, S.K., Ko, D.Ch., Kim, B.M.: Parametric study on mechanical clinching process for joining aluminium alloy and high-strength steel sheets. J. Mech. Sci. Technol. **24**, 123–126 (2010)

8. Nong, N., Keju, O., Yu, Z., Zhiyuan, Q., Changcheng, T., Feipeng, L.: Research on press joining technology for automotive metallic sheets. J. Mater. Proc. Technol. **137**, 159–163 (2003)
9. Oudjene, M., Ben-Ayed, L.: On the parametrical study of clinching joining of metallic sheets using the Taguchi method. Eng. Struct. **30**, 1782–1788 (2008)
10. Oudjene, M., Ben-Ayed, L.: Shape optimization of clinching tools using the response surface methodology with Moving Least-Square approximation. J. Mater. Proc. Technol. **209**, 289–296 (2009)
11. da Silva, L.F.M., Adams, R.D.: Measurement of the mechanical properties of structural adhesives in tension and shear over a wide range of temperatures. J. Adhes. Sci. Technol. **19**, 109–142 (2005)
12. Hergenrother, P.M.: Development of composites, adhesives and sealants for high-speed commercial airplanes. SAMPE J. **36**, 30–41 (2000)
13. da Silva, L.F.M., Adams, R.D.: Techniques to reduce the peel stresses in adhesive joints with composites. Int. J. Adhes. Adhes. **27**, 227–235 (2007)
14. Srinivas, S.: Analysis of bonded joints. NASA TN D-7855 (1975)
15. Geiss, P.L., Koetter, M.P., Presser, M., Raudonat, D.: Hybrid joining with pressure sensitive adhesives. http://www.pstc.org/files/public/TeCH33Papers/ (2010)
16. Brockmann, W., Geiss, P.L., Klingen, J., Schroeder, B.: Adhesive bonding—materials, applications and technology. Wiley-VCH, Weinheim (2008)
17. Balawender, T., Sadowski, T., Golewski, P.: Experimental and numerical analysis of hybrid: clinched–adhesive joint. In: ACE-X conference, Paris 8–9 July 2010
18. Abe, Y., Matsuda, A., Kato, T., Mori, K.: Plastic joining of aluminum alloy and high strength steel sheets by mechanical clinching. Met. Form. **1**, 649–656 (2008)
19. Moroni, F., Pirondi, A., Kleiner, F.: Experimental analysis and comparison of the strength of simple and hybrid structural joints. Int. J. Adhes. Adhes. **30**, 367–379 (2010)
20. Abe, Y., Kishimoto, M., Kato, T., Mori, K.: Joining of hot-dip coated steel sheets by mechanical clinching. Int. J. Mater. Form. **2**(Suppl 1), 291–294 (2009)
21. He, X.: Recent development in finite element analysis of clinched joints. Int. J. Adv. Manuf. Technol. **48**, 607–612 (2010)
22. Saberi, S., Enzinger, N., Vallant, R., Cerjak, H., Hinterdorfer, J., Rauch, R.: Influence of plastic anisotropy on the mechanical behaviour of clinched joint of different coated thin steel sheets. Int. J. Mater. Form. **1**(Suppl. 1), 273–276 (2008)
23. Varis, J.P.: The suitability of clinching as a joining method for high-strength structural steel. J. Mater. Proc. Technol. **132**, 242–249 (2003)

Science and Technology of Bolt-Adhesive Joints

Jan R. Weitzenböck and Dag McGeorge

Abstract This chapter addresses bolt-adhesive joints to transfer both loads and moments. All our examples are taken from maritime applications. Experience has shown that bolt-adhesive joints in the maritime industry are not designed for hybrid action where one joining method improves the performance of the other. Rather they are used in a fail-safe-mode where one joining method takes over should the other fail. Three different applications will be discussed: (1) composite superstructures and composite to steel joints on large ocean going vessels, (2) adhesive bonding of windows, also known as "direct glazing" and (3) cellular sandwich concept for large ocean going ships such as bulk carriers. The sandwich is composed of two steel faces and a lightweight concrete core.

1 Introduction

This chapter addresses bolt-adhesive joints to transfer both loads and moments. All our examples are taken from maritime applications. This chapter does not include thread-locking—using adhesives to stop nuts and bolts from coming loose (see Chapters Science of friction-adhesive joints and Technology of friction-adhesive joints for more information about thread-locking).

Experience has shown that bolt-adhesive joints in the maritime industry are not designed for hybrid action where one joining method improves the performance of the other. Rather they are used in a fail-safe-mode where one joining method takes

J. R. Weitzenböck (✉) · D. McGeorge
Det Norske Veritas AS, 1322 Høvik, Norway
e-mail: Jan.Weitzenboeck@dnv.com

D. McGeorge
e-mail: Dag.Mcgeorge@dnv.com

Adv Struct Mater (2011) 6: 177–199
DOI: 10.1007/8611_2011_54

Published Online: 22 March 2011

over when the other fails. As a consequence, design for adhesive bonding and bolting is usually done independently from each other.

Based on the applications we have seen, we interpret bolt-adhesive connections as *adhesively bonded joints combined with vertical members that limit potential movement in the in-plane direction*. This may include bolts, studs or metal strips that are welded or bolted in place.

This chapter will not address selection of materials, including adhesives. Interested readers are referred to [52, 55]. Having said that, the examples discussed here use common shipbuilding materials, e.g. steel or composite sandwich. This helps to minimise changes needed to the support structures and the yard building processes.

2 Motivation for Using Bolt-Adhesive Joints

Concerns over long-term performance of adhesively bonded joints prevent in many cases the use of adhesive bonding as a joining method for load bearing connects [51]. Hence hybrid joints have received more attention recently. Under which circumstances is it beneficial to use hybrid joints?

- The joint is subject to loading from different directions: Anon [2] the adhesive carries shear stresses while the bolts carry transverse loading. A special case is composite pultrusions or aluminium extrusions which may have interlocking features that are used to connect the profiles.
- As mentioned by Anon [2], bolts may help make a bonded joint survive exposure to fire. Bolts are usually not loaded during service. However, in the event of a fire adhesives soften eventually and the bolts will carry the load.
- A further advantage in using bolt-adhesive hybrid joints is that the issue of predicting long-term performance of the bonded joint is avoided since bonding is being combined with a proven joining method [51].

There are situations where hybrid joints are not effective:

- As Anon [2] point out, structures that are operated at higher temperatures and are required to be stiff may not be suitable for bolt-adhesive joints. The adhesive will soften at elevated temperatures thus reducing the joint's stiffness.
- A further potential disadvantage according to [2] is the fact that electrical continuity is interrupted by the adhesive layer.
- Furthermore, by using two joining methods fabrication and inspection costs are increased. Hentinen and Hildebrand [25] quantified the cost penalty for fabrication of bolt-adhesive joints compared to bonded only solutions.

While "adhesion" is essential for corrosion protection to ensure the corrosion prevention coating adheres to the steel surface, there are few applications where adhesively bonded joints are used in safety critical applications. The most prominent example is the joint between the steel deck and the composite superstructure of the French La Fayette class frigates [34]. Hybrid joining is probably the application that has been developed furthest. It is used for joining of composite superstructures to steel main structures and for the joining of windows.

Recent changes by the International Maritime Organisation [28] open up the possibility for using risk based design to demonstrate equivalent levels of fire safety. This has opened up the possibility for using composites on large ocean going vessels as previously only non-combustible materials were allowed. With it comes the challenge of joining the composite structure to the main steel structure. An early study by Hentinen and Hildebrand [25] demonstrated the feasibility of joining composite to steel and they developed possible joint solutions. McGeorge et al. [37] presented results from research into the use in a frigate-type naval ship of a bonded tuning-fork type of joint offering some degree of mechanical interlocking. They concluded that the bonded joint would offer the required integrity and will not be the weak link in the structural assembly. Bohlmann and Fogarty [7] report results from the prototype development for a US Navy destroyer and the joining methods they had developed. Finally Speth et al. [49] and Ritter et al. [46] report on the superstructure to deck joint for a large yacht.

Adhesive bonding of windows, also known as “direct glazing”, has become standard practice on passenger and cruise ships with more than 10 years of accumulated experience in the shipbuilding industry. Wacker et al. [50] reported the bonding of window panes and their design. Weitzenböck [53] reports on the experience gained so far with bonded windows.

Bergan et al. [6] introduced a cellular sandwich concept for large ships such as bulk carriers. The sandwich is composed of steel faces and a lightweight concrete core. The main motivation was to simplify shipbuilding (fewer parts) and make it cheaper (concrete is cheaper than steel). During material testing it was discovered that the sandwich configuration had to be changed. Initially the load transfer between the concrete core and the steel face relied entirely on the adhesion between concrete and steel. Later on it was discovered that one needs additional mechanical fasteners, such as shear studs to improve adhesion also in the long-term.

All the examples presented in this section will be discussed in more detail later on.

3 How to Gain Confidence

Adhesive bonding has been used for many years in transport applications. However, there are still very few examples where adhesively bonded joints are used in safety critical applications. One of the reasons is that it is difficult to predict the long-term effects of the (marine) environment on such a joint.

Considerable media attention has been given to the Boeing 787 and Airbus A350 aircraft development programmes. Both aircraft make extensive use of composites in the fuselage leading to significant reduction in structural weight and number of components and fasteners. However, what is receiving less attention is the fact that the fuselage sections are still joined using fasteners. In fact they are the main critical item in the assembly process as discussed by Peck [44]. Another interesting detail is that the lack of suitably certified fasteners was one of the main contributing factors for the considerable delays the 787 Dreamliner has experienced to date [36]. Hence, also for the most modern commercial airplanes adhesion in critical load bearing joints is not relied on alone. However, at the same time there is evidence of the long-term performance of bonded (aluminium) aircraft structures. Beevers [5] reports on a study using bonded joints that had been recovered from a scrapped Nimrod/Comet 4C aircraft which was built in 1963. The aim was to establish the condition of the adhesive in a 30-year-old structure. The tests showed that tensile strength and fracture toughness of the adhesive still met or exceeded the manufacturers' original specification. Another aircraft that is making extensive use of adhesives is the Fokker Fellowship F-28. Bond [8] reports that the Fellowship F-28 contains more than 900 bonded assemblies. At the time of writing of Bond's report these aircraft had been in service for up to 25 years without any reported failures or damage of the bonded connections. An example to illustrate the time scale it takes to build confidence in hybrid joining methods is the development of weld bonding (spot welding and adhesive bonding). Studies of the properties of weld bonded joints by Menges and Schmidt [42] and Reinhardt [45] were the basis for the design of a loading ramp for a truck (see also [47]). These East German developments were at least partly based on initial studies carried out earlier in the Soviet Union. Drain and Chandrasekharan [18] report that use of adhesives in hem flanges of, e.g. doors of cars had become common practice. However, structural bonding was still very much a topic of research with impact, fatigue and durability being the key issues. Today weld bonding is a standard joining process known for its superior crash performance. For example Keller et al. [32] report results from a project to use weld bonding for a vehicle subassembly. The crash performance was better than a spot welded assembly. It was noticed that the numerical models to model material behaviour needed further development to be able to make better predictions during the product development process. Another interesting development is that weld bonding had also been developed for aircraft manufacture. Croucher et al. [11] reported on a study to develop weld bonding for fighter planes. More information about weld-bonding can be found in Chapters Science of weld-adhesive joints and Technology of weld-adhesive joints.

Weitzenböck and McGeorge [51] outlined an approach to introduce adhesive bonding into marine applications despite the lack of documented long-term performance. This approach uses sound risk management principles and comprises the following three steps:

1. Hazard identification.
2. Risk assessment.
3. Adoption of suitable risk control measure,

Three risk control measures are proposed:

1. use state of the art in design, material selection and fabrication of bonded connections,
2. make it possible to detect failure early on, before it becomes critical and

3. develop a qualified repair procedure in case the bonded connection fails and requires repair.

This approach implies that not all possible joint configurations on a ship or offshore structure can be bonded as some of the risk control options are difficult or impossible to implement in critical parts. For example, the onset of failure of the bonded assembly may be difficult to detect or there may not be enough time to qualify a repair procedure. However, by introducing a further risk control option, namely bolting, many more application suddenly become feasible.

While adding bolts or similar items to the joint may increase confidence, there is a trade-off to be considered. Fabrication cost will increase as Hentinen and Hildebrand [25] point out. They compared adhesively bonded and bolt-adhesive joints between steel and FRP sandwich material. While there was no difference in cost for the fabrication of the edges, joining to the ship structure was twice as expensive compared to the bonded only solutions. It is interesting to note that for quasi static tensile test both joint concepts obtained almost identical values while for flexural loading the bonded only solutions was clearly superior. However, it is worth pointing out that for certain types of loading, e.g. high temperature, it is advantageous to use hybrid joints, as an adhesive may fail due to high temperature.

There are two main conclusions that can be drawn from these discussions. Adhesives are not usually used alone for major safety critical joints. Use of failsafe design, such as adding bolts or other mechanical fasteners, may help overcome the lack of confidence in the long-term performance of adhesives.

4 Design and Analysis of Bolt-Adhesive Joints

4.1 Introduction

Strictly speaking bolt-adhesive joints are not the preferred option for joining two components. Using two joining methods simultaneously incurs a cost penalty. However, there are situations where it is an advantage to use hybrid joints, e.g. loading from different direction or exposure to high temperatures. Still, the main reason for using bolt-adhesive hybrid joints for marine applications is the uncertainty about the long-term performance of bonded joints and how to document it. By combining adhesive bonding with a well proven and tested joining method, one can accept some un-quantified uncertainty about the durability of the adhesive.

4.2 Science of Bolt-Adhesive Joints

Kelly [33] carried out experimental and numerical studies of the load transfer in hybrid bolt-adhesive composite single-lap joints. He studied how different kinds of adhesives affect the way adhesives and bolts interact. Very stiff adhesives result in joints where the bolts contribute very little to the load transfer. However, adhesives with lower strength and modulus and increased ductility resulted in more flexible joints allowing the bolts to transfer a larger share of the loading. Kelly [33] sums up the results of his study as follows:

- The load transferred by the bolt increases with increasing adherend thickness.
- The load transferred by the bolt increases with increasing adhesive thickness.
- The load transferred by the bolt decreases with increasing overlap length.
- The load transferred by the bolt decreases with increasing pitch distance.
- The load transferred by the bolt decreases with increasing adhesive modulus.

Note: the bolt "pitch distance" is defined as the distance between bolts in a row. Kelly points out that if bolts are to make a significant contribution to the load transfer, the joints need to be flexible, either by selecting a suitable adhesive or joint design. Shenoi and Hawkins [48] provide an introduction and guidance to joint design for composite materials, including bolted connections. Broughton et al. [9] presented a general design procedure for bolt-adhesive joints. They discuss tools and procedure for design and analysis of joints. da Silva et al. [12, 13] review and benchmark all known analytical formulas for bonded lap-shear joints. da Silva et al. [13] selected some approaches from the literature survey and subjected them to a systematic benchmark study. The first conclusion was that no single model works equally well in all cases. The authors recommended candidates for: (1) long overlaps, brittle adhesives and elastic adherends (any linear analysis), (2) ductile adhesives and elastic adherends (Hart-Smith), (3) adherends that yield (Adams, Comyn and Wake) and (4) joints with composites (Adams and Mallick).

Adams [1] proposed a simple bondline strength criterion assuming that the bondline strength is proportional to the overlap length. This criterion is appropriate for short overlaps. But such short overlaps are discouraged because the short overlaps tend to compromise the bonded joint's long term performance (creep, fatigue) and its defect and damage tolerance. For longer overlaps used for structural joints, Adams' simplified approach is unsafe as it ignores that the efficiency of the joint decreases with bondline length. This may mislead the designer to think that increasing bondline length will increase capacity. For structural adhesives, nonlinearity must be taken into account. In some cases this can be done without finite element (FE) analysis. Hart-Smith assumes that the strain to failure is a material property. That works if the real case is similar to the tested case from which the strain to failure was determined (see also [43]). However, generally speaking, Hart-Smith's assumption is incorrect and can lead to problems if the real case differs significantly from the tested case from which the strain to failure was determined, e.g. much stiffer adherends. This phenomenon is documented by Echtermeyer et al. [19], Guthu [23] and McGeorge [41]. FE analysis helps if geometry does not lend itself to use of simple formulas and makes it easy to predict strains at given loads assuming no failure. To be reliable, failure prediction must account for both nonlinear material behaviour and the fracture process that causes failure. This can be achieved by simple formulas for some simple geometry [40] or using FE analysis with plasticity and cohesive elements for more general geometries [23]. For flexible adhesives, time-dependent failure modes govern. However, this is not quite true for very low stresses where linear elastic methods can be used. To explicitly account for creep and time dependent failure is complicated both with regard to modelling and testing to challenge/validate models.

4.3 Practical Design of Bolt-Adhesive Joints

Experience has shown that bolt-adhesive joints in the maritime industry are not designed for hybrid action where one joining method improves the performance of the other ("$1 + 1 \geq 2$"). Rather they are used in a fail-safe-mode where one joining method takes over when the other fails ("$1 + 1 = 1$"). Based on the applications we have seen, we interpret bolt-adhesive connections as adhesively bonded joints combined with vertical members that limit potential movement in the in-plane direction. This may include bolts, studs or metal strips that are welded or bolted in place. Provided a suitable adhesive is used, bolt-adhesive joints can also be part of the corrosion protection strategy by providing an electrical insulation layer against galvanic corrosion.

Broughton et al. [9] provide an overview of the design process for bolted and bonded joints. There are different types of bolted connections. Eurocode 3 [20] divides bolted connections into three categories:

- Category A: bearing type
- Category B: slip-resistance at serviceability limit state
- Category C: slip-resistance at ultimate limit state

Serviceability limits are typically excessive vertical or horizontal deflections. Slip-resistance for bolt-adhesive joints is provided by the adhesives not the compression forces of the bolts. This means that category A would be equivalent to bolt-adhesive joint with a highly flexible adhesive or sealant. The bolts always carry the load. For category B type joints slip should not occur at the serviceability limit state. The design serviceability shear load should not exceed the design slip resistance. This implies that under normal operational loads the adhesive is providing all the strength. And finally for category C type joints slip should not occur at the ultimate limit state. The design ultimate shear load should not exceed the design slip resistance. This implies that the adhesive should carry all the loads—always. Basically, this is an adhesive joint. This is only possible for adhesives with documented long-term performance.

The main long-term concerns for the bolts are probably corrosion and not fatigue as the bolts are not loaded during service [4]. Corrosion of bolts can be minimised by selecting suitable corrosion resistant materials or coatings and ensuring that the design avoids exposure of the joints to water. More information about corrosion can be found in the forthcoming Det Norske Veritas Recommended Practice on "Specification, design and installation of fasteners for the offshore oil and gas industry".

Bolted connections need to pass a number of criteria. Some of the most basic ones are that the bearing strength of the plate and the shear strength of the bolts are sufficient. The bearing strength of composite plates can be determined for example from DNV [17]

$$\sigma_{\text{bear}} = \frac{R_{\text{bear}}}{3\gamma} \tag{1}$$

with σ_{bear} is the shear load divided by $d \times t$, t the thickness of structural laminate, d the bolt diameter, $\gamma = 1.0$ for holes with a difference between bolt and hole diameter <10 mm, $\gamma = 1.6$ for holes with a difference between bolt and hole diameter of <1.0 mm, R_{bear} is the elect value for different laminates from table in DNV [17].

A formula for determining design shear resistance for the bolt may be found in, e.g. Eurocode 3 [20]:

$$F_{\text{v,Rd}} = \frac{\alpha_{\text{v}} f_{\text{ub}} A}{\gamma_{\text{M2}}} \tag{2}$$

with A = tensile stress area of bolt A_{s}, α_{v} between 0.5 and 0.6 (see Eurocode 3 [20]), γ_{M2} = partial safety factor for resistance of bolt (=1.25), f_{ub} = ultimate tensile strength for bolt, $F_{\text{v,Rd}}$ = design shear resistance per bolt.

Bonded joints are strongest when they are loaded in shear. Hence, hybrid bolt-adhesive joint will also mainly be loaded in shear. Furthermore, as the bondline surface is usually quite large, bonded joints tend to be lightly loaded. This is partly also because adhesives that exhibit good gap filling characteristics have much lower shear strength than aerospace grade adhesives with very thin bondline thickness. Now, depending on other requirements such as ability to compensate for thermal miss match, the joint could be designed such that under normal service conditions, the adhesive carries all loads (non-slip condition) while accidental loading or failure of adhesive due to excessive ageing is carried by the bolts (bearing condition). This means that for service conditions the bolt-adhesive joint behaves like a bonded joint while in accident scenario the joint is a bolted joint.

Bolt and adhesive joints are designed independently. Hence there is no universal design basis and requirements. They are different from case to case. Table 1 sums up the requirements or design basis for the application examples discussed later in this chapter. For adhesives, there are no strict requirements as they are not usually allowed for safety critical connections, mainly because their long-term performance is not known precisely. However, for bolts a number of codes or rules are available. For example DNV [17] provides guidance on how to design bolted connections in composite materials. The steel–concrete–steel (SCS) sandwich material was designed on the basis of Eurocode 4 [21] and DNV [14]. For the

Table 1 Design basis for bolt-adhesive joints for application examples

	Adhesive	Bolt
lightweight composite superstructures for ships	No requirements, but low strains or stresses recommended Use closed form solutions or FE	Established codes, e.g. classification society
Direct glazing of windows	Strain/stress limits Simple formulae or FE analysis for larger windows	engineering judgement—no design code
Steel–concrete–steel sandwich plates	Adhesion is not considered	Shear studs—established design codes, e.g. Eurocode 4

windows, the number and location of the bolts are selected on the basis of engineering judgement.

5 Application Examples

5.1 Introduction

The following sections describe known application examples of hybrid bolt-adhesive joints used in marine structures. The first main group of examples are related to lightweight composite superstructures for ships, both large ships and high speed craft. Next, we will be presenting direct glazing of windows, probably the best established hybrid joining case. Lastly we will be outlining an application of a novel material and structural arrangement—the SCS sandwich material.

The case descriptions follow a common structure:

What is the challenge?

- What is the (joining) problem?
- Why hybrid joints?
- What are the design challenges?
- What are the materials technology and fabrication challenges?
- What are the requirements?

Description of solution addressing:

- Analysis and design
- Fabrication and inspection
- Long-term performance

Status?

- What is the (service) experience if any?

5.2 Composite Superstructures

5.2.1 What is the Challenge?

The use of composites in superstructure modules has been investigated in a number of research projects showing promise of considerable weight-saving in excess of 50% of the weight of a steel design for the module enabling considerable fuel savings and emission reductions. Some details on the structural design can be found in for example [7, 25] while fire safety is discussed in [39]. For large ships combustible materials are not allowed. Recent changes in safety regulation permit the use of combustible materials provided one can show equivalent levels of fire safety (SOLAS Regulation 17, see also [28]). McGeorge et al. [39] demonstrate equivalent fire safety via a rigorous risk assessment. To be able to use composite superstructures, one needs to devise methods for efficient and reliable joining of the composite to steel or sometimes aluminium. However, the ultimate challenge for using this new technology in practice is to convince the relevant approval authority or flag state that the safety of this solution is equivalent to steel.

Hentinen and Hildebrand's [25] research was motivated by trying to increasing the use of fibre reinforced plastic (FRP) sandwich structures in superstructures to save weight. Their aim was to develop joints that can easily be installed in a traditional (steel) shipyard. Hence they developed joints with integrated steel substructure that can be installed by welding without the need for bolting, adhesive bonding or lamination at the shipyard. Hentinen and Hildebrand [25] investigated solutions where the metal structure is bolted and/or bonded to the FRP structure (see also Fig. 1).

Bohlmann and Fogarty [7] report on a project to demonstrate the feasibility of building part of the superstructure of a US Navy destroyer in composite. To be able to fit the composite module, it was necessary to join two different materials. However, since adhesive bonding on its own was not acceptable to the Navy officials, a hybrid bolt-adhesive joint was chosen. Many different joint configuration and concepts were evaluated. The one that was selected for further study is shown in Fig. 1. Its main benefits are that it has a smooth outside, and the lower steel plate can be welded directly to the deck and bolted and bonded to the sandwich laminate.

McGeorge et al. [37] developed a bonded tuning-fork type joint offering some mechanical interlocking (see Fig. 1). By testing prototype joints, they showed that the joints were stronger than the elements they were designed to join. A comprehensive ageing programme was undertaken to improve the confidence in the long term performance of the joints [38]. Nevertheless, there appears to remain a challenge concerning confidence in the long term performance of bonded joints among decision-makers.

Ritter et al. [46] and Speth et al. [49] report on a very recent project to convert two old Dutch frigates to super yachts. One of the many changes is the fitting of new lightweight composite superstructures to these vessels. The two papers focus

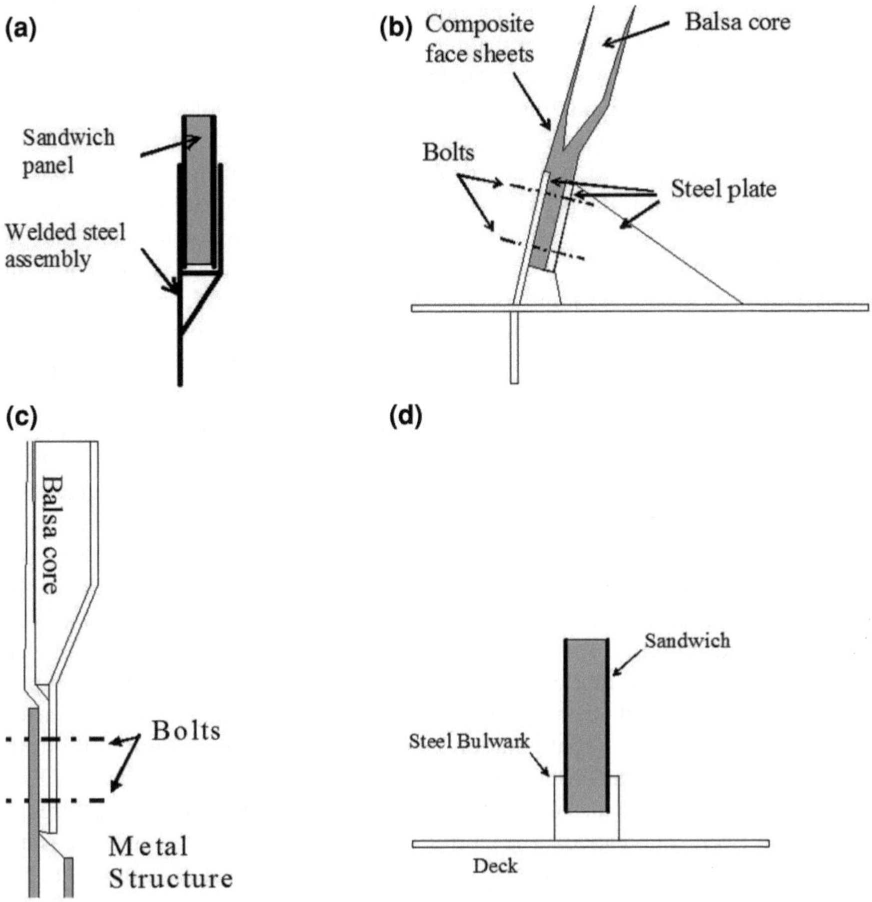

Fig. 1 Hybrid joint concepts for connecting composite superstructures to a steel deck: **a** McGeorge et al. [37]; **b** Bohlmann and Fogarty [7]; **c** Hentinen and Hildebrand [25]; **d** Speth et al. [49]

mainly on the design and verification of the composite-steel joints between superstructure and deck.

5.2.2 Description of Solution

The joint designs by Bohlmann and Fogarty [7] and Hentinen and Hildebrand [25] rely on tapering the sandwich faces into a solid laminate as shown in Fig. 1. Different solutions are chosen for the tapered sandwich core—extruded aluminium or higher density core material. Bohlmann and Fogarty [7] sandwich the laminate between two steel plates. One of the plates is welded to the steel deck and secured by adhesive bonding and bolting. The back plate is bolted to the laminate.

Any gaps are compensated for using liquid shims. This joint design has to meet stringent requirements. The hybrid joint aligns with and connects to the (steel) support structure of the lower deck. This is to minimise impact on the steel design and fabrication of the frigate. Furthermore, this allows the steel part of the joint to be fitted and welded using existing steel ship yard skills. The new superstructure module is made of sandwich material. A requirement from the Navy is that the outside is smooth and follows existing lines—presumably to avoid any changes in radar cross-section. As a consequence, the bolts had to be countersunk into the metal flange resulting in the metal flange becoming quite thick. In fact it is the heaviest item of the whole joint according to [7].

The joint was designed such that the bolts or the adhesives could take all the critical loads without help from the other joining method. Static tests of the prototype joints confirmed failure at more than twice of the design load. Furthermore, there was good agreement between the 2D plane strain FE predictions and experimental results. Bohlmann and Fogarty [7] concluded that the new joint is strong enough to survive all critical loading conditions without performance degradation. In addition the authors proposed further improvements to reduce fabrication costs.

Hentinen and Hildebrand [25] had identified decks and bulkheads in the superstructure of large ships as a possible application for use of FRP sandwich. Being able to join panels to the aluminium or steel main structure is a major challenge. Hentinen and Hildebrand proposed four different joint configurations where the sandwich-metal joint could be prefabricated enabling the installation on board the ship by welding only. This was then compared to an adhesive-bolted joint developed earlier. The most promising joint configuration had an extruded aluminium profile in the tapered section which was further studied using FE analysis and experiments. Two-dimensional plane strain analysis was carried out and tensile and flexural strength experiments were conducted. The correlation between predicted and measured load–displacement was good. However, even though the load at failure was predicted quite well, the deflection was over-predicted almost by a factor of 2. The comparison of bonded only and bolt-adhesive joints revealed that in-plane strength is almost identical. However, in bending, the bonded only solutions were two and a half times stronger than a bolt-adhesive hybrid joint.

Ritter et al. [46] and Speth et al. [49] carried out double lap-shear tests. The test results were used to calibrate their FE model for the double lap shear joint. The model was then extended to a full scale joint model where the composite was modelled with its faces only and finally the whole assembly. Furthermore, the authors used the test results to calculate factors of safety (FOS). Unfortunately no full scale joint tests were performed to validate the predictions.

The load cases investigated in the studies mentioned above were (1) distributed pressure on the outside of the panel [7, 25] and (2) in-plane tensile loading [25, 49]. Failure was predicted by Speth et al. [49] using von Mises stresses. They predicted failure to initiate in the laminate at the base of the joint on the “outside”. Hentinen and Hildebrand [25] also calculated von Mises stresses for the adhesive and interlaminar shear stresses in the composite faces. They identified critical stresses both in laminate and adhesive at the wedge shaped core at the lower end of the panel.

Bohlmann and Fogarty [7] did not predict failure; rather they modelled load and deflection of the joint and compared it with experimental results. They achieved very good agreement between the predicted and measured deflections until immediately prior to failure. They observed that the final failure was initiated by shear failure of the balsa core just outside the joint region.

5.2.3 Status and Outlook

Most authors indicate that adhesively bonded joints are more efficient to build than hybrids. However, Bohlmann and Fogarty [7] identify three reasons that could delay or stop the development of bonded only solutions:

1. Need for more extensive adhesive characterisation,
2. Change of design of supporting structure required, best all composite structure! (this may not be the case for other joint designs)
3. Reluctance of customer to adopt "high risk" approach for critical component

Le Lan et al. [34] discussed the joint design process and selection of the final configuration of the composite steel joints for the French La Fayette-class frigates. Even though hybrid bolt-adhesive joints achieved a higher tensile strength during qualification testing, the bonded joints were selected due to ease of manufacture of the bonded connection. Anon [3] came to the same conclusion.

None of the above studies addressed fire protection. However this is essential for getting approval on large passenger and cargo ships. McGeorge et al. [39] outline a risk based approach for a composite superstructure on a RoRo Ferry by using Regulation 17 from the International Convention for the Safety of Life at Sea (SOLAS). Most studies, discussed here are feasibility studies. There is very limited industrial production experience. However, as Bohlmann and Fogarty [7] point out, fabrication is key to reliability of the joints and the possible acceptance of bonded joints later on.

5.3 Direct Glazing

5.3.1 What are the Requirements?

Adhesive bonding is used extensively in outfitting of ships mainly in non-critical joints, for example flooring compounds. However, there are few examples where adhesives are used to transfer loads. Bonding of windows, also known as direct glazing, has become standard practice on passenger and cruise ships (see also Fig. 2). Typical windows are shown in Fig. 3.

There are international regulations that govern the design and fabrication of windows. The IMO Load Line convention regulates where and what size window or side scuttle can be use on board a ship. This is further explained in Unified

Fig. 2 Cruise ship

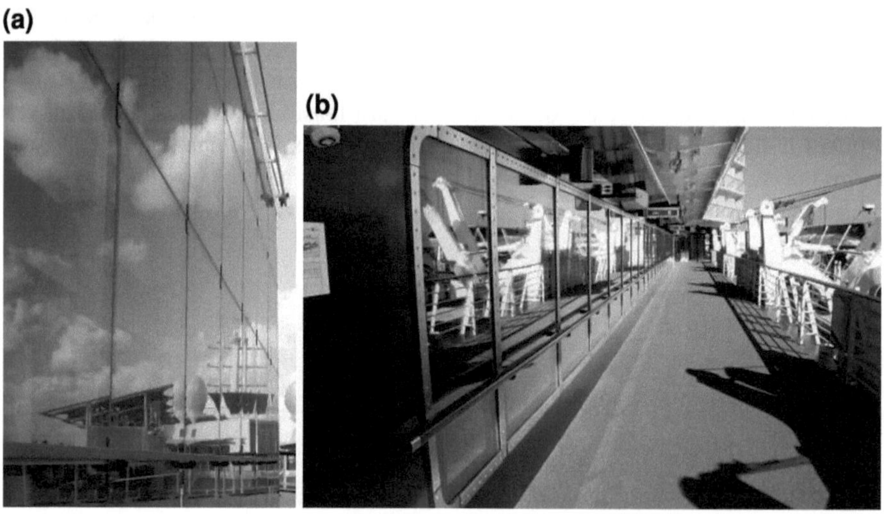

Fig. 3 Bonded windows for cruise ships: **a** glass façade showing (small) retaining frames, **b** fire protective glazing with substantial retaining frames (with permission from Brombach + Gess)

Interpretation (UI) LL62 by the International Association of Classification Societies [26]. The main concern is damage stability and to maintain the watertight integrity of the ship's hull in case of an accident. There are strict requirements on what kind of window can be used for openings below the freeboard deck, the first tier of enclosed superstructures and deckhouses. However, higher up on the superstructure, it is quite feasible to realise (very) large windows or glass facades. LL62 also states that *Side scuttles and windows together with their glasses, deadlights and storm covers, if fitted, shall be of approved design and substantial construction in accordance with, or equivalent to, recognised national or international standards. Non-metallic frames are not acceptable.* Window frames are typically designed such that if the window has to sustain water pressure the glass will be compressing the frame and transferring the load directly to the ship side shell. Only internal pressure or suction loads will act on the bond line directly. Usually, there are concerns about the durability of adhesives and to avoid the need for extensive testing; retaining frames and bolts are installed to secure the window pane against premature failure for the adhesive.

5.3.2 How are Windows Designed?

The design and fabrication of windows and side scuttles is well established. Rectangular windows are designed and built according to ISO 3903 [31], while side scuttles follow ISO 1751 [30]. The glass is tested and marked according to ISO 614 [29]. However, there are no IMO or IACS regulations for large panorama windows. Hence their design approval and fabrication follow-up is based on regulations from classification societies. Most of the windows are made of toughened glass. However, due to safety concerns it is becoming more common to use laminated glass. The polymer interlayer retains the fractured glass thus improving passenger safety in accident scenarios. Large window panes are bonded directly to the ship structure. The function of the glass is not to add stiffness to the ship's superstructure (which is common for cars) but to shield the inside of the superstructure from the elements. While the supporting structure carries the global loads the elastic adhesive layer needs to be able to compensate for the ship structure displacements due to load and thermal mismatch to minimise the loading onto the window panes. The glass panes are to carry wind and water pressure, wind suction and impact loads due to falling objects. The environmental loads are usually specified in class rules, see for example Det Norske Veritas Rules for Ships [15], while displacements due to superstructure deflection have to be calculated using global FE analysis of the ship structure.

The hybrid joint between the glass pane and the steel structure is designed in two stages. First the adhesive joint is specified using some simple formula. For example Burchardt et al. [10] provide a diagram for dimensioning the joint geometry. The minimum bondline thickness is 6 mm while the recommended overlap length is between 12 and 20 mm. According to [16] the minimum joint width can be calculated as follows:

$$d = \frac{bP_{\mathrm{w}}}{2,000\sigma_{\mathrm{t}}}\ (\mathrm{mm}) \tag{3}$$

with d is the joint width, not <0.02 b (mm), b the length of shortest window side (mm), P_w the wind suction load $= 1.25 \times 10^{-3}$ $(50 + 0.5\ V)^2$ (kN/m^2), V the speed in knots, σ_t is the allowable tensile stress for the adhesive (MPa). Normally to be taken as the stress at 12.5% elongation.

The thickness of the bondline is to be not less than [16]:

$$t = kl \times 10^{-3} \tag{4}$$

where $t =$ is the minimum 6 mm and not to be less than $d/2$ for $d > 12$ mm, $k = 1.5$ for glass, $l =$ length of longest window side (mm).

Furthermore, it is important to protect the bondline of external windows against UV radiation by using ceramic screens printed onto the glass. Moreover, the use of sealants is important to avoid having cavities in the joint where water may collect which may in turn triggers corrosion. The adhesive used for direct glazing is usually a 1-part polyurethane adhesive while the sealant is silicone based. Alternatively, the adhesive could also be used for filling any remaining cavities.

DNV [15] provides formulae for calculating the minimum thickness of the monolithic glass pane of windows and side scuttles. The following formula is used to determine the minimum thickness of laminated glass. It is based on the thicknesses of the individual glass panes to ensure the combined glass is as strong as monolithic glass panes [15]:

$$t = \sqrt{t_1^2 + t_2^2 + \cdots + t_n^2} \tag{5}$$

where n is the number of laminates, t_1 to t_n the thickness of each glass in the laminate, t is the equivalent thickness of laminated toughened safety glass.

The thickness of the polymeric interlayer is not taken into account. The resulting glass pane is usually noticeably heavier than an equivalent monolithic glass. High performance interlayers may lead to considerable reduction in total thickness for the laminated glass thus reducing the weight of the windows.

The need to use bolts in addition to the bonded joints finds its basis in IMO [27] requiring that "*..., all windows and side scuttles in bulkheads separating accommodation and service spaces and control stations from weather shall be constructed with frames of steel or other suitable material. The glass shall be retained by a metal glazing bead or angle*". Furthermore UI LL62 [26] specifies *substantial construction* and not allowing *non-metallic frames.* Today, all windows are fitted with additional mechanical fasteners. There are no guidelines on the size and spacing of these fasteners.

5.3.3 Practical Experience

Despite the widespread use of direct glazing, assessing the long-term performance of bonded joints has not been possible yet. There is a lack of documented long-term performance of the bonded windows. Fortunately, DNV keeps systematic records of the inspection reports from DNV classed ships. They are stored in a database called Nauticus Production System (NPS) [53].

A preliminary analysis of the records in NPS related to windows of passenger and cruise ships gave some surprising insights. There were no reports of adhesive failures of windows. The only reports of failures or comments related to windows can be summed up as follows:

- Use of non-certified glass—when fire rating was required
- Window wipers defective on lifeboats or the bridge
- Broken glass due to heavy weather damage
- Crack in side shell of hull that started at the corner of a big window

Discussions with DNV surveyors confirmed that failure of the bondline for windows was not known to be an issue. However, glass breakage due to heavy weather damage, such as green seas, means that further improvements are needed concerning the design and proper dimensioning of the windows, in particular the thickness of the glass panes.

5.4 Steel–Concrete–Steel Sandwich Panels

5.4.1 What is the Challenge?

A feasibility study with respect to strength and weight of an innovative bulk carrier design as shown in Fig. 4 was carried out by Bergan et al. [6]. The innovative bulk carrier design utilises a sandwich concept with steel faces and a concrete core with a density of 900 kg/m^3. The sandwich concept eliminates the need for traditional transverse frames, longitudinal or transverse stiffeners and corrugated bulkheads. Therefore, all surfaces are smooth.

The strength calculations performed showed that the concept was feasible with respect to strength and the weight of the cargo hold region. Furthermore, it was competitive with a traditional steel design at that time. The total weight was similar to a conventional design. A later study by Weitzenböck and Grafton [54] identified a need for further validation of the material properties and joining methodologies in order to confirm the concept.

Experiments carried out in the early phases of the project by Bergan et al. [6] indicated that there was a problem with the initial concept of gluing concrete to steel. Panels that had been produced for experimental studies got damaged during transit—in some cases the steel faces delaminated completely. Gerwick and Venututi [22] report of a study were fatigue performance of SCS beams was monitored. The SCS beams had additional shear connectors, so called stirrups, installed on the inside faces of the steel sheets. The study clearly indicated that after a few thousand cycles only the stirrups were carrying the load. The bond strength between the concrete core and the steel plates had been lost. In line with this, DNV's standard for offshore concrete structures requires that the forces are

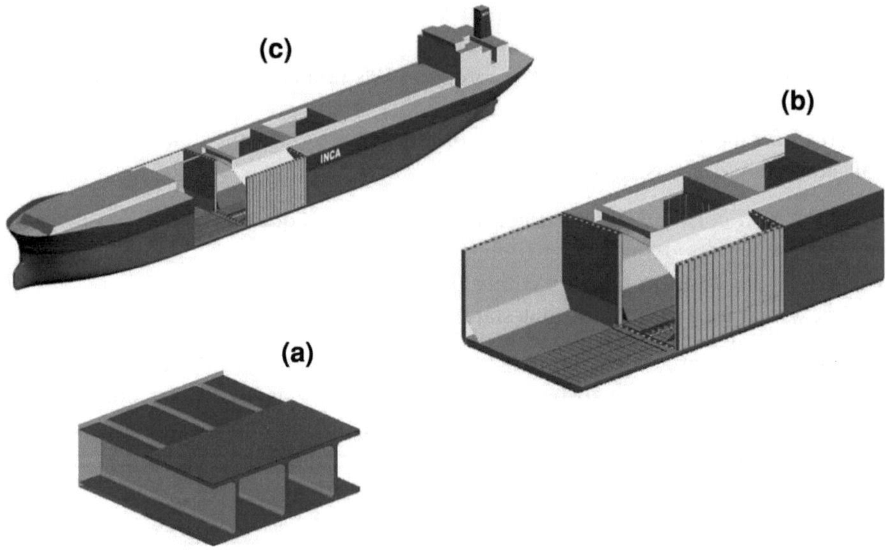

Fig. 4 Steel–concrete–steel (SCS) sandwich concept applied to a bulk carrier: **a** cellular structure assembled from SCS panels, **b** cargo hold made up of SCS structures, **c** bulk carrier assembled from SCS panels

transferred between the members by reinforcements, shear keys or other devices [14].

These observations were the basis for a systematic study of the SCS sandwich panels, the results of which will be reported below.

5.4.2 Solution to Improving Bond Strength of SCS Panels

A benchmark study was carried out as reported by Weitzenböck and Grafton [54]. Two different configurations of the SCS sandwich panel were considered. The original concept consisted of two steel plates with a concrete core and is shown in Fig. 5. Load transfer between the steel plates and the concrete core relies entirely on adhesion between the steel interface and the concrete core. Later on this was modified by the introduction of shear studs as adhesion alone was not considered reliable enough. A typical example is shown in Fig. 6.

The results of the parametric study by Hayman [24] show that for most cases there is a clear advantage in using welded studs to improve the shear strength of an SCS panel. Even if a solution without studs appears to give an approximately equal weight to one with studs, the need to ensure an adequate shear connection between the face sheets and the core makes it advisable in practice to use studs in all cases. Steel–concrete composites with studs have been successfully used for many years in the construction of buildings and bridges; the latter are subjected to fatigue loading. Therefore, steel–concrete composite with studs can be considered a proven

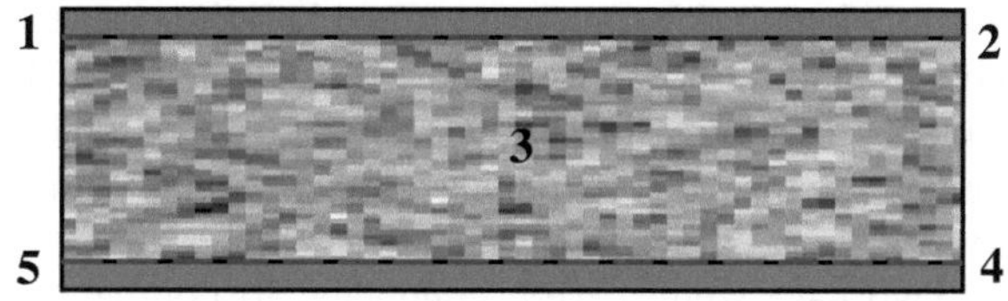

Fig. 5 SCS sandwich—"no studs" (*1* face 1: steel plate; *2* bondline 1 between face 1 and core: adhesion of concrete to steel; *3* core: lightweight concrete; *4* bondline 2 between core and face 2: adhesion of concrete to steel; *5* face 2: steel plate)

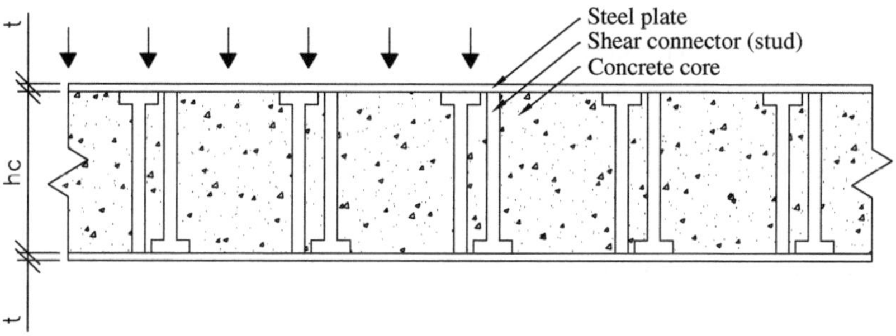

Fig. 6 Typical SCS panel—"with studs"

technology. The novelty of the proposed panels is that, in addition to ensuring the shear connection between sandwich elements, the studs are also used to reinforce the core, avoiding shear tension failure and limiting cracking of the concrete core.

The benchmark study by Lervik and Mayorca [35] focused on assessing SCS-panels with studs from the structural point of view and comparing their performance with typical steel solutions. The study looked into the application of panels to ships and offshore facilities, particularly horizontal elements such as decks. Based on the review of existing design standards and recommendations, the basis for design of SCS was proposed. Ultimate and fatigue limit states are considered. Service limit state was only briefly discussed as there are no specific design requirements for control of deflections in ships and offshore facilities. It is assumed that water tightness is provided by the steel skin. A parametric study was carried out in order to identify SCS-panel (with studs) dimensions and spans for various loads and boundary conditions. The results showed that SCS is capable of covering relatively long spans. Because studs were designed to transfer the horizontal shear force between plates and core and to prevent shear tension failure, design was governed by either shear compression or bending capacity.

The challenge that SCS panels are faced with for most applications in conventional ship applications is its self weight. In the parametric study two different concrete cores were considered, light weight concrete (LWC) and high strength concrete (HSC). The aim was to investigate if the tenfold increase in compressive strength could give any benefit in spite of the almost threefold weight increase. It was concluded that for panels in which the bending capacity limits the maximum feasible length, LWC-cores resulted in lighter panels for a given span. It should be noted that it is possible to increase the shear and bending capacity of the HSC-panels by pre-stressing the panels or increasing the steel plate strength. However, this option makes HSC-panels competitive only for high imposed loads, which are several times larger than the self-weight.

Two case studies were conducted to compare SCS panels with typical steel solutions [35]. The comparison only considered replacing some decks of an existing ship with SCS. The basic conclusion was that it was difficult to match the weight of the existing steel decks with SCS panels. The limitation of the studies is that the main structural system is the result of an optimization process for a traditional steel structure. Steel structures are good at resisting tensile forces but may buckle under compression forces. The SCS-panel is capable to efficiently carrying both axial compression and tension. Taking into consideration that SCS behaves differently than steel, it is foreseen that a better utilization of SCS can be achieved if a main structural arrangement was developed that exploits the SCS advantages.

5.4.3 Status of SCS Sandwich Technology

SCS is still at a prototype stage awaiting commercial exploitation. By adding shear studs to the steel interfaces the SCS can be designed and produced using existing knowledge and practices. However, the main stumbling blocks are not technical but of economical nature. SCS panels are best suited to solutions which are not weight critical and where other benefits of the technology can be utilised to offer additional advantages.

6 Conclusions

The reoccurring message from the discussions of the maritime applications is that hybrid bolt-adhesive joining is used mainly because there is a lack of confidence in the durability of adhesively bonded joints. This is also why existing designs do not usually utilise the composite action of the two joining methods. Rather, the two joining methods are used in parallel where one joining method takes over when the other fails. While there are many good technical solutions, most decision makers in industry and regulators are risk-adverse and tend to prefer traditional design solutions and material choices.

Bolt-adhesive joints are usually used to join dissimilar materials. The application cases discussed here illustrate this. While direct glazing of windows is an accepted way for making large-scale windows, composite superstructures and the necessary hybrid joints have yet to be established as an industrial standard for large ocean going vessels. However, polymer sandwich structures have been used on small and larger vessels for many years. And finally, SCS sandwich requires further development in order to industrialise this material.

References

1. Adams, R.D.: The design of adhesively-bonded lap joints: modelling considerations. In: 46th SAMPE Symposium, pp. 402–414 (2001)
2. Anon: Guide to the Structural Use of Adhesives. The Institution of Structural Engineers, UK (1999)
3. Anon: Committee V.5 naval ship design. In: Proceedings of the 16th International Ship and Offshore Structure Congress, Southampton, UK (2006)
4. Bickford, J.H.: An Introduction to the Design and Behavior of Bolted Joints, 3rd edn, pp. 840–871. Marcel Dekker, NY (1995)
5. Beevers, A.: Forensic studies of adhesive joints, part 2—bonded air-craft structure. The performance of adhesive joints (PAJ)—project 3: environmental durability. http://www.adhesivestoolkit.com/ (1995). Accessed Sept 2010
6. Bergan, P., Bakken, K., Thienel, K.C.: Analysis and design of sandwich structures made of steel and lightweight concrete. In: III European Conference on Computational Mechanics: Solids, Structures and Coupled Problems in Engineering (ECCM-2006), Lisbon, 5–8 June 2006
7. Bohlmann, R.E., Fogarty, J.H.: Demonstration of a composite to steel deck joint on a Navy destroyer. In: 9th International Conference on Marine Applications of Composite Materials: paper L, Melbourne, FL, USA, pp. 19–21 (2002)
8. Bond, A.E.: A review of industrial case histories. MTS programme on the performance of adhesive joints—project 3: environmental durability. http://www.adhesivestoolkit.com/ (1995). Accessed Sept 2010
9. Broughton, W.R., Crocker, L.E., Gower M.R.L.: Design Requirements for Bonded and Bolted Composite Structures. National Physical Laboratory, NPL Report MATC(A)65, January (2002)
10. Burchardt, B., Diggelmann, K., Koch, S., Lanzendörfer, B.: Elastic Bonding. Verlag Moderne Industrie, Germany (1998)
11. Croucher, T., Vanaman, A., Lee, S.: Computer controlled weldbonding of aluminium on the Fairchild A-10. Metal Prog. (1981)
12. da Silva, L.F.M., das Neves, P.J.C., Adams, R.D., Spelt, J.K.: Analytical models of adhesively bonded joints—part I: literature survey. Int. J. Adhesion Adhesives **29**, 319–330 (2009)
13. da Silva, L.F.M., das Neves, P.J.C., Adams, R.D., Wang, A., Spelt, J.K.: Analytical models of adhesively bonded joints—part II: comparative study. Int. J. Adhesion Adhesives **29**, 331–341 (2009)
14. DNV: Offshore concrete structures. Offshore Standard—DNV-OS-C502 (2007)
15. DNV: Rules for Ships. Sect. 6, Part 3, Chap. 3. "L. Side Scuttles, Windows and Skylights" (2010)
16. DNV: Rules for High Speed, Light Craft and Naval Surface Craft. Sect. 1, Part 3, Chap. 6. G. Windows, 8 pp (2010)
17. DNV: Rules for High Speed, Light Craft and Naval Surface Craft. Sect. 9, Part 3, Chap. 4. Bolted Connections, 28 pp (2010)
18. Drain, K., Chandrasekharan, S.: Automotive applications for adhesives. In: ASM Engineered Materials Handbook. Adhesives and Sealants, vol. 3, pp. 551–557 (1990)
19. Echtermeyer, A.T., McGeorge, D., Sund, O.E., Andresen, H.W., Fischer, K.P.: Repair of FPSO with composite patches. Presented at the fourth international conference on composite materials for offshore operations (CMOO-4), Houston, TX, USA, 4–5 October 2005
20. Eurocode 3: Design of steel structures, part 1–8: design of joints, EN 1993-1-8 (2005)
21. Eurocode 4: Design of composite steel and concrete structures—part 2: general rules and rules for bridges (2005)
22. Gerwick, B.C., Venututi, W.J.: Fatigue Behaviour of Prestressed Concrete in a Ship Hull. Technical and Research Bulletin 2-24, The Society of Naval Architects and Marine Engineers (1979)

23. Guthu, G.O.: Modelling of mechanical response and fracture in adhesively bonded joints. Master thesis, Faculty of Mathematics and Natural Sciences, University of Oslo (2010)
24. Hayman, B.: Benchmark study of INCA panels without studs. In: Weitzenböck, J.R., Grafton, T. (eds.) Assessment of the INCA Steel-Concrete-Steel Sandwich Technology—Public Report. DNV Technical Report No. 2010-1284 (2010)
25. Hentinen, M., Hildebrand, M.: How to efficiently join FRP structures to metal ships? In: 3rd International Conference on Fast Sea Transportation, FAST'95, Lübeck-Travemünde, Germany, 25–27 September 1995
26. IACS: Side Scuttles, Windows and Skylights (Regulation 23). Unified Interpretation LL62, 2 pp (2010)
27. IMO: SOLAS Chapter II-2 Part B, Regulation 33: Windows and Side-Scuttles (1992)
28. IMO: SOLAS Chapter II-2 Part F, Regulation 17: Alternative Design and Arrangements (2002)
29. ISO 614: Shipbuilding and marine structures—toughened safety glass panes for rectangular windows and side scuttles—punch method of non-destructive strength testing (1989)
30. ISO 1751: Shipbuilding and marine structures—ships' side scuttles (1993)
31. ISO 3903: Shipbuilding and Marine Structures—Ships' Ordinary Rectangular Windows, 2nd edn (1993)
32. Keller, H., Howard, M., Hoyer, J.: Weldbonding an Improvement, Theoretically and Practically. Swissbonding, Rapperswil (2006)
33. Kelly, G.: Load transfer in hybrid (bonded/bolted) composite single-lap joints. Compos. Struct. **69**, 35–43 (2005)
34. Le Lan, J.Y., Livory, P., Parnelix, P.: Steel/composite bonding principle used in the connection of composite superstructure to a metal hull. In: Proceedings of SANDWICH2, Gainesville, USA (1992)
35. Lervik, A., Mayorca, P.: Benchmark study of INCA panels with studs. In: Weitzenböck, J.R., Grafton, T. (eds.) Assessment of the INCA Steel-Concrete-Steel Sandwich Technology—Public Report. DNV Technical Report No. 2010-1284 (2010)
36. Marsh, G.: Boeing's 787: trials, tribulations, and restoring the dream. http://www.reinforcedplastics.com/view/4282/boeings-787-trials-tribulations-and-restoring-the-dream/ (2009). Accessed 31 Aug 2010
37. McGeorge, D., Lilleborge, J., Høyning, B., Eliassen, G.: Survivable composite sandwich superstructures for naval applications. In: Proceedings of the SANDWICH'6, Fort Lauderdale, USA (2003)
38. McGeorge, D.: On the long term properties of steel-composite interfaces for marine applications. Presented at the fourth international conference on composite materials for offshore operations, Houston TX, USA, 2005
39. McGeorge, D., Høyning, B., Nordhammar, H.: Lightweight composite sandwich RoPax superstructure. In: Papanikolau, A. (ed.) Risk-Based Ship Design. Springer, Berlin (2009)
40. McGeorge, D.: Inelastic fracture of adhesively bonded overlap joints. Eng. Fract. Mech. **77**, 1–21 (2010). doi:10.1016/j.engfracmech.2009.07.002
41. McGeorge, D.: Predicting failure of bonded structural joints. In: Weitzenböck, J.R. (ed.) Using Adhesives in Marine Engineering (2011) (to be published)
42. Menges, G., Schmidt, P.: Untersuchungen zum Langzeitverhalten kombinierter Kleb-Punktschweisßerbindungen (Investigations into the Long-Time Behaviour of Combined Bonded and Spot Welded). Schweissen und Schneiden, 19 pp (1967)
43. Osnes, H., McGeorge, D.: Experimental and analytical strength analysis of double-lap joints for marine applications. Compos. Part B Eng. **40**(1), 29–40 (2009)
44. Peck, S.J.: Streamlining automated processes for composite airframe manufacturing. JEC Compos. Mag. June–July(58), 34–36 (2010)
45. Reinhardt, K.G.: Festigkeitseigenschaften und Gesichtspunkte für die Bemessung von Punktschweiß-Klebverbindung. ZIS Mitteilungen **10**, 1609–1628 (1968)
46. Ritter, G.W., Speth, D.R., Yang, Y.P.: Qualification of adhesives for marine composite-to-steel bonded applications. J. Ship Prod. **25**, 198–205 (2009)

47. Ruttge, H.: Vorteile durch Einsatz des Punktschweißklebeberfahrens beim Herstellen von Ladebordwänden. Schweißtechnik **19**, 167–169 (1969)
48. Shenoi, R.A., Hawkins, G.L.: Practical design of joints and attachments. Compos. Mater. Marit. Struct. **2**, 28 (1993)
49. Speth, D., Yang, Y.P., Ritter, G.: Qualification of adhesives for marine composite-to-steel applications. Int. J. Adhesion Adhesives **30**, 55–62 (2010)
50. Wacker, G., Brügge, D., Fach, K., Franzelius, W.: Theory and practice: requirements for adhesive bonding in marine structures. Proc. Inst. Mech. Eng. Part M J. Eng. Marit. Environ. **218**(M4), 217–226 (2004)
51. Weitzenböck, J.R., McGeorge, D.: The designer's dilemma: how to deal with the uncertainty about the long-term performance of adhesively bonded joints. Proc. Inst. Mech. Eng. Part M J. Eng. Marit. Environ. **218**(4), 273–276 (2004)
52. Weitzenböck, J.R., McGeorge, D.: BONDSHIP Project Guidelines, 254 pp (2005)
53. Weitzenböck, J.R.: Adhesive Bonding in the Maritime and Offshore Industry. Swissbonding, Rapperswil (2009)
54. Weitzenböck, J.R., Grafton, T.: Assessment of the INCA steel-concrete-steel sandwich technology—public report. DNV Technical Report No. 2010-1284 (2010)
55. Weitzenböck, J.R. (ed.) (2011) Using adhesives in marine engineering (to be published)

Science of Friction–Adhesive Joints

Eugenio Dragoni and Pierfranco Mauri

Abstract This chapter addresses the fundamental properties of hybrid friction–adhesive joints which combine any form of mechanical tightening (stimulus for friction forces) with anaerobic adhesives. By filling the voids around the microareas of true metal-to-metal contact between the mating parts, anaerobic adhesives allow the full area involved by the engagement to be usefully exploited. Advantages ranging from sealing action, fretting suppression, noise reduction and enhanced strength derive from this combination. The focus of the chapter is on predicting the mechanical strength of these joints. The literature covering the static and the fatigue strength is reviewed showing that proper choice of the adhesive can increase the overall strength of the joint well above the strength of the purely mechanical joint based on friction only. Simple equations are also provided for the strength calculation of practical engineering assemblies.

1 Introduction

Purely mechanical joints based on the frictional forces arising at the pre-loaded contact between metal parts have become a common assembly solution in machine constructions since a long time. Anaerobic adhesives (anaerobics) are an effective

E. Dragoni (✉)
Department of Sciences and Methods for Engineering, University of Modena and Reggio Emilia, Via Amendola 2, 42122 Reggio Emilia, Italy
e-mail: eugenio.dragoni@unimore.it

P. Mauri
Henkel Italia S.p.A., Via Amoretti 78, 20157 Milan, Italy
e-mail: Piero.Mauri@it.henkel.com

Adv Struct Mater (2011) 6: 201–225

DOI: 10.1007/8611_2010_34

Published Online: 23 July 2010

means to improve the performance of these joints because they tolerate being trapped between closely matching surfaces subjected to substantial clamping forces. Threaded connections, flanged couplings, taper or interference fits (Fig. 1) are typical examples taken from the wide range of applications which benefit from these machinery adhesives [12]. A distinguished property of anaerobic adhesives is the low demand that they put on the design of the joint which in most cases can be efficiently bonded as it was conceived for the standard assembly. This characteristic represents a remarkable advantage over virtually all other types of adhesive and makes them ideal candidates to solve many problems of mechanical design.

Anaerobic adhesives belong to the large family of thermosetting acrylic polymers and are kept stable as one-part liquids by the oxygen in the air. Easily applied to the adherends by manual or automatic means, anaerobics cure when confined in the airless pockets (hence the name) left by the roughness of the mating metal surfaces upon tightening of the joint (Fig. 2). Advantages of the adhesive-augmented friction joints include added sealing action, improved corrosion resistance, reduced fretting wear and increased mechanical strength. The strength enhancement brought forth by the adhesive is comparable to (often greater than) the strength of the purely mechanical (dry) joint. This improvement results in more compact (thus cheaper) constructions when a new design is undertaken to meet a specified load-carrying capacity. Alternatively, existing designs can be upgraded at marginal costs whenever the strength expectations of the market become more demanding with time.

This chapter reviews the scientific literature dealing with the static and the fatigue strength of bonded joints based on this hybrid technology. The focus of the review is on the fundamental properties of the friction–adhesive interface rather than on applications. The technological aspects of anaerobics usage together with several representative case studies are presented and discussed in the next chapter of this book.

(a) (b) (c)

Fig. 1 Typical applications of anaerobic adhesives to complement mechanically tightened friction joints: **a** threaded connections, **b** flanged couplings, **c** cylindrical fits

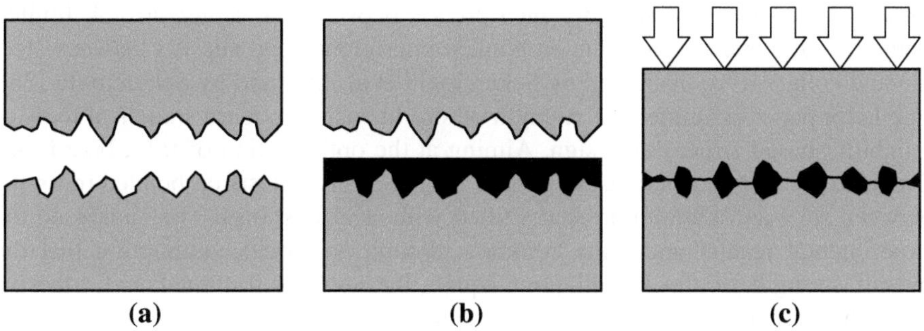

Fig. 2 The three stages of formation of a hybrid friction–adhesive joint: **a** dry joint, **b** adhesive application, **c** joint closure and tightening

Systematic research on anaerobics can be traced back to the paper by O'Reilly [19] covering static testing and design of bonded cylindrical interference-fits, a mainstay of anaerobic technology. Mahon [17] and Sawa et al. [24] demonstrated how anaerobics can be used to enhance the strength of flanged couplings. A review of anaerobics application to axisymmetric joints is given by Romanos [23] and Bartolozzi et al. [2]. Dragoni [6] showed that anaerobic threadlockers can improve the fatigue life of threaded connections by reducing the fretting damage (thus delaying crack initiation) between nut and bolt threads. The effects of bolt preload, bolt diameter, thread pitch and surface treatment on breakaway, breakloose and prevailing torques in commercial nut-bolt connections bonded with an anaerobic threadlockers were measured by Sekercioglu and Kovan [29]. Yoneno et al. [32–34], Sawa et al. [25] and Kawamura et al. [14] presented an analytical model for the stress analysis in the adhesive layer of bonded interference fits between hollow round shafts and annular hubs undergoing push-off and torsional loading. Their predictions are far on the conservative side when compared with the outcome of static tests. Dragoni and Mauri [7] investigated the intrinsic properties of friction–adhesive interfaces by torsional testing of annular butt joints subjected to controllable normal pressure. Their measurements on a single adhesive supported the conclusion that the cumulative strength of the hybrid interface equals the sum between the strength of the purely adhesive interface and the strength of the purely frictional interface for any contact pressure. By extending the investigation to an assortment of anaerobics tested both in ideal conditions (butt interface) and in real working conditions (threads, cylindrical fits and overlaps), Dragoni and Mauri [8] observed that the strength of the hybrid joints increases with the contact pressure but the superimposition of strengths (purely frictional + purely adhesive) is not always verified. More recently, a paper by Dragoni [9] evaluated the fatigue behavior of hybrid friction–adhesive tapered joints, proposing an empirical relationship between the fatigue limit and the static strength. The result can help the designer in predicting the behaviour under cyclic stresses starting from the more readily available static failure stress. Canyurt [3] used a genetic algorithm to estimate the fatigue strength of bonded tubular joints including the effect, based on third-party results, of assembly interference between the adherends. A limited range of static and fatigue tests on bonded interference and slip fits between steel pin and collars were discussed by Sekercioglu et al. [27] and by Sekercioglu [28]. The latter paper examines the strength of slip fits using modern statistical tools of reliability-based structural design. Aiming at the optimization of front bike forks, Croccolo et al. [4] carried out static and fatigue push-off tests on bonded press-fits between steel and aluminium shafts fitted with steel bushings. They analyzed the experimental results under the common, though empirical, assumption that the overall strength of the hybrid joint equals the sum of frictional and adhesive strengths, calculated independently of each other.

By examining the above literature and particularly the papers by Dragoni [9] and Dragoni and Mauri [7, 8], this chapter has three main purposes. First, summarize the static strength of hybrid interfaces, formed from a variety of adhesives and showing both ideally regular (annular butt joints) or realistically irregular (threaded, cylindrical and double-lap joints) stress distributions over the bondline. Second, correlate the fatigue strength of the hybrid interface to the corresponding static strength from tests on bonded taper fits. Third, provide a micromechanical model of the hybrid interface and supply engineering formulae for the strength calculation of typical industrial joints.

2 Experimental Details

2.1 Static Strength of the Ideal Interface

An ideal friction–adhesive interface is a bonded interface submitted to homogeneous (uniform) contact pressure (upon joint formation) and homogeneous shear stresses (upon breakaway). Testing of these particular interfaces aims at understanding the intimate interaction between friction and adhesive forces exchanged by the bonded parts. Experiments on nearly ideal interfacial conditions were performed by Dragoni and Mauri [7, 8] using the simple device shown in Figs. 3, 4a.

The equipment in Fig. 3 realizes axisymmetric butt joint conditions according to ASTM E229. The primary engagement occurs at *x*-*x* between the annular end faces (∅30 × 37 mm) of the steel bushings 4 and 5 (quenched and drawn 39NiCr Mo3 to EN 10083-3, with ultimate tensile strength $S_u = 1{,}100$ N/mm^2 and yield strength $S_y = 850$ N/mm^2). The axial contact force is provided by the centre bolt 2 (M14 × 120, property class 12.9), tightened through the flanged nut 8 (property class 10.9). The friction between the threads of bolt and nut is kept to a minimum by smearing them with low-friction lubricating paste (Molykote G-n plus). The bolt head, angularly secured by setting screw 1, pushes on bushing 4 through quenched steel washer 3, whereas nut 8 presses on bushing 5 through collar 7 and thrust roller bearing 6 (INA 81206). The bearing serves the double purpose of limiting the torque on the annular interface upon tightening and reducing the resisting torque from outside the interface upon breakaway (see below).

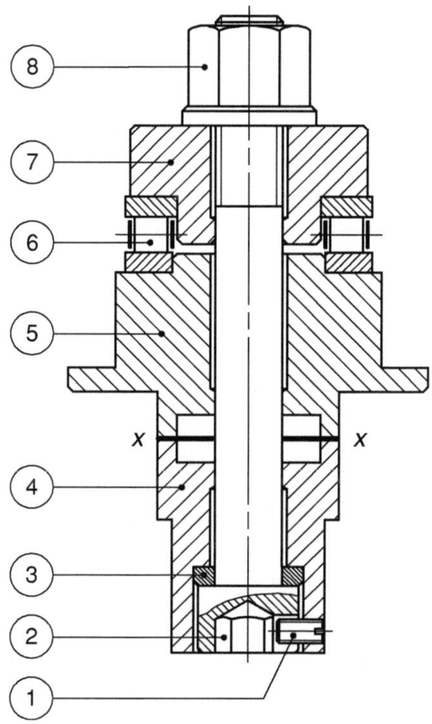

Fig. 3 Section view of the device used to produce uniform interfacial stress conditions (at *x–x*): (*1*) setting screw, (*2*) centre bolt, (*3*) washer, (*4*) lower bushing, (*5*) upper bushing, (*6*) roller bearing, (*7*) collar, (*8*) flanged nut

(a) **(b)** **(c)** **(d)**

Fig. 4 Overview of the specimens used for static characterization of the hybrid friction–adhesive interface (in *white* the *bondline*): **a** ideal interface, **b** threaded connection, **c** cylindrical fit, **d** double-lap joint

The tightening and the breakaway operations are simplified by lateral flats machined on bushings 4 and 5 to allow fixation in a vice and torque application with a wrench (Fig. 4a). Although designed to be cheap (no expensive load cells) and simple to use (hand operated), the described device ensures ideal working conditions at the interface. First of all, the flat active surfaces facilitate cleaning, treatment and closure of the joint. In the second place, the narrow annular shape transfers constant pressure under the closing force and nearly uniform shear stresses under the breakaway torque. In particular, the bondline is free from stress singularities which are an inherent feature at the edges [1] of more common specimens like the overlap joint [21], the pin-collar coupon [11] and many others [16].

The tests with the device in Fig. 3 were carried out in four steps as follows:

1. *Surface preparation*: removal of adhesive residuals (if any) and manual regeneration of surface roughness to $Ra = 0.8–1.2\ \mu m$ on flat sanding table (grit size P60), followed by check of the planarity of the surface (hand adjustment with file, if necessary) and final degreasing with liquid trichloroethane.
2. *Assembly and preloading*: fixation of bushing 4 in a vice, application of anaerobic adhesive (if necessary), assembly of specimen and tightening of nut 8 under prescribed torque (electronic wrench) to apply a specific contact force according to an experimentally determined torque-preload calibration curve (see [7]).
3. *Consolidation of interface*: permanence in oven at 40°C for 8 h (curing of adhesive) and subsequent acclimatisation in still air at room temperature (20°C) for about 12 h for complete growth of metal junctions between roughnesses [30]).
4. *Breakaway*: blocking of bushing 5 in a vice and gradual application of torque (electronic wrench) to bushing 4 up to fracture of the interface. The maximum (breakaway) torque recorded by the wrench less the small friction torque of the roller bearing 6 at the applied axial load (see [7]) gives the torsional strength of the interface.

The investigation by Dragoni and Mauri [7, 8] involved the products Loctite 638, 601, 243, 242 and 222. These products span the full range of anaerobics in terms of mechanical strength and field of application. Loctite 638 is a high-strength retainer, particularly suited for permanent cylindrical fits. Loctite 601 is a medium/high-strength retainer, tailored for cylindrical joints and general purpose bonding. Loctite 243, 242 and 222 are medium- and low-strength threadlockers, used to prevent rotation and conserve preload in threaded joints.

Along with the bonded interfaces, the tests included a number of unbonded assemblies, tightened under a variety of loads, to gain information on the strength of the dry interface. All tests were performed on twelve identical devices as in Fig. 4a, which were used to carry out sequential runs of twelve different test conditions, randomly allocated to the specimens.

2.2 Static Strength of Real Joints

For the tests carried out under realistic working conditions, Dragoni and Mauri [8] examined three joint geometries: threaded connections (Fig. 5a), cylindrical fits (Fig. 5b) and double-lap joints (Fig. 5c). Pictures of these specimens are shown in Fig. 4b–d, next to the device used to achieve ideal interface conditions. A description of the joints and of the testing procedure follows.

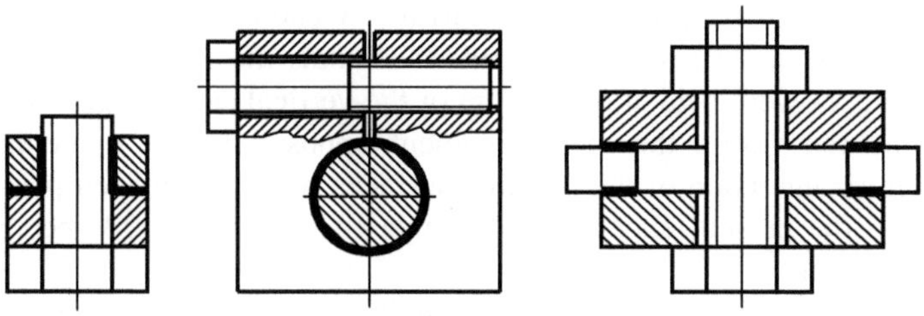

(a) (b) (c)

Fig. 5 Joint geometries providing realistic interface conditions: **a** threaded connection, **b** cylindrical fit, **c** double-lap joint (*bondline* shown with *heavy lines*)

1. *Threaded connections*. The threaded connection in Fig. 5a comprises a commercial M12 × 30 steel bolt (grade 8) and two commercial M12 × 10 steel nuts (grade 8). Bolt and nuts were initially degreased by repeated soaking in liquid trichloroethane. After that, each sample was assembled by driving the lower nut against the head of the bolt followed by controlled tightening of the upper nut, with adhesive applied (if any) to both thread and bearing face. Permanently engaged with the incomplete thread turns underneath the bolt head, the first nut provides a neat countersurface for the second nut which embodies the actual test piece. Details on the conversion of the tightening torque applied to the upper test nut into effective preload induced in the bolt are provided by Dragoni and Mauri [8]. The strength measurements were performed by recording with an electronic wrench the peak torque required to unscrew the top nut after consolidation of the interface. Due to the difficulty in restoring the working surfaces of the parts and in view of their low cost, each specimen was tested only once and test replications were performed on new pieces.
2. *Cylindrical fits*. The cylindrical fit in Fig. 5b, fabricated in 33 samples, includes a ∅16 mm steel pin (head and shank of a commercial M16 hexagonal bolt of grade 8), a square (40 × 15) steel clamp (steel C40 to EN 10083-2) and a transverse commercial M8 steel screw (grade 12.9). The clamp receives the pin in a centre hole (∅16 mm) and is made flexible by a through radial slot, perpendicular to the screw axis. The parts were prepared by manual sanding the cylindrical surfaces of pin and clamp with emery paper (grit size P60), degreasing with liquid trichloroethane and smearing with lubricating paste (Molykote G-n plus) the threads and the bearing head surface of the transverse screw. Lubrication aimed at achieving a low (about 0.05) and repeatable frictional coefficient throughout the tests. After application of the adhesive (if any) to the cylindrical surfaces, the pin was inserted into the hole of the clamp and the parts were tightened at the requested torque by means of the transverse screw. The clamping force on the pin can be related to the tightening torque on the screw by means of textbook formulae. The strength measurements were made by recording with an electronic wrench the torque needed to break the interface under twist. The hexagonal head of the pin and the square outline of

the clamp helped in fitting the parts to the wrench and to a vice during this operation. To avoid stressing of the interface under the closing force of the vice, the test pieces were held at the bottom of the vertical sides (Fig. 5b) so that the force line ran within the metal ligament opposite to the tightening screw. Although regeneration of the working surfaces would have been possible after testing, in the present experimental campaign each of the 33 specimens was used only once.

3. *Double-lap joints*. The double-lap joint in Fig. 5c, manufactured in eleven pieces, is composed of two hollow steel (C40 to EN 10083-2) discs (∅14 × 50 × 15 mm), three peripheral steel (C40 to EN 10083-2) cylinders (∅20 × 10), a commercial (M14 × 60) centre steel bolt (grade 12.9), a hardened steel washer (HRC = 60) and a commercial (M14 × 12) steel nut (grade 10). The specimens were prepared by manual sanding the flat surfaces of discs and cylinders with emery paper (grit size P60), degreasing all parts by repeated soaking in liquid trichloroethane and smearing bolt and nut threads with lubricating paste (Molykote G-n plus). Upon assembly, the three cylinders (each of exactly the same height) were interposed between the (coaxial) discs in such a way that their centres lay on the rim of the discs with 120° angular spacing. The equal thickness and the regular spacing ensured an equal share among the cylinders of the axial preload imposed on the bolt by a controlled torque. The adhesive (if any) was applied to the bases of each cylinder prior to formation of the specimen. The strength measurements were made by recording the maximum load (applied by a hydraulic testing machine) needed to displace radially by a small amount (1 mm) each cylinder with respect to the discs. In this way, each specimen supplied three readings involving the same assembly conditions. The surfaces of discs and cylinders were regenerated once (and the parts tested a second time) by chemical removal of cured adhesive residuals and mechanical flattening on a plane grinding machine.

The real joints tested by Dragoni and Mauri [8] were bonded with adhesives Loctite 638 and 243, the most representative products (in commercial terms) within the categories of high- and medium-strength anaerobics of the Loctite product range. A number of unbonded joints were again included in each test run so as to establish a reference baseline for the strength values at different levels of the contact pressure.

2.3 Fatigue Strength

Fatigue failures in the mating parts of a clamped connection are a common occurrence in mechanical constructions. Although the degradation of rubbing surfaces (fretting) is often a stimulus to fracture initiation [6, 13], these failures can be classified as a stress concentration problem [31] and is well documented in the technical literature (see [15] for the shaft-hub connection).

The work by Dragoni [9] examined the intimate strength of the interface (be it dry, slip bonded or hybrid) which represents the weakest link of the joint once classical stress concentrations have been eliminated. In order to achieve failure at the interface and gain information on its properties, the test specimens shown in Fig. 6 were adopted. The specimen comprises a tapered (1:50) pin (∅10 × 30 mm) and a collar (∅10 × 30 × 9 mm) with tapered (1:50) center hole. The pin is a commercial item made of alloy steel, hardened to $S_u = 1.500$ N/mm^2 and ground on the conical surface. The collar, obtained by cutting, boring and reaming a ground round bar, is made of carbon steel, quenched and drawn to $S_u = 800$ N/mm^2. The specimen dimensions are similar to those prescribed for cylindrical pin and collars by ISO 10123, the main standard addressing the static strength of anaerobics. Cylindrical specimens of this type (pin ∅12.7 × 50 mm, collar ∅12.7 × 25.4 × 11 mm, both of mild steel) were used in Dragoni's [9] work (see below) for exploratory fatigue tests on slip bonded (no interference) assemblies.

Use of tapered specimens for systematic fatigue testing has many advantages: inexpensive fabrication and easy preparation of the parts (Fig. 6a); simple adjustment of the contact pressure through control of the axial push-on force (Fig. 6b); straightforward axial loading by push-off compression on reverse (Fig. 6c); certain accomplishment of failures at the interface; clear detection of failure by neat ejection of the pin from the collar. The main limitation of the specimens lies in the irregularity of the contact pressure (upon assembly) and of the shear stress (upon loading) over the interface, including singularities at the edges of the contact [1]. As a consequence, although significant for real joints with the same proportions as the specimens, the measured strength values are probably lower than the true properties of the interface.

Dragoni [9] manufactured the specimens in homogeneous batches of 30 (cylindrical specimens) and 100 samples (tapered specimens), all showing a roughness of the mating surfaces in the range $Ra = 1.2$–1.6 μm. Before assembly, the parts were cleaned by repeated soaking in liquid trichloroethane followed by wiping with cotton cloth and final drying (solvent evaporation under hood cabinet). When applicable, the parts were bonded with the anaerobic adhesive Loctite 638, the high grade retaining product used extensively in the static testing of the previous sections. The tests were performed according to a factorial scheme which

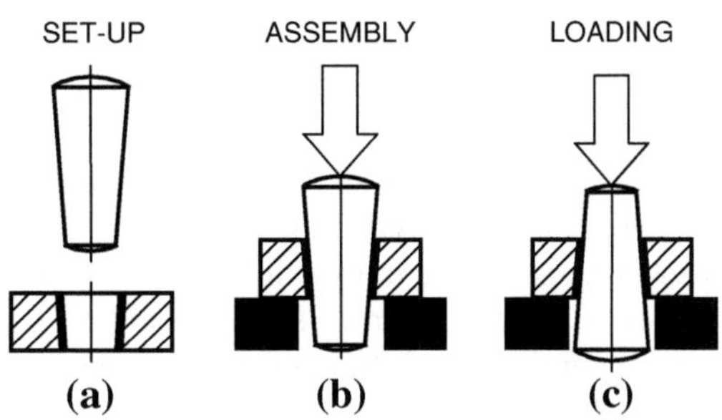

Fig. 6 Geometry of the tapered specimen for fatigue testing showing set-up (**a**), assembly (**b**) and loading arrangements (**c**). *Bondline* shown with *heavy lines*

combined in all relevant ways the three following experimental factors, each variable over two levels: set-up, dry (no adhesive) or bonded (Loctite 638); assembly, slip fit (no contact pressure) or press fit (mean pressure of about 150 N/mm^2); loading, quasi-static (monotonic) or fatigue (repeated stress).

The coefficient of friction, needed to convert the axial assembly force into (mean) contact pressure, was retrieved (for both dry and adhesive joints) from two parameters. One parameter was the slope of the force–displacement diagram recorded at assembly. The second parameter was the ratio between maximum assembly force and corresponding hoop strain in the collar as captured by an electrical-resistance strain gauge applied on the outer surface (see Sect. 3.3). Elaborated by means of the formulas for taper connections and thick-walled cylinders [9], the above parameters provided the coefficient of friction between the parts.

The fatigue tests were carried out under cyclic loading (stress ratio $R = F_{min}/F_{max} = 1/19$) at a constant frequency of 110 Hz. The load was applied axially to the thinner end of the pins as shown in (Fig. 6c). Despite the high frequency, Dragoni [9] detected no appreciable temperature increase in the adhesive. The criteria adopted to terminate the single fatigue test were either the achievement of joint failure (testified by gross disengagement of the pin) or the attainment of a fatigue life of 10 million cycles (run-out). The fatigue limit was identified using a reduced form of the staircase method [5, 22]. For one run-out (within each specimen type) that had survived the maximum stress amplitude, loading was prolonged to either failure or completion of 200 million cycles. The other survived specimens were fractured statically and the residual ultimate strength was recorded.

3 Experimental Results

3.1 Static Strength of the Ideal Interface

Figure 7 shows the typical relationship between applied torque and twist angle measured at breakaway of friction and hybrid interfaces. The assembly contact pressure was 150 N/mm^2 in both cases and the hybrid interface was bonded with Loctite 638 [7]. The curves were obtained by means of an electronic wrench fitted with an encoder for measuring the rotation angle of bushing 4 with respect to the vice fastening bushing 5 (see Fig. 3). Slopes *OA* and *OB* of the curves in Fig. 7 reflect the behaviour of the rig (comprising take up of backlashes and elastic deformation) up to yielding of the interface (peaks *A* and *B*). After achieving the peak values, both curves exhibit a drop. For the purely friction interface, the drop is due to switching from static to sliding friction. For the hybrid interface, it is consequent upon failure of the adhesive bonds. By increasing rotations, the residual strength of the hybrid interface remains stationary, whereas that of the

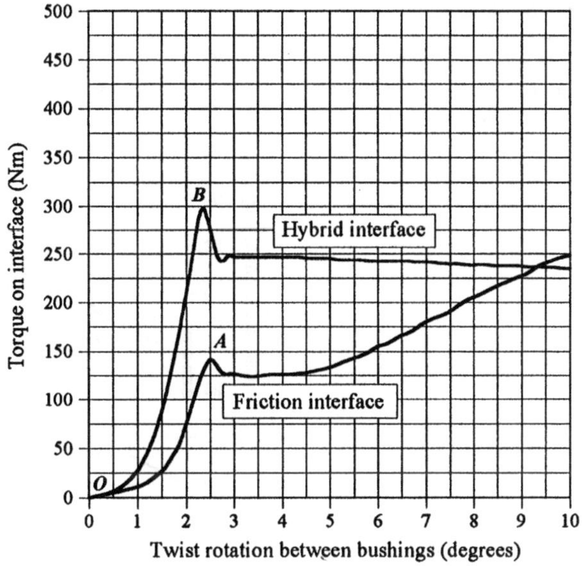

Fig. 7 Relationship between applied torque and relative twist of bushings 4 and 5 in the device of Fig. 3 (adhesive = Loctite 638; contact pressure = 150 N/mm^2; [7]

friction interface grows steadily. Close observation of the unbonded surfaces after the test spotted large areas of galling, which explains such an increase. Output curves as in Fig. 7 were collected for several preliminary trials showing a quite reproducible response. In the actual tests, only the peak torque values (*A* or *B*) encountered anywhere during a bushing rotation of about 4° was recorded by the wrench.

For the product Loctite 638, Fig. 8 presents the results by Dragoni and Mauri [7] in terms of ultimate shear strength (vertical axis) plotted against the mean contact pressure (horizontal axis). It is seen that the strength of both bonded (solid circles) and unbonded (hollow circles) interfaces steadily builds up linearly with the contact pressure and that the slopes of the two interpolating lines are almost the same (about 0.17). For any given contact pressure, the cumulative strength of the hybrid interface can thus be calculated by adding the strength of the purely adhesive interface (about 28 N/mm^2 at zero contact pressure) to the strength of the purely frictional interface at that particular pressure. This result is consistent with the physical explanation that the adhesive is squeezed out from the small spots where the crests of the mating rough surfaces touch each other and is accumulated in the relatively larger surrounding volumes. According to this model, the inter-metallic junctions provide the same frictional strength (proportional to the contact pressure) as in the dry joint, while the surrounding adhesive bonds provide a strength contribution (independent from the pressure) equal to that of the adhesive alone.

However, when the strength of hybrid interfaces bonded with other anaerobic adhesives is considered [8] a different picture emerges. Figure 9 collects the strength results from all the adhesives tested by Dragoni and Mauri [7, 8] under ideal conditions. Superimposed on the data for Loctite 638 (solid circles) and for

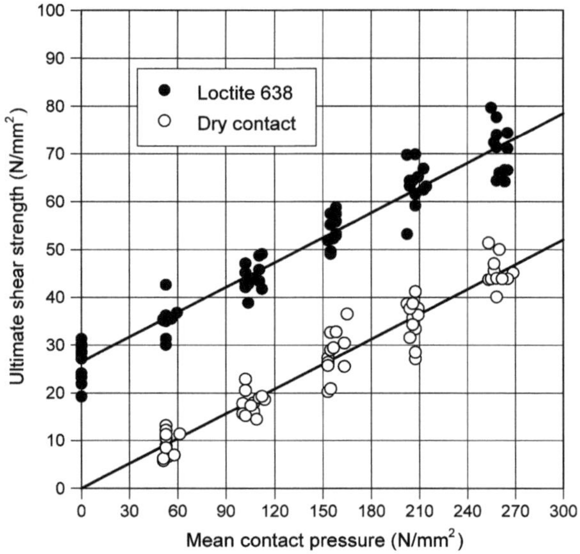

Fig. 8 Static strength of a strong anaerobic (Loctite 638) and of dry contact measured under ideal conditions with the device in Fig. 3 [7]

the unbonded interface (dashed line), already shown in Fig. 8 [7], displayed in Fig. 9 are also the results for products Loctite 601, 243, 242, and 222 [8]. Figure 9 confirms that the cumulative strength of all bonded interfaces steadily increases with the applied contact pressure, but the rate of the build-up is not the same for all products. For the adhesives Loctite 243 (examined under four pressure levels) and 601 (examined under three pressure levels) the increase is also remarkably linear. For the adhesives 222 and 242, nothing can be said on linearity since the only available data involve the maximum pressure (260 N/mm^2) deliverable by the

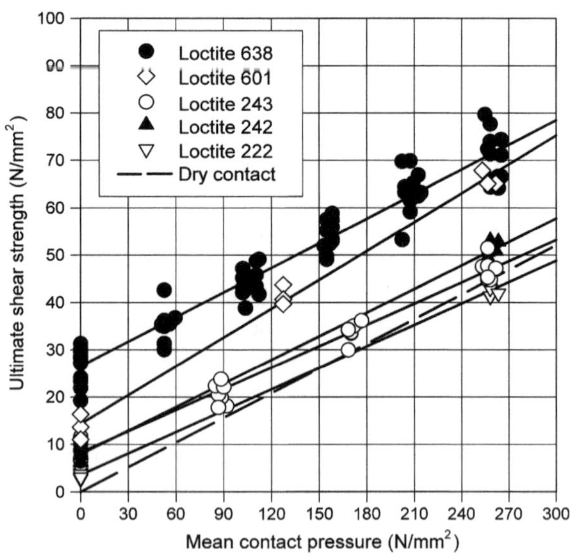

Fig. 9 Static strength of several anaerobics and of dry contact measured under ideal conditions with the device in Fig. 3 [8]

equipment in Fig. 3 or no pressure at all. Although the trend of adhesive 242 (solid triangles) and 638 (solid circles) exhibit the same slope characterising the strength of the unbonded interface (dashed line), different strength gradients than the dry parts are displayed by the interfaces formed with adhesive 601 (steeper slope) and adhesives 243 and 222 (flatter slope). In the case of product 222, the line of the bonded interface crosses over that of the dry interface within the range of contact pressures examined. Overall, the observed behaviour cannot support the superposition criterion surmised above and calls for a different interpretation (see Sect. 4).

3.2 Static Strength of Real Joints

The strength results for the real joints of Fig. 5 are collected in Fig. 10 (threaded connection), Fig. 11 (cylindrical fit) and Fig. 12 (double-lap joint). Unlike Figs. 8 and 9, where the local parameters (contact pressure and shear strength) of the interface are plotted, Figs. 10, 11, 12 display the macroscopic properties of the joints (preload or tightening torque on the horizontal axis and breakaway force or torque on the vertical axis). The choice is due to the uneven working conditions over the real interfaces, particularly irregular in terms of contact pressure, that spoil the meaning of local variables (compare, for instance, the pressure between active and inactive flanks of the engaging threads in the bolted joint in Fig. 5a). Figures 10, 11, 12 provide both the individual experimental points (different symbols in relation to the particular interface) and the linear regressions (straight

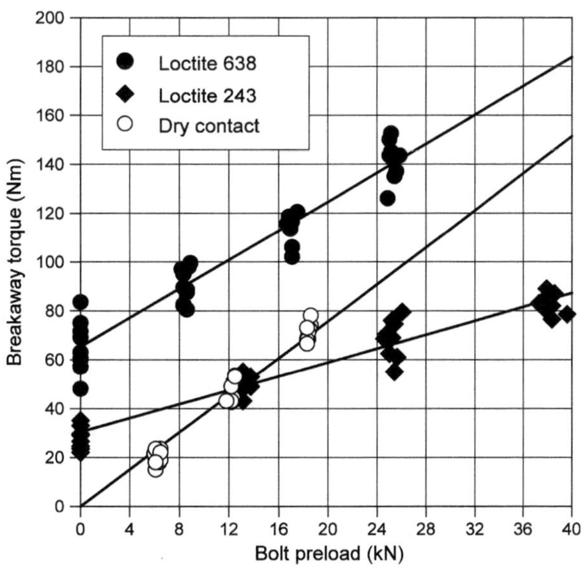

Fig. 10 Static strength under different bonding conditions of the threaded connections in Fig. 5a [8]

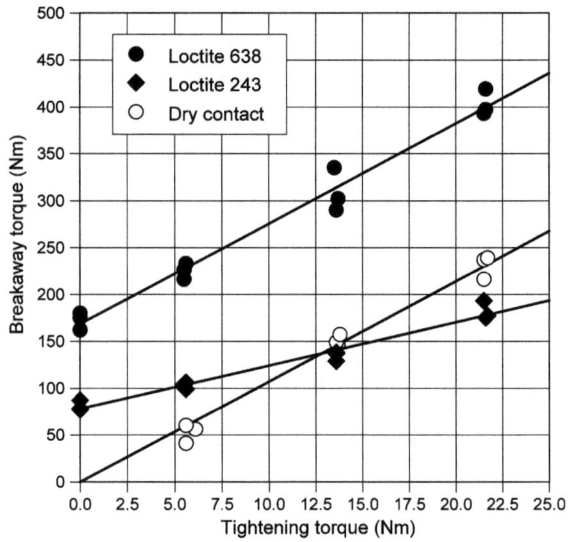

Fig. 11 Static strength under different bonding conditions of the cylindrical fits in Fig. 5b [8]

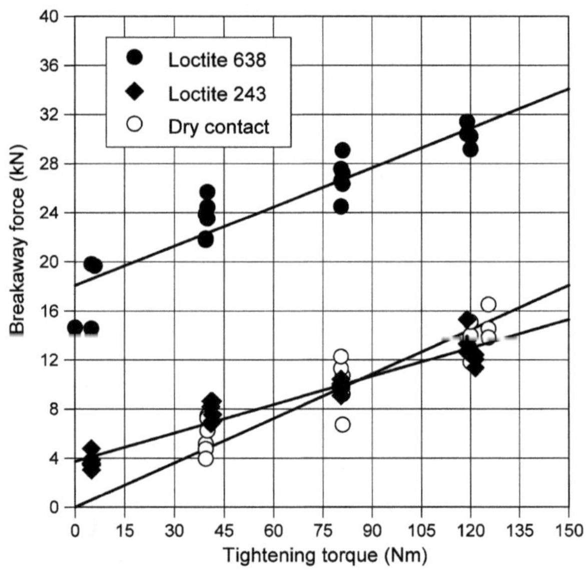

Fig. 12 Static strength under different bonding conditions of the double-lap joints in Fig. 5b [8]

lines) of each family of interfaces. For the dry joints, the interpolating lines were forced to start from the origin.

The data for the threaded connections (Fig. 10) derived from a total of 110 tests, parted in two runs of 55 fresh (not recycled) samples. Each run comprised eleven assembly conditions (four pressure levels for the bonded joints and three levels for the unbonded joints) with five repetitions. The preloads appearing in Fig. 10 were calculated from the tightening torques actually applied to the specimens during the tests. The conversion is needed because of the different frictional coefficients affecting the three interfaces (unbonded and bonded with Loctite 243 or Loctite 638) during tightening, when the two adhesives behave as (dissimilar) lubricants with respect to the dry contact. The relationship between torque and preload was measured directly on several specimens, modified by replacing the lower nut (Fig. 5a) with an annular load cell. The details on the torque-preload conversion procedure are provided by Dragoni and Mauri [8]. The conversion explains the different range of preloads covered by the experimental points in Fig. 10, originated by different frictional coefficients between threads in response to fixed torque levels applied to the upper nut in Fig. 5a. No such conversion was required for the cylindrical fits (Fig. 5b) and the double-lap joints (Fig. 5c). Supplied by an independent bolt, in both cases the contact force is proportional to the tightening torque through a constant (though unknown) coefficient irrespective of the particular working conditions at the active interface.

The strength data of the cylindrical fits (Fig. 11) embrace a total of 33 tests, performed on a single batch of all available samples. The batch included eleven assembly conditions (four pressure levels for both bonded joints and three for the dry joint) with three replications.

The results for the double-lap joints (Fig. 12) were obtained from two batches of eleven specimens (all available rigs shown in Fig. 5c). Each batch encompassed eleven different assembly conditions (four pressure levels for both bonded joints and three for the dry joint) with three repetitions (the three small cylinders) per assembly.

As for the ideal interface (Figs. 8, 9), the behaviour of the real joints in Figs. 10, 11, 12 indicated a similarity between the dry joints and the joints bonded with Loctite 638 in terms of strength build-up with the contact pressure. Likewise, the characteristic lines of the joints bonded with Loctite 243 show (more markedly than the ideal interface) a lower gradient than that of the dry joint. Though sensitive to the contact pressure, the strength of product 243 intersects that of the dry parts in the range of measurements for all three real couplings. Again, the simple model based on the superimposition of dry friction strength and adhesive strength fails to predict the cumulative strength of the hybrid joints, especially if bonded with a medium-strength anaerobic (Loctite 243). A micromechanical model describing the observed macroscopic behaviour of the hybrid interface, regardless of the bonding adhesive used, is proposed in Sect. 4.

3.3 Fatigue Strength

Figure 13 provides an example of the force–displacement diagram at insertion of the pin into the collars (see Fig. 6b). The ratio $\Delta F/\Delta s$ in Fig. 13 measures the slope of the linear steady-state portion of the diagram that follows the nonlinear run-in arc. Using $\Delta F/\Delta s$ and the equations for thick-walled taper fits, Dragoni [9] calculated the coefficient of friction and the mean contact pressure between pin and

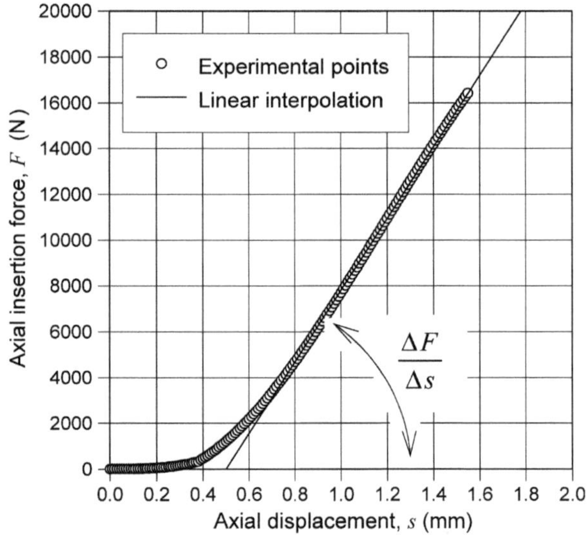

Fig. 13 Typical plot of insertion force against axial displacement at assembly for the tapered specimen in Fig. 6 [9]

collar for each joint. As an alternative, the friction coefficient was also calculated by measuring the circumferential expansion of the outer surface of the collar by means of electrical-resistance strain gauges (Fig. 14).

Figure 15 displays an assortment of tapered specimens after test, comprising a dry press fit (Fig. 15a), a bonded slip fit (Fig. 15b) and a bonded press fit (Fig. 15c). It is seen that, according to expectations, all specimens failed neatly at the interface, without any sign of damage in the mating parts.

The results by Dragoni [9] derived from the reference static tests on bonded and unbonded taper press fits are displayed in Fig. 16, which plots the ultimate unit shear strength against the mean contact pressure. The unit shear strength is simply

Fig. 14 Close-up of a tapered specimen during assembly showing the collar instrumented with an electrical-resistance strain gauge (*circled*)

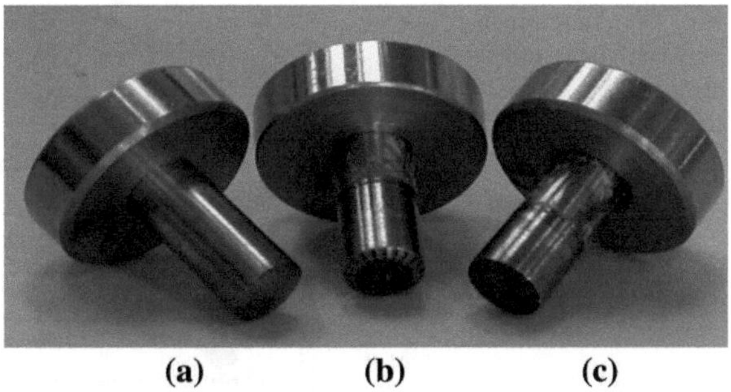

Fig. 15 Tapered pin-collar specimens after testing: **a** dry press fit, **b** bonded slip fit, **c** bonded press fit

given by the ratio of the ultimate load over the area of engagement (291 mm^2) between pin and collar. In Fig. 16, the linear interpolations for both dry and bonded joints are superimposed (solid lines) on the set of experimental points (circles). For the dry joints, the origin is used as a virtual point since no contact pressure implies no mechanical strength. Figure 16 supports the main finding of the previous section that the static strength increases with the contact pressure but with different gradients for dry and bonded joints. The heuristic principle of superimposition of effects (independent action of friction and adhesive) is again negated as a rational tool for calculating the static strength of the hybrid joint.

Also addressing specific quantities (unit shear strength and contact pressure), Fig. 17 contrasts the static strength (solid lines) with the (mean) fatigue strength (dashed lines) measured in the tests by Dragoni [9]. The values of the fatigue strength correspond to the loading amplitude at the 50% of probability of failure. The static strength results (taken from Fig. 16) are accompanied by the confidence interval (plus or minus one standard deviation). The scatter is not provided for the

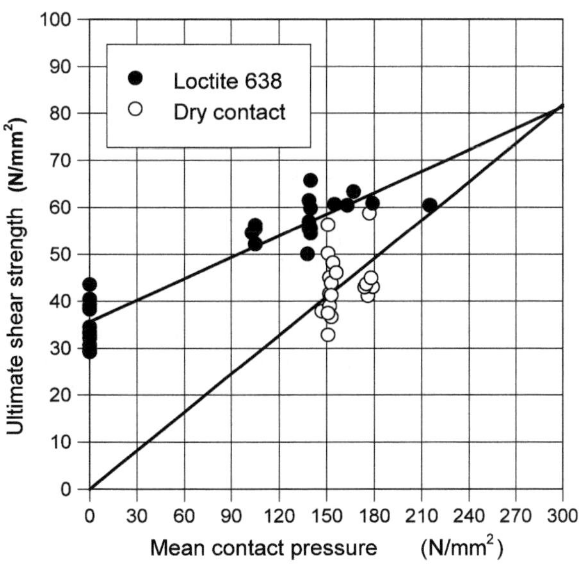

Fig. 16 Variation of ultimate shear strength with contact pressure for dry and bonded taper fits of Fig. 6 [9]

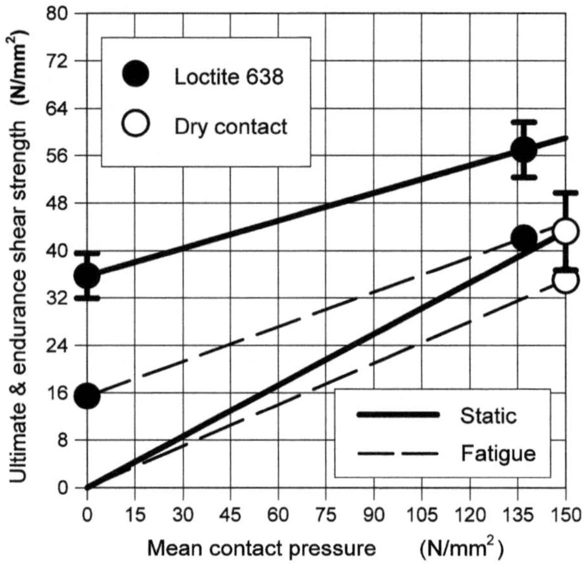

Fig. 17 Comparison between static and fatigue strength for dry and bonded taper fits [9]

fatigue strength results because the reduced staircase method adopted does not supply it [5]. The high scatter affecting the response of the dry press fits can be explained by the uncertainty that accompanies the friction forces between unlubricated metals.

Most interesting in Fig. 17 is the comparison of the fatigue strength with the static strength, from which distinct behaviors for dry and bonded joints emerge. The fatigue strength of the bonded specimens (dashed line with end solid circles) shows a definite decay with respect to the static strength. In absolute terms, the decay is approximately independent of the contact pressure and equals roughly one half of the static strength of the adhesive alone. This observation suggests a simple rule that can be usefully applied in design practice (as far as a life of 10 million cycles is concerned).

In comparison to Fig. 17, the fatigue tests of Croccolo et al. [4] indicate that when exposed to cyclic loading fluctuating from the static strength with amplitude of one half the adhesive strength, bonded press fits lose entirely their retaining capacity. This lower performance can be explained with the adhesive filling less efficiently the interface in the cylindrical press fits by Croccolo et al. [4] than in the taper fits by Dragoni [9].

When the attention is moved in Fig. 17 to the fatigue strength of the dry joints, quite another conclusion is drawn. First of all, it must be remarked that all failures taking place in this case occurred at the launch of the test with the specimens not even undergoing the very first load cycle. In fact, the failures were of static nature and the fatigue tests were actually static tests performed with the (reduced) staircase method. All specimens that could survive the first load cycle also survived the chosen threshold life of 10 million cycles. From this outcome it can be surmised that the dry interface is virtually unaffected by fatigue damage.

This explains why the fatigue data point of the dry press fits in Fig. 17 falls almost within the scatter interval of the static strength for the same joint. The difference between the mean values of the two strengths can be explained by the fact that the static and the fatigue data derive from sets of specimens prepared and tested separately from each other.

Dragoni [9] also reports that the residual static strength of the dry press fits surviving the fatigue testing coincides with that of the fresh joints. By contrast, a decay of the residual static strength affected the bonded joints which survived fatigue testing (10 million cycles), an indication of the absence of a true fatigue limit for the category of bonded press fits. This speculation is corroborated by the behavior of the run-outs (one for each type of fit) for which the fatigue loading was prolonged beyond the threshold of 10 million cycles. The run-outs dry press fit endured a life of 200 million cycles with no signs of failure (after which the testing was terminated). By contrast, the two bonded specimens failed after a (total) life of 90 million cycles (slip fit) and of 105 million cycles (press fit). Whether a true fatigue limit for bonded slip fits and bonded press fits actually exists remains an open question and represents a fruitful topic for future research.

4 Joint Modelling

4.1 Micromechanical Model of the Hybrid Interface

Consider the simple model depicted in Fig. 18a of a hybrid interface between two massive adherends, bonded and tightened under the normal force P. In Fig. 18a the roughness of the contacting surfaces has been condensed into two protrusions

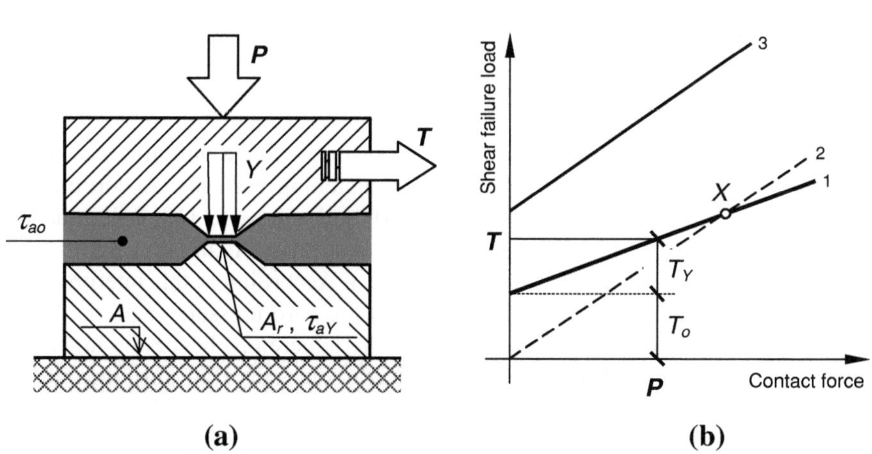

Fig. 18 Micromechanical model of the hybrid interface (**a**) and corresponding predictions (**b**) for its macroscopic shear strength [8]

that touch each other over a fraction, A_r, of the nominal contact area, A. It is also assumed that the adhesive fills the voids surrounding the protrusions, where it sustains no pressure, and forms a thin layer between the protrusions, subjected to the yield pressure, Y, of the softest adherend [10].

In order to break the joint, the area, A_e ($= A - A_r$), of the adhesive around the asperities and the area, A_r, of the adhesive entrapped between the asperities must be fractured. For very stiff adherends (as typical of anaerobics) loaded statically, those areas are likely to break simultaneously. If τ_{ao} is the unit shear strength of the adhesive at zero pressure and τ_{aY} the shear strength of the adhesive at pressure Y, the shear failure load, T, of the joint amounts to

$$T = \tau_{ao} A_e + \tau_{aY} A_r = \tau_{ao} (A - A_r) + \tau_{aY} A_r \tag{1}$$

From the equilibrium condition of each adherend in the vertical direction, $YA_r = P$, the true area of contact, A_r, is calculated as

$$A_r = P/Y \tag{2}$$

Combination of (1) and (2) followed by some rearrangements yields

$$T = \tau_{ao} A + (\tau_{aY} - \tau_{ao}) P/Y = T_o + T_Y \tag{3}$$

Equation (3) shows that the macroscopic strength of the joint, T, is the sum of a constant term, $T_o = \tau_{ao} A$, and a variable term, $T_Y = [(\tau_{aY} - \tau_{ao})/Y]\, P$, proportional (whenever $\tau_{aY} > \tau_{ao}$) to the contact force. This produces the generic diagram of Fig. 18b (curve 1).

For a dry joint ($\tau_{ao} \equiv 0$, $\tau_{aY} \equiv \tau_Y =$ shear strength of the metal junctions), Eq. 3 predicts $T_o = 0$ and $T_Y = (\tau_Y/Y)\, P$ in accordance with Coulomb's law (Fig. 18b, curve 2).

For a bonded joint where the adhesive would be squeezed out of the junctions upon tightening ($\tau_{aY} \equiv \tau_Y \gg \tau_{ao}$), Eq. 3 would predict the same constant term $T_o = \tau_{ao} A$ as in the purely adhesive joint and a variable term $T_Y = [(\tau_Y - \tau_{ao})/Y]$ $P \approx (\tau_Y/Y)\, P$ with the same slope as the dry joint. This is the rationale behind the criterion of superimposition of effects, stated in the Introduction and recalled throughout the paper, that must be dismissed on experimental grounds.

A possible explanation of the experimental results is that a thin film of adhesive is actually formed between the crests of the matching surfaces. According to a mechanism well documented for polymers [20] and adhesives [18, 26], under the high local pressure (Y) this film could achieve a shear strength (τ_{aY}) significantly greater than the shear strength at no pressure (τ_{ao}). According to this interpretation, it would be expected that the higher the strength at zero pressure (τ_{ao}), the higher the strength (τ_{aY}) under the yield pressure of the adherends, hence the higher the slope, $(\tau_{aY} - \tau_{ao})/Y$ in Eq. 3, of the characteristic curve in response to the clamping force (Fig. 18b, curves 1 and 3). Observation of Figs. 8, 9, 10, 11, 12 indicates that this is the case, with the stronger anaerobic (Loctite 638) outperforming the medium one (Loctite 243) in both respects (greater strength at zero pressure and greater strength gradient under pressure). When a weak anaerobic is used (low value of $\tau_{aY} - \tau_{ao}$ in Eq. 3), the slope of the characteristic line for the hybrid interface becomes lower than the slope for the dry joint and the two curves cross each other (point X in Fig. 18). In this case, the adhesive between the crests of the roughness behaves as a solid lubricant.

The proposed micromechanical model points out the role of the shear strength of the pressurised adhesive for the rational characterisation of the hybrid interface under steady stresses. Research aimed at confirming this mechanism would help understanding the intimate behaviour of these joints and would also give directions for the development of new adhesives tailored to specific applications.

4.2 Strength Calculation of Actual Joints

Although elaborate models exist for the stress distribution at the interface of hybrid adhesive–friction joints [14, 25, 32–34] their usefulness for actual strength prediction are very limited. This is because the formulae are quite elaborate and because the geometries examined so far are essentially restricted to pin and collar engagements under axial or torsional loading.

By assuming that the metal adherends are rigid (not deformable) and that the hybrid interface behaves as an elastically deformable layer, the strength of the joint can be estimated with elementary strength of material models. Although extreme, the above assumption leads to simple equations, easy to use and accurate enough for most engineering problems.

This section presents the strength equations for five common geometries of hybrid friction–adhesive joints: (1) threaded connection, (2) overlap joint, (3) cylindrical fit, (4) taper fit, (5) flanged coupling. For all equations, the joint type and dimensions are defined in Fig. 19a–e and the following material properties are used:

τ_{ao} = static shear strength of the adhesive (e.g. measured on pin-collar slip fits)
τ_T = cumulative static shear strength of the hybrid interface (measured on hybrid joints under the same contact pressure as the final joint)

1. Threaded connection (Fig. 19a)
   Maximum axial force: F = not applicable
   Maximum torque: $M_t = (\tau_{a0} + \tau_T)\pi H(D - 0.65\,a)^2 + 0.65 f\,VD$ where a is the pitch of the thread (to ISO 68), V is the preload in the screw and f is the coefficient of friction between nut and plate.
2. Overlap joint (Fig. 19b)
   Maximum axial force: $F = \tau_T A$
   Maximum torque: M_t = not applicable
3. Cylindrical fit (Fig. 19c)
   Maximum axial force: $F = \tau_T \pi D L$
   Maximum torque: $M_t = \tau_T \frac{\pi D^2 L}{2}$

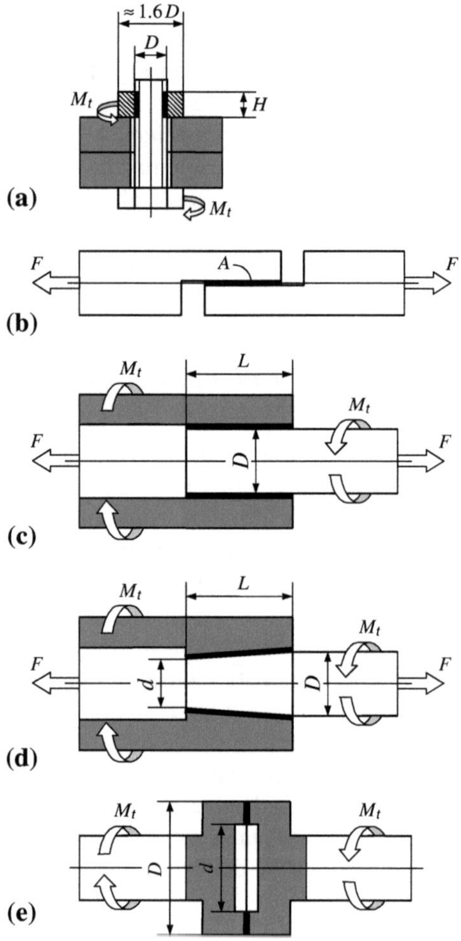

Fig. 19 Typical configurations of hybrid friction–adhesive joints encountered in engineering practice: **a** threaded connection, **b** overlap joint, **c** cylindrical fit, **d** taper fit, **e** flanged coupling

4. Taper fit (Fig. 19d)
   Maximum axial force: $F = \tau_T \frac{\pi(D+d)\sqrt{(D-d)^2+4L^2}}{4}$ (use only if $D - d \leq 0.3L$)
   Maximum torque: $M_t = \tau_T \frac{\pi(D+d)\left[(D+d)^2-2dD\right]\sqrt{(D-d)^2+4L^2}}{16D}$
5. Flanged coupling (Fig. 19e)
   Maximum axial force: $F =$ not applicable
   Maximum torque: $M_t = \tau_T \frac{\pi\left(D^4-d^4\right)}{16D}$

Use of the above equations is reliable for the prediction of the static strength of the joints. Due to the uncertainty of fatigue data, adaptation to cyclic loading should be performed with care. For example, based on the fatigue results presented in Sect. 3.3, the cyclic strength under repeated stresses can be estimated by replacing τ_{ao} with $0.5\tau_{ao}$ and τ_T with $\tau_T - 0.5\tau_{ao}$.

5 Conclusions

Anaerobic adhesives (anaerobics) are a simple and very effective means of enhancing the performance of classical friction joints in which retaining forces are promoted by mechanical tightening. Entrapped in the airless environment arising between closely mating machinery parts, anaerobics cure thanks to intimate contact with active metals (for example iron and copper) and to the absence of oxygen. Advantages of anaerobic-augmented friction joints are extra sealing action, lower fretting wear, better corrosion resistance, greater stiffness and higher mechanical strength. These benefits can be achieved with minor or no changes to the design of the parts with respect to the classical assembly. This chapter has reviewed the literature on friction–adhesive joints with the focus on their mechanical strength, both static and fatigue. Data were gathered from experiments on ideal interfaces (annular butt joints under uniform stress conditions) and actual joint geometries (threaded connections, cylindrical and taper fits, double-lap joints under realistically irregular stresses). The results can be summarised as follows:

- the overall static strength of both dry (friction only) and hybrid joints (friction and adhesive) steadily builds up with the contact pressure (clamping force),
- the static strength build-up of the joints bonded with strong anaerobics (retainers) is similar to that of the dry joint (same strength rate with contact pressure),
- when low- or medium-strength anaerobics (i.e. threadlockers) are used, the static strength build-up is much lower in the hybrid than in the dry joint,
- for high contact pressures, the dry joint can achieve a mechanical strength much higher than the joint bonded with low- or medium-strength anaerobics (solid lubrication),
- predicting the cumulative static strength of the hybrid joint as the sum of adhesive strength and friction strength, calculated independently from each other, can result in gross errors,
- the fatigue strength of all joints (dry and bonded) increases with the contact pressure induced at assembly,
- the fatigue strength of the dry joint is comparable to its static strength at equal pressure,
- the fatigue strength of the bonded joint is significantly lower than its static strength; the decay seems to be independent (in absolute terms) of the contact pressure.

A micromechanical model explaining most of the experimental evidence was presented which spurs further research for the correct interpretation of the basic mechanisms underlying this hybrid joining technology. Simple equations were provided to help the designer predict the strength of virtually all types of friction–adhesive joints occurring in engineering applications.

References

1. Akisanya, A.R.: On the singular stress field near the edge of bonded joints. J. Strain Anal. **32**, 301–311 (1997)
2. Bartolozzi, G., Croccolo, D., Chiapparini, M.: Research on shaft-hub adhesive and compression coupling. ÖIAZ **144**, 198–201 (1999)
3. Canyurt, O.E.: Fatigue strength estimation of adhesively bonded tubular joint using genetic algorithm approach. Int. J. Mech. Sci. **46**, 359–370 (2004)
4. Croccolo, D., De Agostinis, M., Vincenzi, N.: Static and dynamic strength evaluation of interference fit and adhesively bonded cylindrical joints. Int. J. Adhes. Adhes. **30**, 359–366 (2010)
5. Dixon, W.J., Massey, Jr F.J.: Introduction to Statistical Analysis, 4th edn, pp 434–438. McGraw-Hill, New York (1985)
6. Dragoni, E.: Effect of anaerobic threadlockers on the fatigue strength of threaded connections. Int. J. Mat. Prod. Technol. **14**, 445–455 (1999)
7. Dragoni, E., Mauri, P.: Intrinsic static strength of friction interfaces augmented with anaerobic adhesives. Int. J. Adhes. Adhes. **20**, 315–321 (2000)
8. Dragoni, E., Mauri, P.: Cumulative static strength of tightened joints bonded with anaerobic adhesives. Proc. Inst. Mech. Eng. Part L **216**, 9–15 (2002)
9. Dragoni, E.: Fatigue testing of taper press fits bonded with anaerobic adhesives. J. Adhes. **79**, 729–747 (2003)
10. Halling, J.: Principles of Tribology. McMillan, London (1975)
11. Harrigan, T.P., Kareh, J.E., Harris, W.H. (1990) The influence of support conditions in the loading fixture on failure mechanisms in the push-out test: a finite element study. J. Orthop. Res. **8**, 678–684 (1975)
12. Haviland, G.S.: Machinery Adhesives for Locking Retaining and Sealing. Marcel Decker, New York (1986)
13. Hoeppner, D.W., Chandrasekaran, V., Elliot, C.B.: STP 136—Fretting fatigue: current technology and practices. ASTM (2000)
14. Kawamura, H., Sawa, T., Yoneno, M., Nakamura, T.: Effect of fitted position on stress distribution and strength of a bonded shrink fitted joint subjected to torsion. Int. J. Adhes. Adhes. **23**, 131–140 (2003)
15. Kollmann, F.G.: Welle-Nabe-Verbindungen. Gestaltung, Auslegung, Auswahl. Springer, Berlin (1984)
16. Liechti, K.M., Hayashi, T.: On the uniformity of stresses in some adhesive deformation specimens. J. Adhes. **29**, 167–191 (1989)
17. Mahon, F.: Use of anaerobic adhesives to enhance strength and capacity of flanged couplings. Paper 950125, SAE Intl Congress and Exposition, Detroit, 27th February–2 March (1995)
18. Mengel, R., Haeberle, J., Schlimmer, M.: Mechanical properties of hub/shaft joints adhesively bonded and cured under hydrostatic pressure. Int. J. Adhes. Adhes. **27**, 568–573 (2007)
19. O'Reilly, C.: Designing bonded cylindrical joints for automotive applications. Paper 900776, SAE Intl Congress and Exposition, Detroit, 26th February–2nd March (1990)
20. Raghava, R.S., Cadell, R.M.: The macroscopic yield behaviour of polymers. J. Mater. Sci **8**, 225–232 (1973)
21. Renton, W.J.: The symmetric lap shear test—what good is it? Exp. Mech. **16**, 409–415 (1976)
22. Rice, R.C.: Fatigue data analysis. In Metals Handbook: Mechanical Testing, vol. 8, pp. 703–704. American Society for Metals, Metals Park, Ohio (1985)
23. Romanos, G.: Strength evaluation of axisymmetric bonded joints using anaerobic adhesives. Int. J. Mat. Prod. Technol **14**, 430–443 (1999)
24. Sawa, T., Sasaki, R., Yoneno, M.: An analysis of pipe flange connections using epoxy adhesives/anaerobic sealant instead of gaskets. ASME J. Press Vessel Technol. **117**, 298–304 (1995)

25. Sawa, T., Yoneno, M., Motegi, Y.: Stress analysis and strength evaluation of bonded shrink fitted joints subjected to torsional loads. J. Adhes. Sci. Technol. **15**, 23–42 (2001)
26. Schlimmer, M.: Anstrengungshypothese für Metallklebverbindungen. Materialwissenschaft und Werkstofftechnik **13**, 215–221 (2004)
27. Sekercioglu, T., Gulsoz, A., Rende, H.: The effects of bonding clearance and interference fit on the strength of adhesively bonded cylindrical components. Mater. Design **26**, 377–381 (2005)
28. Sekercioglu, T.: Strength based reliability of adhesively bonded tubular lap joints. Mater. Design **28**, 1914–1918 (2007)
29. Sekercioglu, T., Kovan, V.: Torque strength of bolted connections with locked anaerobic adhesive. Proc. Inst. Mech. Eng. Part L **222**, 83–89 (2008)
30. Werthm, S.: (1938) Kräfte an Längspre*β*sitzen. VDI-Z **82**, 471–475
31. White, D.J., Humpherson, J.: Finite element analysis of stresses in shafts due to interference-fit hubs. J. Strain Anal. **4**, 105–114 (1969)
32. Yoneno, M., Sawa, T., Shimotakahara, K., Motegi, Y.: Axisymmetric stress analysis and strength of bonded shrink-fitted joints subjected to push-off forces. JSME Int. J. Ser. A **40**, 362–374 (1997a)
33. Yoneno, M., Sawa, T., Shimotakahara, K., Motegi, Y.: Push-off tests and strength evaluation of joints combining shrink fitting with bonding. Proc. SPIE **2921**, 193–198 (1997b). doi: 10.1117/12.269815
34. Yoneno, M., Sawa, T., Motegi, Y.: Axisymmetric stress analysis and strength of bonded shrink-fitted joints of solid shaft subjected to torsional loads. JSME Int. J. Ser. A **41**, 517–524 (1998)

Technologies of Threadlocking and Interference-Fit Adhesive Joints

Felix Kleiner and Wolfgang Fleischmann

Abstract Adhesives can sometimes replace, but in most of the cases improve the properties of classical joining techniques. Two examples are presented in this chapter: the use of chemical threadlockers which can reliably protect bolted joints against loosening under all operational conditions. Additionally they offer sealing and corrosion protective forces. In case of cylindrical joints, where loads have to be transmitted, the use of adhesives can give considerable advantage in strength under static and dynamic loading conditions over classical shrink- or press-fits and allows for technical and economical optimisation of the application. This chapter provides practical information to handle aspects of design, production and operation.

1 Threadlocking

1.1 Introduction

Threaded assemblies today are still the most important detachable parts used in machinery construction, installation and repair. However, for a reliable application, the bolts must be protected against undesired loosening under service conditions by "threadlocking" methods.

F. Kleiner (✉) and W. Fleischmann
Henkel AG & Co. KGaA, Standort München, Gutenbergstrasse 3,
85748 Garching, Germany
e-mail: felix.kleiner@henkel.com

W. Fleischmann
e-mail: wolfgang.fleischmann@henkel.com

Adv Struct Mater (2011) 6: 227–255

DOI: 10.1007/8611_2010_39

Published Online: 16 November 2010

The two main causes for the failure of a threaded assembly are:

- relaxation of tension, and
- self-loosening.

A threaded assembly 'relaxes' when a permanent change in the length of the bolt occurs in the direction of its axis; the substrate itself relaxes as in gasketed surfaces, or temperature changes. This reduces the bolt tension and thus also reduces the residual clamping force. Permanent changes in length may be produced by:

- *Settling*: The rough faces of the contiguous parts (e.g. nut, washer) become smoother under the pressure of the bolt tension.
- *Creeping*: The surface pressure at the bearing surface of the bolt or nut exceeds the compressive strength of the material.

If the elasticity of the assembly can be increased, so that the expected amount of settling and creeping can be compensated for, then the drop in pre-stress force can largely be prevented. This is made possible by:

- bolts with a high l/d ratio (l = bolt length, d = bolt diameter);
- collar bolts and collar nuts as well as hardened and tempered washers which reduce surface pressure and thus the settling on the bearing surfaces;
- bolts and nuts with pressed-on spring head washer or concave bearing washer;
- conical spring washers or cup springs.

After the tightening procedure the clamp load is maintained by the preload in the bolt. This arises because the bolt has been stretched like a spring and the tension acts so as to pull the nut towards the head, thus compressing the clamped members. Once the tightening torque is removed, the tension then tends to unwind the nut from the bolt. The friction in the threads and under the head of the bolt and nut oppose this force and maintains the tension in the bolt. A mathematical representation of this system is as follows:

$$T_L = \frac{F_V d \mu_{th}}{2 \cos \rho} + \frac{F_V d_h \mu_h}{2} - \frac{F_V d \tan \phi}{2} \tag{1}$$

where T_L, resultant torque on the bolt after tightening; F_V, tension in the bolt; d, pitch diameter of the thread; d_h, effective head diameter; ϕ, helix angle of the thread; ρ, thread half angle (30° for ISO threads); μ_{th}, coefficient of thread friction; μ_h, coefficient of friction between the bolt head and the flange assuming the nut is stationary.

If the clamped system is subjected to alternating loads or vibration, the locking effect caused by the friction becomes less effective, allowing the nut to rotate on the threads releasing the tension. The vibrations can be in the longitudinal or transverse direction or a combination of both. The transverse vibrations due to alternating horizontal loads are much more detrimental and can rapidly loosen a standard untreated fastener. Longitudinal forces due to pulsating axial loads will lead to loosening to a lesser degree.

These measures can prevent uncontrolled loosening of correctly loaded bolts:

- The use of high-tensile bolts may allow pre-stress forces that are high enough to prevent relative movement.
- Design that increases the l/d ratio (l = bolt length, d = bolt diameter) increases the elasticity of the assembly. (Historically, l/d ratio >6 have been optimum.)
- Friction can be increased by influencing the surface finish and structure on the bearing surfaces of the bolt and nut.
- By applying adhesive, the degree of freedom for lateral movements is eliminated as the gaps between the threads are completely filled, and at the same time thread friction is increased after the adhesive has cured.
- By creating a positive connection (i.e. interference-fit bolts, welding spots), slip in the thread can be limited.

1.2 Threadlocking: Types and Methods

Threadlocking types and methods can be subdivided into three groups:

1.2.1 The Settling Method

Increasing the elasticity of the assembly compensates for settling after assembly. The pre-stressed force is thus largely retained and relaxation of the threaded assembly is prevented. This method, however, cannot prevent self-loosening of the threaded assembly if relative movements exist between the stressed parts. Examples of settling methods are conical spring washers or cup springs of high rigidity. The locking effect of other elements, for example, spring washers, elastic washers, toothed and fan-type washers, are inadequate. They are unsuitable for securing bolts of property 8.8 (grade 5 is the US equivalent) or higher.

1.2.2 Loss Prevention Devices: (Prevailing Torque-Type Devices)

Loss prevention devices allow partial relaxation or loosening, but prevent the threaded assembly from falling apart. Examples: crown nuts, wire retainers, bolts with thread inserts made of metal or plastic. These techniques often avoid loss of the fixture but are ineffective in maintaining clamp load.

1.2.3 Self-Loosening Prevention Devices

Self-loosening prevention devices prevent threaded assemblies from loosening themselves. These include:

- bolts and nuts with locking teeth
- ribbed flange bolts
- adhesives

1.3 Industrial Threadlockers

Devices to prevent self-loosening must meet the highest standards concerning threadlocking. The industry has developed single-component liquid adhesives that completely fill the microscopic gaps between interfacing threads. They cure to a tough, solid, thermoset plastic when they come into contact with metal in the absence of air. The adhesive creates an interfacial connection—keying to the surface roughness to prevent any movement of the threads. Hence the problem is solved where it arises: in the threads. This is why liquid threadlockers are among the most effective means for locking fasteners.

It is important that the total length of the thread be wetted, and that there is no restriction to the curing of the adhesive (certain oils or cleaning systems can impede or even completely prevent the adhesives from curing by anaerobic reaction).

The liquid adhesive may be applied by hand or with the help of special dispensing devices. Proper wetting of a thread is dependent on the following parameters: the size of the thread, the viscosity of the adhesive and the geometry of the parts. If the parts are of large dimensions, then wetting both faces ensures an adequate application of the adhesive. With blind hole threads, it is essential that the adhesive be applied to the bottom of the threaded hole. The quantity must be such that after assembly, the displaced adhesive should fill the whole length of the thread.

Some industrial threadlocking products cured by anaerobic reaction help, when in a liquid state, to control the coefficient of friction in the thread. The values are comparable with those of oiled bolts. This property allows threadlocking products cured by anaerobic reaction to be integrated into automatic production lines, using existing assembly equipment. Note, however, that pre-stress and installation torque need to be defined.

When someone buys a nut and bolt, they buy with few exceptions just one thing, and that is clamping force. They want to be able to predict what the force is going to be and how long it will stay at that value. In addition, at the end of some time, they may wish to remove the clamping force. Nuts and bolts fill this function well, but must be "engineered" properly to give satisfactory long-term results.

We tighten a screw or bolt by applying torque to the head. A clockwise torque makes the bolt-to-nut distance shorter. If a resistance is met (such as clamping a flange), the bolt will continue to rotate until a balance is obtained between the torque applied to the head and the reaction torque of the assembly. Three components contribute to the reaction torque friction under the head of the bolt, friction between the mating threads and an induced torque due to the bolt tension.

Fig. 1 Typical K values—lubricating threadlockers on various materials

Typical* K Values - Lubricating Threadlockers on Various Materials

Substrate	Oil	Lubricating Threadlocker
Steel	0.15	0.14
Phospate	0.13	0.11
Cadmium	0.14	0.13
Stainless 404	0.22	0.17
Zinc	0.18	0.16
Brass	0.16	0.09
Silicon bronze	0.18	0.24
A1. 6262.TA	0.17	0.29

All specimens were dipped in 5% solution (95% water) and dried before the threadlocker was applied.

(Heat Bath Corp., Lab Oil 72D)

*Range of values for any lot of fasteners was ±15%; however, different fastener lots can increase the variation to ±20%

The equilibrium relationship is often expressed mathematically:

$$T = KdF \tag{2}$$

where T, torque; d, nominal diameter of bolt; F, induced force or clamp load; K, an empirical constant, which takes into account all friction and the variable diameter under the head and in the threads where friction is acting. (It is not the coefficient of friction although it is related to it.)

Values of K can be determined experimentally (see Fig. 1 for typical values).

The variation in friction and, therefore, K is wide since it is the result of extremely high pressure between surfaces that may be rough, smooth, oxidised, chemically treated, and/or lubricated. Oily steel has a K that varies between 0.11 and 0.17 or $\pm$ 20%. Total friction absorbs 80–90% of the tightening torque. Therefore, it is prudent to test a particular combination in a torque-testing device to determine proper torque values for assuring good control of bolt tension. Technical data for lubricants and other thread-treating materials will often have the K values plotted in torque tension curves. These values were obtained on 3/8 $\times$ 16 nuts and bolts (metric approximate: M10) where the nut was turned. Both the threads and the nut face were lubricated. An unlubricated thrust surface, either nut or bolt head can almost double the K value. Threadlocking agents yield controlled torque tension curves and therefore are excellent for in-line production.

Factors that influence K value are: fastener and flange materials, insertion speed, bolt quality, product selected, thread tolerance, and surface finish.

For a required clamp load F the designer must choose the appropriate fastener. The clamp load is related to the stress in the bolt S by the following equation:

$$S = \frac{F}{A} \tag{3}$$

where A is the cross sectional area of the bolt or more usually the sum of the areas of several bolts used to provide the clamp load. If the bolt material is a pre-condition, the optimum value of S is known (75% of proof load is often used) and therefore A can be calculated. The size and number of bolts can then be determined from the torque tension Eq. 2.

The tightening torque will create a torsional shear stress in the bolt. If this exceeds the stress rating for the bolt, the number of bolts must be increased thereby reducing the demands on the individual fastener. The procedure may run in reverse if spatial requirements dictate the bolt configuration beforehand, so that the above equations can be used to determine what fastener material should be used.

In practice K value is application specific and must be determined empirically for each situation. The torque tension Eq. 2 should be used as a guideline only, with the actual value being measured experimentally.

1.4 Fatigue Effects

The fatigue strength of a bolted joint must be evaluated in two ways: fatigue of the bolt, and fatigue of the bolted material. The properly tightened bolt will not fail in fatigue in a rigid joint. Initial bolt tension will stay relatively constant until the external tension load on the joint exceeds the bolt load. If designers do not permit the calculated service load to be greater than the bolt pre-load, the bolt will experience no appreciable stress variation, and without stress variation, there can be no failure by fatigue, regardless of the number of load cycles on the joint.

This is not the case where considerable flexibility is present. Variable stress in screw or bolt fastenings increases with the flexibility of the connected parts. If flexibility is too great, the variable stress present may be high enough to cause eventual fatigue failure of the fastener regardless of the initial bolt pre-load.

The greatest single factor that can eliminate cyclic stress variation due to cyclic loading is proper pre-tensioning or pre-loading of the fastener. Test results indicate that rigid members bolted together by relatively elastic bolts offer the best method to prevent fatigue failure.

However, a liquid threadlocking product will fill all the voids inside the thread, and will therefore give no room for micro-movement of the threaded fastener.

Designers are well versed in designing for fatigue resistance. A properly clamped joint is designed so that it will not endure forces often enough to cause

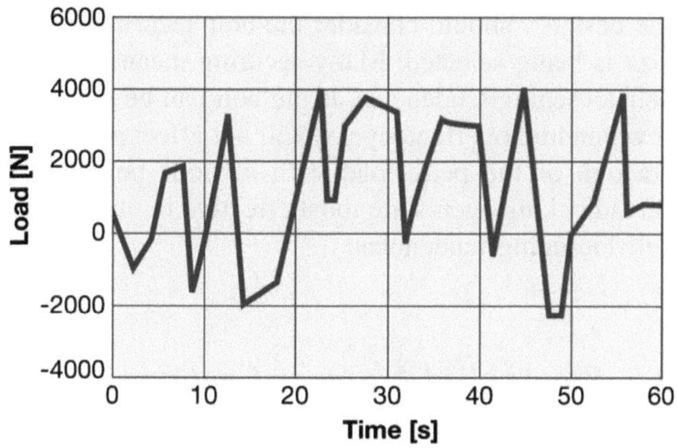

Fig. 2 Fatigue load spectrum (Falstaff, Fighter Aircraft Load Spectrum for Fatigue)

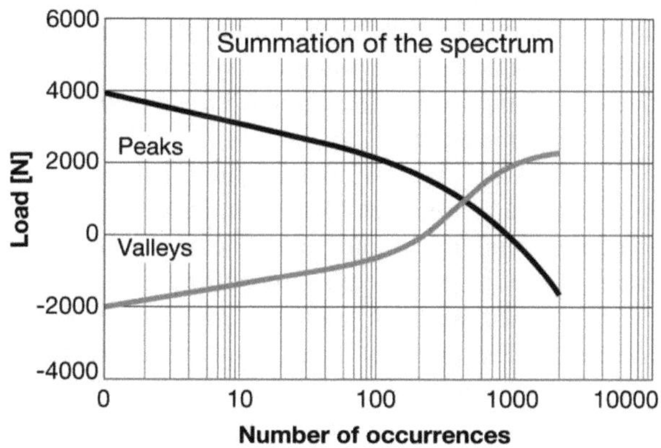

Fig. 3 Summation of the spectrum. The *black line* represents the number of positive load peaks the joint has seen (e.g. 10 times 750 N), the *grey line* shows the distribution of negative peaks

fatigue failures. The problem becomes apparent when a force spectrum of an application is studied. Figure 2 shows a small segment of an aircraft joint strain gauge readout.

The immediate question is, what should be the design load of the joint? We know that a joint clamped tighter than applied loads will not experience cyclic loads on the bolts. Should the clamp load be so high that the bolt never sees the maximum peaks at 4,000 N, or can a calculated risk be made to allow a few hundred peaks, which surely could not cause fatigue damage below the fatigue limit?

The answer is twofold. First, the spectrum should be analysed for number of occurrences. Figure 3 is the result of such an analysis, showing how often the joint has seen the different load peaks in positive (tensile) or negative (pressure) direction.

Second, another consideration as serious as fatigue must be dealt with. Overloading of a joint will cause self-loosening of the fasteners. Only 50 side slips of a joint will cause 20% clamp load loss. As clamp load is lost, lower and more frequent loads cause slipping and failure becomes precipitous. Therefore, the designer should consider the bolt-securing means at the same time as the bolt size is being selected. Many securing means affect the clamp load as well as the self-loosening tendencies. If the bolt can be secured against self-loosening, then a few hundred overload cycles will not affect performance. The design load can be a fraction of the peak load with obvious benefits of cost and weight. Industrial threadlocking agents are most effective in maintaining clamp load and eliminating self-loosening tendencies.

1.5 Pre-Applied Fastener Coatings

If hand application of adhesive is not desired on a continuous assembly line, or if dispensing units cannot be used, then any fastener can be pre-coated with adhesives as an alternative. These pre-applied products are very well suited to high volume production line assembly systems, without the need for extra operations, and can be applied to most threaded male or female fasteners. This results in lower costs and time saving.

The locking systems contain micro-capsules within the dry-to-touch coating. When the parts are assembled, the micro-capsules will crush and release the active ingredient and the liquid threadlocking adhesive will cure similar to a liquid anaerobic, with good sealing properties to fluids and gases.

Pre-applied products coated on any fastener also provide advantages for quality assurance. The quantity of adhesive dispensed for the coating is very consistent due to constant quality checks performed by specialised coating companies. Existing assembly equipment can usually be adapted to use pre-coated bolts without tooling changes.

Most pre-applied products are water based which means no organic solvents are used, they are non-flammable and non-hazardous. They are all brightly coloured for easy identification and offer a long storage on-part life.

All threaded components normally can be re-used by applying a threadlocking liquid anaerobic product.

1.6 Comparison of Threadlocking Methods

The best way to evaluate a threadlocked connection is to test its behaviour under load cycles in a dynamic test machine. The lower the loss in bolt tension, the more reliable the assembly.

To determine clampload retention curves for various locking devices, a bolt test machine was used. The braced bolt is stressed vertically to the bolt axis by a displacement unit that is adjusted with a cam.

As can be seen in Fig. 4, liquid threadlocker has the best clampload retention performance among those tested. Most mechanical methods fail this test. This does

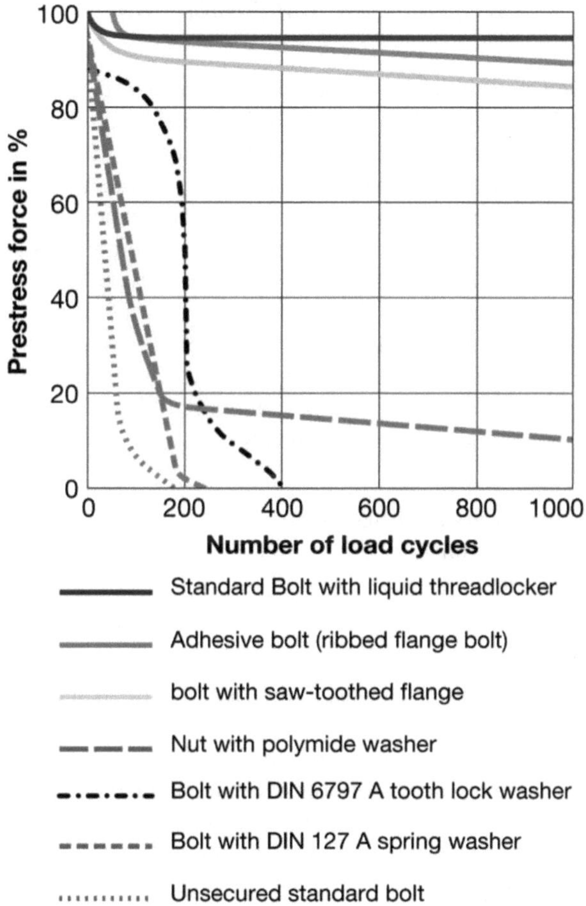

Fig. 4 The characteristic clamp load retention performance of various threadlocking systems as tested on a transverse shock and vibration machine

not mean that these methods are not useful up to a certain degree. However, when their functional operation is compared with their cost, they are difficult to justify. With the liquid threadlocking system, there is no further effort or expense required for mechanical locking elements because "one size" fits all. Most threadlocking problems are thus solved in an economical way.

A similarly favourable load cycle performance is shown only by the surface-compacting ribbed flange bolt. Its disadvantages include: high costs, the relatively large amount of space required for flange-bearing surface, and the unavoidable damage to the surface of the clamped parts around the bolt-bearing surface. Teeth on bolts with a saw-tooth flange penetrate the bearing surface of the clamped material. Bearing surfaces of the head and the nut are damaged during loosening, limiting their possible applications. Parts with hardened surfaces cannot be reliably connected.

1.7 Threadlocking and Production

Bolt reusability is important in maintenance and repair work. Correctly selected industrial threadlockers can be loosened by ordinary tools without damage to bolts.

In a comparative test, breakloose torque may be used as a measure for wetting, adhesion and degree of curing. In general, there is no direct relation between breakloose torque and self-loosening resistance. In practice, it is often required that threaded assemblies be capable of disassembly using normal tools.

The industry has developed products with low and high break-loose torques to meet these needs. For distinction, the definitions ‘may be disassembled using normal tools’ and ‘difficult to disassemble’ were chosen. Because adhesive fills the joint, the fastener cannot seize due to corrosion. A bolt that has been locked with adhesive may be reused by removing old adhesive before applying new adhesive.

Breakloose strength depends on:

- length of thread
- thread material match
- surface finish
- pre-stress torque
- bolt diameter

Bolts should not be reused if they have been stressed to the apparent limit of their elasticity. If re-used, the risk of a breakage exists when the same pre-stress force is applied.

A major advantage of threadlocking products is that the liquid adhesive is effective regardless of bolt sizes and diameters. While special locking bolts require large inventories, Industrial threadlocking adhesives can be used on all standard and special bolts. Searching for or ordering special locking systems or bolts is eliminated.

Threadlockers guarantee more than functional reliability: the cured liquid that fills the voids in the threads not only prevents relative movement—it also seals the joint. Excellent chemical resistance allows threadlockers to be used as sealants against most gases and fluids used in industry. They also seal out moisture and corrosives that can cut the life of the assembly. This sealing effect allows through holes to be drilled instead of blind holes, an easier and cheaper assembly method.

1.8 Adhesive Application

Application to one surface is sufficient to provide the locking effect of liquid threadlockers. When applied by the drop to threads, enough adhesive must be applied so that the adhesive is distributed evenly over several threads when the screw or bolt is inserted. Industrial threadlockers can be applied to both male and female threads. The important factor is that the inner space is completely filled.

Dispensing systems allow accurate control of quantity. The industry offers special equipment to meet customers' needs. Manual or automatic application systems may be used depending on the degree of automation desired. Methods in which the dispensing valve is applied to the work piece by hand are described as manual systems.

Automatic systems can be subdivided into 'semi-automatic' and 'automatic'. The dispensing valve or the work-piece is transported automatically to the dispensing position in semi-automatic systems and the dispensing valve is actuated manually. This guarantees exact repeatability of the dispensing location and quantity dispensed. With fully automatic systems, work-pieces are positioned by handling devices (rotary tables etc.), and the dispensing valve is actuated automatically.

The usefulness of an application system comes from its integration into the production process. Individual conditions determine whether manual or automatic dispensing will be most appropriate. Chemical and physical properties of the individual threadlocking adhesives also influence application method.

2 Retaining

2.1 Retaining Metal Cylindrical Assemblies with Adhesives

The term "retaining compounds" describes adhesives used in cylindrical assemblies joined by inserting one part into the other. A typical example is bearings mounted in electric motor housings.

The American Standards Association began to standardise press fit methods in 1920. It was not until 1955 that ASA Std. B4.1 "Preferred Limits and Fits For Cylindrical Parts" was published. This procedure was limited to parts heavy enough to handle the stresses of forced fitting.

A retaining adhesive is a synthesised chemical formulation which is applied as a liquid into a cylindrical joint. The liquid adhesive then undergoes a chemical reaction to change to a thermoset plastic which has good adhesion to the adjoining metal substrates.

There are many different types of adhesives, often categorised according to the chemistry type e.g. silicones, polyurethanes, acyrlics etc. The most suitable and successful type of adhesives used in cylindrical joints are anaerobic adhesives.

For the chemical reaction which changes an anaerobic adhesive from a liquid to a solid to occur, two conditions are necessary—absence of oxygen and the presence of an active metal surface. Bonding of cylindrical non-metallic components such as plastics can be achieved using concepts presented in this chapter.

Popular applications for retaining adhesives:

- Mounting bearings in housings or on shafts
- Fixing rotors, gears, sprockets and pulleys on shafts
- Retaining cylinder linings

- Sealing engine core plugs
- Eliminating keys and set screws
- Avoiding distortion of precision tools or machines
- Assembling drill jig bushings
- Fixing oil filler tubes in castings
- Restoring accuracy to worn machine tools

Cylindrical assembly methods:

Assemblies used to transmit load in torsional, radial or axial directions from a shaft to a hub (or vice versa) can be divided into four categories:

1. Positive-drive (such as splines)
2. Friction-drive (such as press fits)
3. Welding or soldering
4. Adhesive bonding

Categories 3 and 4 are sometimes grouped together and referred to as interfacial drive assemblies.

2.1.1 Positive-Drive Assemblies

The most common positive-drive assembly is the conventional key/keyway. Others include splines, pins and serrations. They are relatively easy to assemble or disassemble and can transmit very high levels of torque. These benefits are offset by disadvantages such as high stresses due to the "notch effect" that occurs in the area of a key, high machining costs, the need for axial restraints, and backlash inherent in most designs.

Keyed assemblies also have uneven distribution of mass, an imbalance that can lead to vibration at high speeds.

2.1.2 Friction-Drive Assemblies

Well-known friction-drive assemblies include clamp rings, press fits, shrink fits, and taper fits.

These assemblies can be very economical, there is no imbalance at high speeds, and additional axial anchoring is not required.

However, these methods have disadvantages: they rely on friction alone to transmit torque, and so they are limited by materials, surfaces and design, and very close tolerances are needed to obtain specific load capacities—leading to high production costs. Except for tapered fits and expansion bushes, the assemblies can also be difficult or impossible to disassemble because of surface galling. Interference fitting creates stresses in the components that can lead to failure, particularly when combined with operational stresses. Positive and friction/drive assemblies are susceptible to fretting; micro-movement that leads to corrosion, accelerated wear and potential failure.

Welding, soldering, and adhesive bonding can create joints where no micro-movements are possible. Correctly designed joints can therefore eliminate the potential for fretting.

2.1.3 Welding or Soldering

Resistance to fretting, very high strength and high thermal and electrical conductivity are advantages of welded or soldered assemblies. However, there are a number of restrictions to the use of these methods. For example, normally only compatible metals can be joined, the parts can be distorted by the high temperatures required, and heating of the material can lead to residual stresses and structural degradation. Disassembly can also be difficult or impossible.

2.1.4 Adhesive Bonding

There are three distinct types of bonded cylindrical assembly, as shown in Fig. 5:

- *Bonded slip-fits*: parts are manufactured with a clearance, and the cured adhesive transmits all the load.
- *Bonded shrink- and press-fits*: the load is transmitted by both the cured adhesive and by friction between the parts due to the interference fit.

In both types, the adhesive is applied as a liquid, completely fills the joint, and cures to a tough thermoset plastic to link the surfaces together.

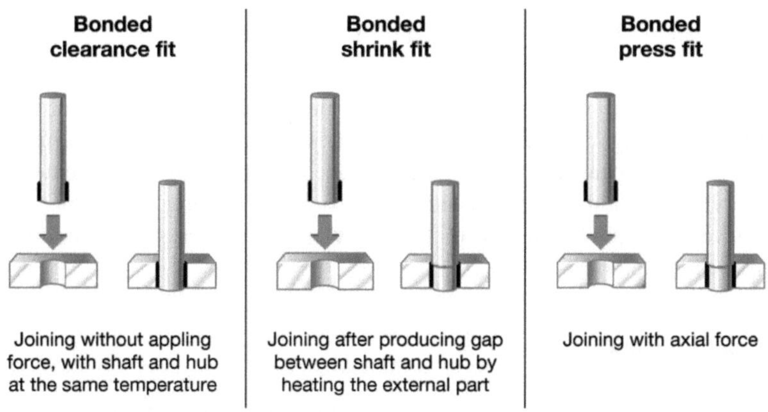

Fig. 5 Various bonded cylindrical assemblies

2.2 Benefits of Bonded Cylindrical Assemblies

The reasons for choosing an anaerobic retaining adhesive depend on each specific application. Some common benefits include:

- Replace mechanical assemblies
- Reinforce cylindrical joints
- Eliminate fretting corrosion
- Produce stronger, more rigid assemblies
- Eliminate backlash in keys and splines
- Reduced need for extra securing components
- Realisation of lightweight constructions
- Reduced tolerance requirement
- Parts can be dismantled by heating to reduce adhesive strength
- Allow the use of dissimilar materials
- Uniform stress distribution, and reduced stress in parts
- Reduced machining costs
- Allow bearings and bushes to self-align
- Hard and soft surfaces can be joined without damage
- Completely sealed joint, eliminating corrosion

The use of retaining compounds is limited by some factors. They are plastic materials and should not be used at high temperatures. Each individual grade will have its own temperature limitations, but generally, continuous operating temperatures above 150°C should be approached with caution.

Retaining adhesives must only be used where the gap between parts is small. At higher gaps, the strength and time to cure for the adhesive will be reduced.

Retaining adhesives work best on metal substrates. They can be used to bond plastic and ceramic materials but performance will be more limited.

2.3 Designing Adhesive Retained Assemblies

Bonding can simplify the design, production and assembly of parts. For example, when fitting bearings the adhesive compensates for misalignment, within certain limits. Shafts and bearings assembled without the stress of misalignment, have longer service life and can be reused after a simple cleaning operation.

Adhesives also allow the use of thinner wall components when compared to interference fit methods.

2.3.1 Loading Conditions

It is important to consider the loading conditions in order to optimise the design of a bonded assembly. Most assemblies carry a combination of: axial, torsional,

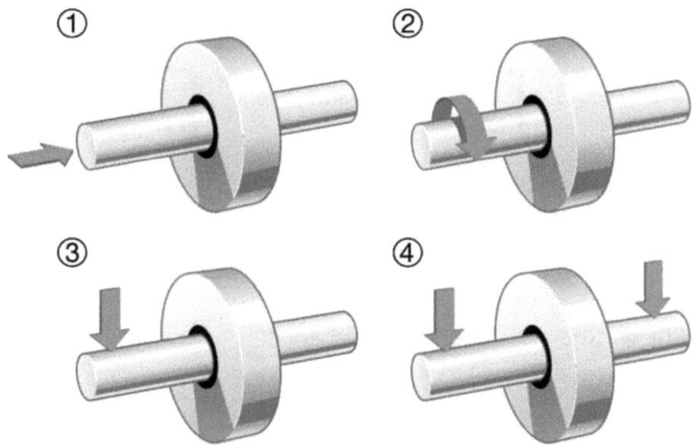

Fig. 6 Loading conditions (*1* axial, *2* torsional, *3* bending, *4* radial)

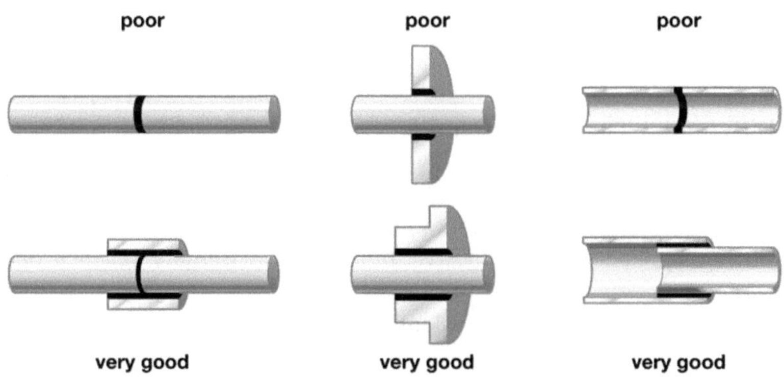

Fig. 7 Examples of design improvements in bonded cylindrical assemblies

radial and bending loads (Fig. 6). Loads may be considered to be static or dynamically applied.

2.3.2 Joint Design

Retained assemblies should be designed to minimise peel and cleavage loads, and maximise compressive and shear loads on the adhesive during use. Relatively small changes in design details can lead to large improvements in ultimate strength (Fig. 7).

2.3.3 Stress Distribution

In practice, stresses in cylindrical assemblies are not evenly distributed over the whole length of the joint (Fig. 8). For instance, stresses will be concentrated at

Fig. 8 Stress distribution across bond

the edge of the joint area, hence increasing the joint length will not lead to a proportionally stronger assembly. Designers can create favourable geometries to deal with such stress concentrations.

2.3.4 Differential Thermal Expansion

In many cases, dissimilar materials with different coefficients of thermal expansion are used in cylindrical assemblies. This can result in large undesirable tensile strains being applied to the adhesive film as the assembly reaches operating temperature.

The thermal expansion can be calculated by the following equation:

$$d = d_0(1 + \alpha\Delta T) \tag{4}$$

where d, diameter after expansion (mm); d_O, original diameter (mm); α, coefficient of linear thermal expansion (mm °C^{-1}); ΔT, temperature differential (°C).

Three techniques can be employed to ensure that the adhesive bond can accommodate any thermal expansion that may occur:

Adhesive Augmented Interference Fit

An adhesive should be used to supplement an interference fit. Provided a small interference is maintained throughout the total operating temperature range, the assembly will operate successfully.

Large Gap Clearance Fit

The relatively low modulus and high coefficient of expansion of industrial retaining products mean that it is possible to eliminate or reduce tensile stress in the joint by specifying tolerances to ensure a substantial film thickness of adhesive is present.

Shrink Bonded Clearance Fit

A shrink bonded clearance fit should be specified (typically 0.05 mm clearance for a 50.0 mm diameter), when the outer part has a higher coefficient of thermal expansion than the inner part, i.e. a steel shaft (or bearing) into an aluminum pulley (or housing). The adhesive should be applied to the male component,the female component heated, then the parts assembled (a temperature differential ΔT of 100°C is typical).

This assembly method results in a compressive stress on the adhesive during static conditions. When differential expansion during service results in an enlargement of the bond gap, the compressive stress on the adhesive relaxes. This allows considerable differential expansion before an unacceptable tensile load is placed on the adhesive.

2.4 Dynamic Loading

When retained assemblies are subjected to dynamic (cyclic) loads, consideration needs to be given to the fatigue resistance of the joint.

To describe the dependency of applied alternating load and number of load cycles to failure, Wöhler curves are used (Fig. 9). After a constant drop the curve typically goes straight, representing the fatigue limit. The generation of those curves takes considerable time as for statistical significance every load level has to be repeated and the test frequency is limited by the available test set up and the properties of the test assembly.

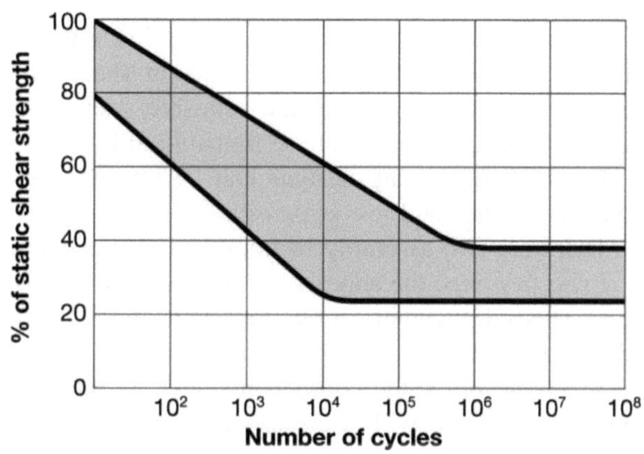

Fig. 9 Wöhler curve for a bonded slip fit assembly undergoing alternating loading at 15–20 Hz. The area between the curves represents the statistical scatter of the results

2.5 Combined Assembly Methods

Adhesives can have a beneficial effect on all conventional cylindrical assembly methods, allowing higher loads to be transmitted and eliminating fretting.

2.5.1 Positive Drive Assemblies with Adhesives

Splines, keys etc. are susceptible to fretting due to the backlash inherent in their design. Retaining adhesives fill all the voids in the assembly. When cured, they eliminate movement between the two parts and thus increase the long-term load carrying capacity.

The axial load capacity of the adhesive means that in some assemblies there could no longer be a need for axial restraints, such as circlips, collars, or pins.

2.5.2 Interference Fits with Adhesive

The strength of a non-bonded interference fit is calculated using the formula:

$$\text{Strength} = \text{Hub pressure} \times \text{coefficient of friction} \times \text{contact area} \qquad (5)$$

The variables are dependent on the following:

- Interfacial hub pressure is dependant on the actual level of interference, modulus of the materials, and the component design. The higher the interference, the higher the pressure.
- Coefficient of friction is dependent on the material, the surface finish, and the surface condition.
- Contact area: studies have shown that the maximum actual metal-to-metal contact between the surface of a shaft and hub is only 25–30% even in the most severe press or shrink fit joint.

When an adhesive is added to the joint, the contact area is raised to 100%, allowing the whole joint area to transmit shear and compressive loads. The dry interference fit strength and the adhesive strength can be added together, which means that higher loads can be transmitted by the same size of joint.

In practice, existing designs can be up-rated, and new designs will be more compact, lighter and less expensive.

There are two assembly techniques for interference fitting with adhesives:

Press fit joints: the adhesive is applied as a film to one or both surfaces, and the parts pressed together. It is important to avoid wiping or pushing away the adhesive during the assembly of parts.

Shrink fit joints: the components have an interference at working temperature, but are assembled with a clearance. The preferred technique for adhesive shrink

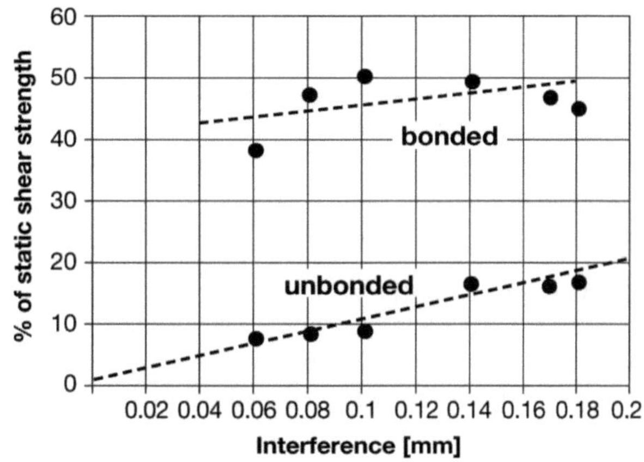

Fig. 10 Comparison of bonded and unbonded shrink fit joints. Pin 120 mm diameter, collar 120 mm inner diameter, 140 mm outer diameter. Joint length 25 mm

fits is to apply the adhesive to the male component, and to heat the female component to obtain the fitting clearance. Shrink fitting techniques give optimum joint strength, as the heat in the assembly helps to create a rapid cure and the contraction of the outer component exerts a compressive load on the curing adhesive (Fig. 10).

The result is a total strength far in excess of the combined adhesive shear strength and the dry interference strength.

An alternative method is to freeze the male component and apply adhesive to the female. This technique is not normally recommended as condensation (frost) on the cold component can affect the adhesive cure and the bond strength.

In cases of very high interferences or small diameters where the thermal expansion is too small, a combination of both methods is sometimes used.

2.6 Detail Design

2.6.1 Surface Finish

A surface finish in the range 0.8–3.2 μm of R_a (arithmetic average of the roughness profile, e.g. DIN EN ISO 4287) is recommended. In general, very smooth surface finishes should be avoided as they reduce the degree of "keying" available.

Conventional machining techniques create a lay in the surface that results in a coarse finish in the axial direction, and a fine finish in the circumferential direction. Note should therefore be taken of the loading direction when specifying surface finish.

2.6.2 Lead-in Chamfers

To minimise the possibility of the adhesive being scraped off the surfaces during assembly, it is advisable to specify chamfers on the leading edge of both male and female components. A 1.0 mm chamfer at 15°–35° to the axis of the shaft is recommended.

2.6.3 Stepped Shafts

When two or more components are to be fitted on to one end of a shaft, the assembly should be designed with steps between each bond area to allow easy assembly (Fig. 11).

2.6.4 Tapered Joints

The use of retaining adhesives with tapered assemblies eliminates potential fretting and increases load carrying capacity. Tapered bonded assemblies ensure a closely controlled concentricity and can be an effective way of ensuring good adhesive coverage on assemblies with long bond areas, that may be subjected to a scraping action during assembly.

2.6.5 Variable Thickness Bond Lines

Bond areas designed with variable stepped gaps can be useful when there is a need for a thin adhesive film over part of the bond area (i.e. for close concentricity control) and a thick adhesive film to allow for thermal expansion.

2.6.6 Sealing with Retaining Adhesives

Retaining adhesives completely seal assemblies, therefore, through-holes can be used in gear casings, water jackets etc. without risk of leakage. This can save machining cost and weight when compared to traditional blind hole designs.

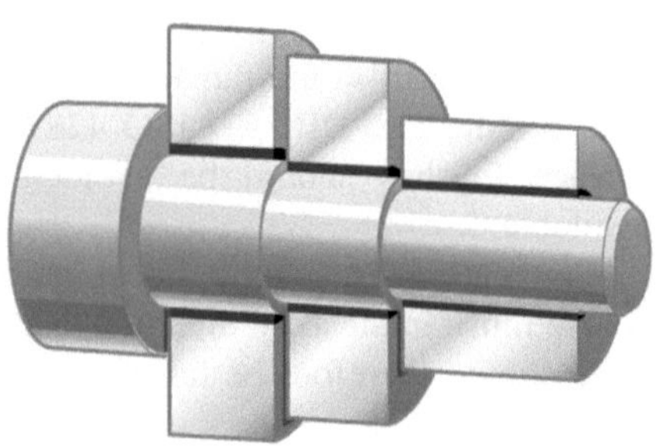

Fig. 11 Stepped shafts. Reducing shaft size to allow assembly without scraping off the adhesive

2.7 Adhesive Application and Assembly

2.7.1 Surface Preparation

Components should be clean and free from contamination such as grease, oil, cutting coolants, protective coatings, etc. Normal practice is to degrease using solvents, or aqueous washing solutions.

When selecting a cleaning process, care should be taken to ensure that no residue remains on the surface, and that no chemical is present which may inhibit the adhesive curing. The most common cause for concern is residual nitrites from some aqueous washing solutions or cutting coolants. These materials are not removed by solvent degreasers, and require a clean water wash to ensure satisfactory cure.

The use of etching techniques, abrasion or grit blasting has a positive affect on assembly strength. However, for comparability all standard data should be generated on degreased surfaces to the relevant specifications.

2.7.2 Oil Tolerant Retaining Adhesives

The industry manufactures oil-tolerant retaining adhesives that can effectively bond "as received" components where moderate "oily" contamination may be present. These are ideal for high volume production.

This technology is particularly useful where oil impregnated sintered bushes are to be retained, when full degreasing would reduce bearing life, and press fitting can 'close in' the bore diameter. Oil tolerant retainers allow bushess to be slip fitted after simply wiping clean the bond surface.

2.8 Production Hints

2.8.1 Where to Apply Adhesive

Application of adhesive over the complete bond area of both parts is optimum. However in practice application by bead to the leading edge of one or both parts should result in a reliable result. It may be necessary to apply two or more such beads for a long bond area. Each assembly configuration will be different, and tests should be conducted to ensure the adhesive application technique results in adequate joint fill after assembly.

When parts are assembled into blind holes, the adhesive can be forced out of the joint by the air being compressed. This can be avoided by the provision of a ventilation hole. If this is not possible, the adhesive should be dispensed into the bottom of the hole. During assembly, hydraulic pressure will force the adhesive back up the hole and coat the total bond area by hydraulic pressure when the parts are assembled.

Retaining products can be applied directly from small volume bottles which have dispensing nozzles that allow manual application.

The industry also designs, builds and commissions special purpose equipment to meet unique dispensing requirements. Special purpose equipment can be either semi automatic, or fully automatic. It can be stand-alone, or integrated into new or existing assembly lines.

2.8.2 Assembly of Components

Slip Fit

Where possible, cylindrical parts should be turned slightly against each other during assembly; this will spread the adhesive, and help achieve full coverage of the bond area. Where it is important to maintain concentricity and alignment between two bonded parts, a fixture should be used during the adhesive's initial cure.

Components must be correctly positioned at the outset. Corrections or movements made during the curing process damage polymer chains that have already formed, thus reducing the final strength of the joint.

Press Fit

Incorporating adhesives into existing press fit design can lead to considerable strength enhancement. However, the assembly force is also significantly increased, therefore, the press should be adjusted accordingly (Fig. 12). Assembling at high speed (typically 1,000 mm/min) will reduce the assembly force by between 30 and 40% when compared to typical assembly press speeds.

Shrink Fit

The female component of a shrink fit assembly must be heated. The maximum recommended shrink fitting temperature when using retaining adhesives is 200°C. A gap of 0.1% of the nominal diameter (0.01 mm minimum) should be achieved for assembly. The adhesive should be applied to the complete surface of the cold part (shaft).

The elevated temperature of the heated component will cure the adhesive rapidly. It is therefore important to align the parts and assemble into the final position in one movement.

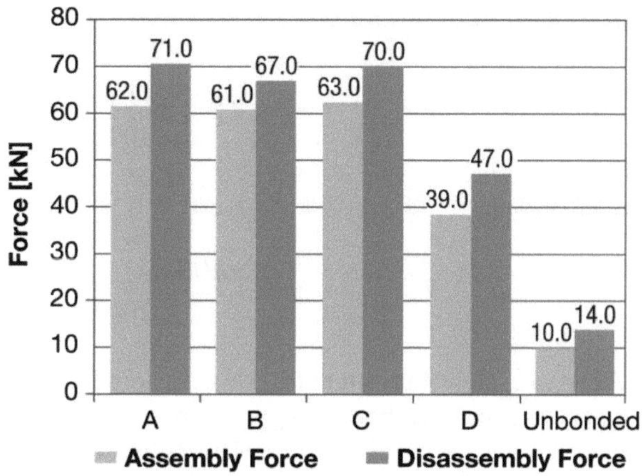

Fig. 12 Comparison of assembly and disassembly forces of press fit joints for different anaerobic adhesives A to D. Pin 20 mm diameter, collar 20 mm inner diameter, 40 mm outer diameter. Joint length 24 mm

2.8.3 Curing of Adhesive

Retaining adhesives cure at room temperature. However each grade has a specific cure profile. It is good practice to ensure no load is placed on the bonded assembly, until sufficient adhesive strength is achieved.

2.8.4 Disassemble and Reassemble

When heat is applied (maximum 250°C) to anaerobic adhesives, the strength is reduced allowing the parts to be disassembled with conventional extractor devices. When regular disassembly of bonded cylindrical joints is a design requirement, anaerobic adhesives with the appropriate strength and/or temperature performance should be selected. Prior to reassembly, any adhesive residue should be removed.

2.8.5 Retaining Adhesives for Maintenance and Repair

Unserviceable worn parts can often be reliably repaired with anaerobic adhesives. For example, loose bearings can be refitted in housings with anaerobic adhesives, allowing worn housings or shafts to be salvaged. To ensure correct alignment, centring may be required when gaps are large. With extremely large gaps that cannot be filled by the adhesive alone, a component can often be saved by a combination of the use of metal shims and adhesive.

2.9 Calculating the Strength of Bonded Cylindrical Assemblies

The attainable strength of a bonded joint and the torque that can be transmitted are calculated with the following formulae:

Static axial push-out force:

$$F = \frac{\pi dl[(\tau_{B2} f_c) + (P\mu)]}{1000} \tag{5}$$

Static Torque:

$$T = \frac{\pi d^2 l[(\tau_{B2} f_c) + (P\mu)]}{2000} \tag{6}$$

where F, press-out strength (kN); T, torque capacity (Nm); d, nominal joint diameter (mm); l, bond length (mm); τ_{B2}, static shear strength of adhesive (in accordance with ISO 10123/MIL-R-46082) (N/mm^2); f_c, product of correction factors, ($f_1 \cdot f_2 \cdot f_3 \cdot f_4 \ldots$..); P, radial contact pressure due to interference fit calculated using classic "thick-walled cylinder" theory (for slip fitted assemblies $P = 0$) (N/mm^2); μ, effective coefficient of friction.

The calculated results should be regarded as approximate values for the strength that can be expected in practice. The exact strength can only be determined by tests on actual parts.

2.9.1 Correction Factors

There are several standard test methods for determining the strength of adhesives in cylindrical joints. Commonly ISO 10123/MIL-R-46082 standard pins and collars are used. Such test results are useful for comparing adhesive grades and for quality control purposes; however, they cannot be used directly for design calculations. It is usual to define a series of correction factors, based on empirical testing and practical experience. These correction factors are used to adjust the adhesive strength value determined from the standard test method, to account for the actual physical and operating conditions of the intended application:

f_1: type of material
f_2: type of assembly
f_3: actual clearance/interference
f_4: geometry
f_5: operating temperature
f_6: heat ageing
f_7: operating environment

For specific applications, it may also be necessary to take account of additional factors, such as:

- Dynamic loading
- Surface roughness
- Differential expansion

2.9.2 Type of Material f_1

The curing of anaerobic adhesives is influenced by the catalytic effect of the surface and the adhesive strength is partially dependent on the surface structure of the material. The following correction factor is to be applied for the material to be bonded:

- Steel 1.0
- Alloy steel 0.9
- Cast iron 0.8
- Stainless steel 0.8
- Aluminium 0.5
- Copper and alloys 0.4
- Plated metal surfaces 0.2

When joining two different materials the smaller value is to be taken in each case. When bonding plastics, the adhesive strength can also be weakened by poor wettability. Cyanoacrylate adhesives may be more suitable when assembling small plastic parts.

2.9.3 Type of Assembly f_2

Depending on whether the assembly is a slip, press or shrink fit, the effective strength contribution from the adhesive will be different:

- Slip-fit 1.0
- Press fit 0.5
- Shrink fit 1.2

These values are conservative and approximate and are influenced by actual design criteria.

2.9.4 Actual Clearance/Interference f_3

For slip fitted joints, the adhesive strength can be affected by large bond gaps. The most favourable gap for bonding is between 0.02 and 0.07 mm diametrical clearance (Fig. 13). For interference fitted joints this factor is 1.

2.9.5 Geometry f_4

Shear stress within a cylindrical assembly is not evenly distributed across the overlap length; maximum stresses are concentrated at the edge of the components. This means that assembly strength does not increase linearly with increasing bond width. The effective bond length to joint diameter ratio (*l/d* ratio) also affects the

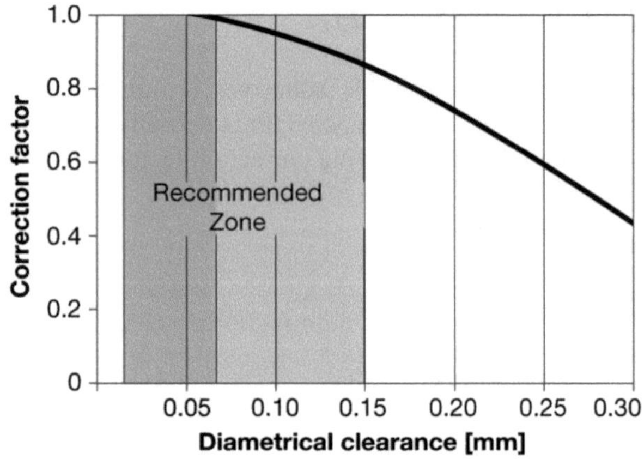

Fig. 13 Relationship between correction factor and diametrical clearance

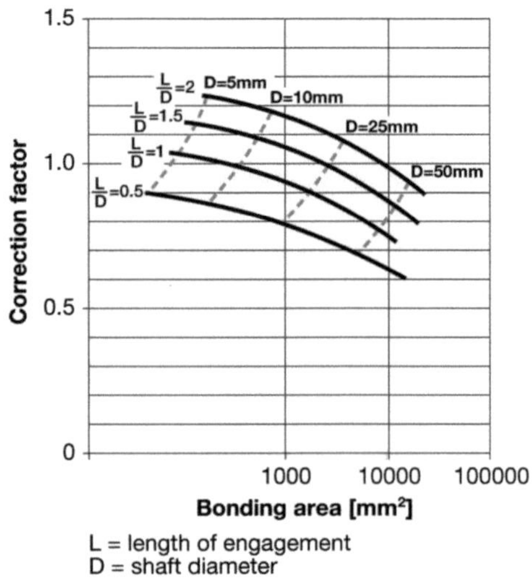

Fig. 14 Relationship between correction factor and bonding area for different length/diameter ratios

final bond strength. Both these effects are combined in the geometry factor (Fig. 14).

2.9.6 Operating Temperature f_5

The strength of an adhesive is reduced at elevated temperature. The degree of reduction is dependent on the specific grade selected. The relative strength of retaining products depending on the operational temperature is normally delivered by the manufacturer in the technical data sheet. In other cases it must be determined by individual tests.

Factor f_5 can be calculated from this data as follows:

$$f_5 = \frac{\text{Relative strength}\,[\%]}{100} \tag{7}$$

2.9.7 Heat Ageing f_6

Heat ageing is the result of long term exposure to high temperatures. This can be a permanent increase or decrease in the adhesive strength when tested at room temperature. The product used, the duration, and intensity of the heat influences the effect. This information is normally delivered by the manufacturer in the technical data sheet. In other cases it must be determined by individual tests.

Factor f_6 can be calculated from this data as follows:

$$f_6 = \frac{\text{Relative strength}\,[\%]}{100} \tag{8}$$

2.9.8 Operating Environment f_7

Industrial retaining products are resistant to most industrial environments. Chemical and solvent resistance values should be delivered by the manufacturer in the technical data sheet. In other cases it must be determined by individual tests.

Factor f_7 can be derived from these tables as follows:

$$f_7 = \frac{\text{Relative strength}\,[\%]}{100} \tag{9}$$

2.9.9 Dynamic Loading

In many applications the loading on the joint is dynamic, i.e. the applied load fluctuates to create an alternating stress on the joint. This stress may be applied in either an axial or torsional sense. The industry has conducted dynamic fatigue studies on bonded cylindrical assemblies to determine *S/N* (Wöhler) curves and endurance limits.

Fatigue tests have focused primarily on torsional loading as this is the most common mode in cylindrical joint applications. From these studies, a series of dynamic correction factors has been established to determine dynamic fatigue load limits as a percentage of the calculated static strength.

$$\text{Dynamic axial load capacity}, F_{\mathrm{DYN}} = Fk_{\mathrm{DA}} \tag{10}$$

$$\text{Dynamic torque capacity}, T_{DY} = Tk_{DT} \quad (11)$$

where k_{DA}, 0.12 for axial bonded slip fit joints; k_{DT}, 0.30 for bonded slip fits, and 0.35 for bonded interference fits.

These values are conservative and approximate, and depend upon actual design criteria.

2.9.10 Surface Roughness

While a surface roughness of approximately 0.2 μm (R_a) is required for press fits, 1.0–3.5 μm (R_a) is sufficient for bonded joints. This cuts the cost of machining. With a surface roughness of more than 3.5 μm (R_a), there is the danger that the whole surface will not be wetted, and the influence of notch effects will become increasingly noticeable.

3 Conclusion

During the last years adhesives have gained an important status in industrial applications. This chapter addresses two classical areas of application: threadlocking and joining cylindrical assemblies (retaining). From an industrial, application oriented perspective support is given to select the most suitable product, to estimate the attainable strength, to consider design aspects and operational influences in order to achieve the full potential of the junction.

Bibliography

1. Loctite Worldwide Design Handbook, 2nd edn. (1998)
2. Habenicht, G.: Kleben, Grundlagen–Technologie–Anwendungen. Springer, Verlag
3. Adams, R.D. (ed.): Adhesive Bonding, Science, Technology and Applications. Woodhead Publishing Limited, UK
4. ESDU—Engineering Data Handling Unit. http://www.esdu.com
5. Brockmann, W., Geiß, P.L., Klingen, J., Schröder, B.: Klebtechnik, Klebstoffe, Anwendungen und verfahren. Wiley-VCH, New York
6. Habenicht, G.: Kleben—erfolgreich und fehlerfrei. Vieweg + Teubner
7. Petrie, E.W.: Handbook of Adhesives and Sealants. McGraw-Hill, USA
8. Fosta-Forschungsberichte p 229, 259, 332, 340, 477, 508, 513, 676 bei http://www.stahl-online.de
9. Romanos, G.: Technologie und Dimensionierungsgrundlagen geklebter und hybridgeklebter Welle-Nabe-Verbindungen. VDI-Berichte 1790. Welle-Nabe-Verbindungen. Gestaltung, Fertigung, Anwendungen. Tagung Wiesloch b. Heidelberg, 19/20 November 2003/VDI-Gesellschaft Entwicklung Konstruktion Vertrieb S. 271–285. VDI Verlag, Düsseldorf (2003)
10. Kreuzer, R., Romanos, G.: Zuverlässigkeit von Flüssigdichtmitteln unter dynamischer Beanspruchung. Konstruktion **36** (H. 6), 60–62 (2002)

11. Kreuzer, R., Romanos, G.: Zuverlässigkeit von Flächendichtungen auf Basis von Flüssigdichtmitteln unterdynamischer Beanspruchung. VDI-Berichte 1579. Zuverlässig Abdichten–Moderne Dichtungssysteme in der Anwendung; Tagung Baden-Baden, 07/08 November 2000/VDI-Gesellschaft Entwicklung Konstruktion Vertrieb S. 69–88. VDI Verlag, Düsseldorf (2000)
12. Romanos, G.: Strength evaluation of axisymmetric bonded joints using anaerobic adhesives. Int. J. Mater. Prod. Technol. **14** (H. 5/6), 430–444 (1999)
13. Romanos, G.: Bruchmechanische Bewertung stoffschlüssiger Welle-Nabe-Verbindungen unter Torsionsbelastung. VDI-Berichte 1384. Welle-Nabe-Verbindungen: Systemkomponenten im Wandel; Tagung Fulda, 28/29 April 1998/VDI-Gesellschaft Entwicklung Konstruktion Vertrieb S. 347–357. VDI Verlag, Düsseldorf (1998)
14. Romanos, G.: Klebetechnologie konstruktiv ausnutzen. Beanspruchungsgerechte Gestaltung geklebter Welle-Nabe-Verbindungen. Der Konstrukteur **28** (H. 10), 20–22 (1997)
15. Romanos, G.: Design and analysis aspects of adhesive joints in drive train applications. Mechanical behaviour of adhesive joints. Analysis, testing and design. In: Adams, R.D. (Hrsg. Proceedings of the EUROMECH Colloquium 358. Nevers, France, 04/06 September 1997). S. 361–375 Pluralis, Paris (1997)
16. Romanos, G., Fischer, M.: Konstruktionsprozeß geklebter Bauteile—Methoden, zeitgemäße Werkzeuge, Erfahrungen. Berichtsband zum 21. Kolloquium Konstruktionstechnik "Zeitgemäßer Produktentwicklungsprozeß". Otto-von-Guericke-Universität Magdeburg, 02/03 July 1997, S. 117–131

Science of Mixed-Adhesive Joints

James G Broughton and Michael D Fitton

Abstract Adhesively bonded joints are an effective means of connecting components, particularly in cases where the adherends are thin and dissimilar to one another, providing joining solutions in numerous automotive, aerospace, marine and increasingly civil infrastructural applications. Indeed, in many cases the bonded joint capacity is often limited by the strength of the adherends, either from yielding or through-thickness failure, e.g. delamination. The mixed-adhesive technique has been proposed as a means of tailoring the bondline stiffness (typically by placing at least two adhesives with contrasting stiffness in the bondline) to improve the joint capacity, i.e. by reducing the stress concentrations that are invariably located at the ends of the bonded overlap. Various analytical and numerical scientific approaches are discussed with an aim to establishing key parameters of mixed-adhesive joint design including experimental validation, which in some cases has demonstrated over 75% improvements in static joint strength.

1 Introduction

Adhesively bonded joints are often the most appropriate means of joining highly stressed thin sheets of dissimilar materials because they offer lightweight, strong and durable properties without any significant alteration or damage to

J. G. Broughton (✉)
Joining Technology Research Centre, School of Technology, Oxford Brookes University, Wheatley Campus, Wheatley OX33 1HX, UK
e-mail: jgbroughton@brookes.ac.uk

M. D. Fitton
Science & Technology Facilities Council, Rutherford Appleton Laboratory, Harwell Science and Innovation Campus, Didcot OX11 0QX, UK
e-mail: michael.fitton@stfc.ac.uk

Adv Struct Mater (2011) 6: 257–281

DOI: 10.1007/8611_2010_52

Published Online: 25 February 2011

the substrates. Ideally, this is accomplished by transferring load from one adherend to the other through shear. One of the simplest, and therefore most common, joint configurations is that of the bonded overlap joint. However, it is well established that the strength of an overlap joint may be severely limited by the effects of differential straining in the adherends and an eccentric load path through the joint, both of which set up stress concentrations at the ends of the overlap, see Fig. 1 [1]. Additional adverse effects resulting from large thermal gradients between highly dissimilar materials can magnify these stresses further [8].

Current methods for relieving these stress concentrations have developed from applications where the substrates are relatively thick and it is possible to modify the end of the overlap, e.g. by tapering the overlap and/or using large fillets of adhesive [1, 17]. Tapering of thin sheet, on the other hand, can be costly or may not even be practical although Fessel et al. [11] showed that by pre-deforming the adherends it is possible to eliminate most of the out-of-plane stresses (peel) caused by adherend bending in single overlap joints. Likewise, large fillets of adhesive are not easily controlled during manufacture and cure, and thin substrates leave little room to develop a reasonable size of fillet. Furthermore, sheet materials that exhibit a significant amount of yielding, such as titanium or aluminium, tend to limit the ultimate load capacity of the joint long before the adhesive exhibits any substantial degradation that leads to fracture. Yet, it is quite desirable to have the adhesive undergo yielding, effectively 'blunting' the stress concentrations at the overlap ends. A method involving tailored bondline properties, which exploits this phenomenon has been proposed by a number of authors though rarely demonstrated practically. Nevertheless, this approach has the potential to reduce component costs, negating the need for tapering, as well as increasing strength in, for example, thin-skinned composite structures. These structures are commonly found in motorsport and aerospace applications where they are often subjected to high stresses and substantial thermal gradients.

Overlap joints, made from thin fibre-reinforced plastic (FRP) composite adherends, often fracture or delaminate in the composite adherend adjacent to the peak stress concentrations due to the comparatively low through-thickness strength of the composite. Clearly, mixed substrate materials, like for example titanium bonded to FRP, may suffer a similar fate, and it is therefore vital to consider these highly localized stress concentrations, in particular peak peel stresses, when designing the joint [14].

The mixed-adhesive technique utilises an adhesive bondline with at least one transition between two adhesives exhibiting substantially differing stiffness.

Fig. 1 Principal stress distribution in the adherends of a lap joint loaded in tension

The intention being to relieve the high stress concentrations at the end regions of the overlap whilst maintaining high shear resistance. As already mentioned, until recently this approach had only been discussed theoretically [15, 21, 23, 26, 27] and only a relatively few published experimental demonstrations have yielded significant improvements in joint performance. For example, Sancaktar and Kumar [24] selectively toughened a single adhesive at the ends of the overlap, whilst Pires et al. [22] employed two highly contrasting adhesives. Both authors found that bondlines exhibiting a graded stiffness increased the strength of the joint, albeit marginally. Notably, the above experimental work was carried out on metallic adherends and did not consider the use of fibre reinforced plastic (FRP) composites. Work by Fitton and Broughton [10] and da Silva and Adams [4] provided some of the first experimental demonstrations of significant improvements in the static strength of epoxy mixed-adhesive joints used in FRP–FRP, FRP–aluminium and FRP–titanium bonded joints.

This chapter describes the general findings of published work on mixed-adhesive joints, focusing on analytical and numerical modeling that has been predominantly employed to establish key mechanical parameters that influence joint capacity. A more practical and applied treatment of mixed-adhesive joints is given in Chap. 11; Technology of mixed-adhesive joints.

2 Adhesive Characteristics

The basic mechanical behaviour and performance of current structural adhesives can be crudely categorised according to their stiffness. Structural adhesives exhibiting high bulk strength do so because they generally consist of a tightly packed and highly cross-linked polymeric network, which in turn gives rise to a relatively stiff mechanical response with little strain capacity. An adhesive exhibiting low stiffness is the result of a more open, less cross-linked polymeric network that inevitably exhibits low bulk strength but relatively greater strain capacity [19]. Tensile properties of a broad range of epoxy adhesives, determined in accordance with BS 2782-3 Method 320C: 1976, have been reported by Fitton [9] as part of a selection trial for mixed-adhesive joints. The tensile modulus and tensile strain capacity of the adhesives are plotted in Fig. 2. The high modulus adhesives clearly demonstrated a low strain at failure, reflecting a brittleness and susceptibility to failure at stress concentrations. On the other hand, the lower modulus adhesives exhibited high strains at failure, reflecting a tougher and more ductile behaviour much more capable of absorbing stress. Fitton [9] also compared adhesive tensile modulus and tensile strength, for which a linear relationship was observed, Fig. 3. This is because despite the tougher and more ductile structure exhibited by low modulus adhesives their more open polymer network results in reduced strength and poor resistance to creep. Ideally, it would be desirable to have both high strength and high strain to failure but the testing by Fitton [9] has shown that an adhesive exhibiting both is not generally possible and it is likely that

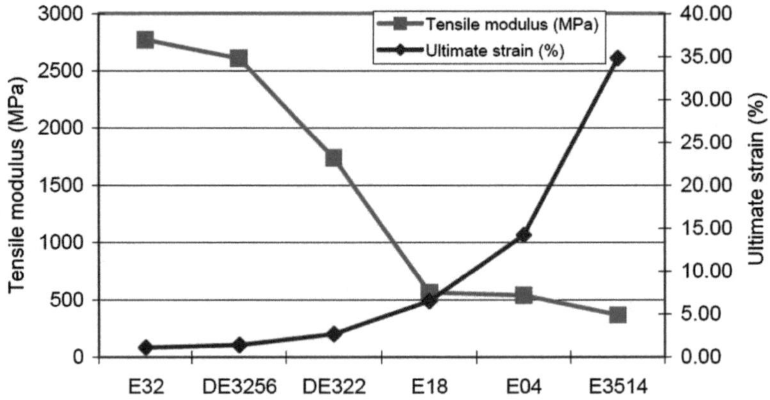

Fig. 2 Tensile Modulus versus ultimate strain of a selection of (Bondmaster) structural epoxy adhesives

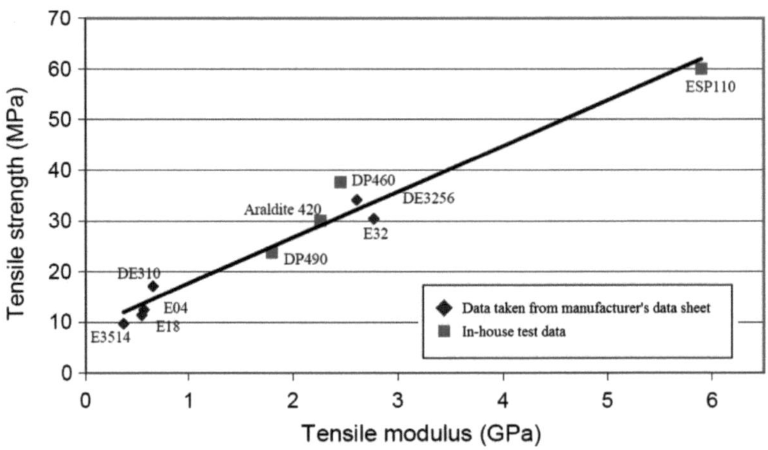

Fig. 3 Tensile strength versus tensile modulus of a selection of structural adhesives

a compromise will need to be sort. Additionally, low modulus adhesives tend to be more susceptible to elevated temperatures, again due to the reduced cross-linking between the polymer chains.

Perhaps an obvious subsequent step would be to combine the most desirable adhesive characteristics of two or more adhesives, thus tailoring the bondline accordingly. However, a number of issues have to be considered; not least selecting two adhesives that exhibit compatible working characteristics, cure regimes, adherend surface pre-treatments, and resistance to environmental exposure. Assuming the above are indeed achievable, it remains necessary to be able to determine the most appropriate mechanical properties of the adhesives. Figure 4 shows the tensile stress–strain behaviour for two of the adhesives that were used in

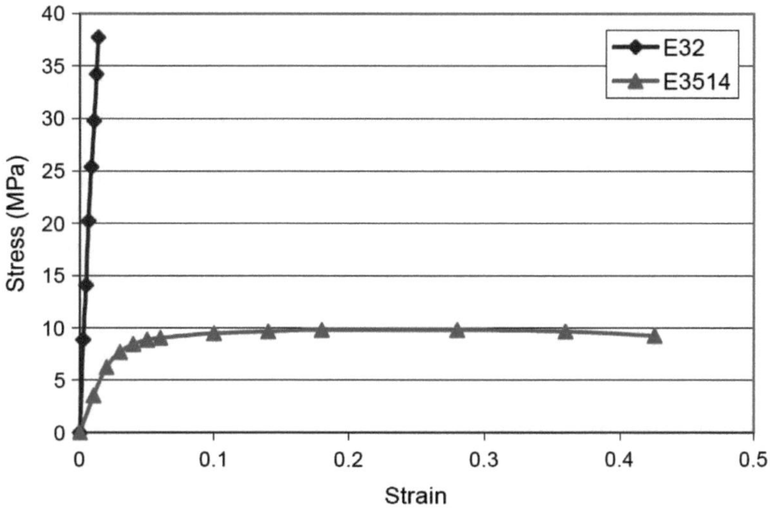

Fig. 4 Tensile stress–strain behaviour of E32 and E3514 (Bondmaster) structural adhesives

Fitton's work (E32 and E3514) for a mixed-adhesive joint. The tensile behaviour of E32 is almost linear to failure, whilst E3514 is highly non-linear. The behaviour of E32 is typical for a relatively high modulus un-toughened epoxy adhesive and has a tensile stiffness over seven times greater than that of E3514. Correspondingly, its strength is almost four times higher but its strain capacity is nearly 15 times less. Although no fracture data was reported in Fitton's work it is well documented that ductile polymeric materials exhibit greater toughness than their stiffer counterparts, which is generally attributed to their ability to better dissipate strain energy at a crack tip [19]. This is particularly relevant to the performance of 'real' joints, in as such that they are generally exposed to fatigue loading and environmental exposure during their service life, and hence the prevention of crack propagation is critical. Comparing the shear strengths of the two adhesives E32 and E3514, which were found to be 28 and 16 MPa respectively (determined in accordance with BS ISO 11003-2: 2001), there was much less of a contrast. This is an important observation, as shear strength data is often used as a baseline for comparing adhesive performance. Although there was no shear strain at failure reported in *Fitton's* work, it would be highly unlikely that the shear strain capacity of the higher modulus adhesive exceeded that of the lower modulus adhesive, and would follow a similar trend found for bulk tensile strain (though clearly not in magnitude). In actual fact, shear strength data is of little use in the prediction of real joint performance, either due to the reason outlined above with regard to service life conditions or because the adherend material often limits the joint capacity. It is for these reasons that the bulk stiffness of the adhesives (either tensile or shear modulus) is seen as a better indicator for improving joint performance when considering mixed-adhesive joints. This is particularly important when considering the location of the adhesives in the bondline, and is discussed later.

Thus, whilst it is important to be able to ascertain the most effective bulk properties of the two adhesives (stiffness, strength and strain capacity both in tension and shear), it is particularly important to understand the interaction between the stiffness of the adhesives and the stiffness and strength of the adherends, and how their combined interaction ultimately dictates the magnitude and the location of the stresses generated in the bond line. da Silva and Adams [3] note that where mixed-adhesive joints are subject to high/low temperature environments it is also essential that the bulk behaviour is ascertained at the respective temperature extremes.

3 Performance

3.1 Single-Adhesive Joint

In practical terms, it is impossible (or unwise) to predict the long-term performance of a bonded joint, based upon the selected adhesive's bulk mechanical performance alone. A comprehensive understanding of the joint design, applied loading and exposure to the intended environment should be considered a minimum. Given that environmental aspects are often unique to the application and generally cannot be altered, joint design is one of the more influential of these considerations. The main principles to consider in the design of a bonded joint are well documented and include:

- obtaining maximum bond surface area (e.g. stepped-lap joint, scarf joint)
- avoiding tensile (peel) forces (e.g. balanced or co-axial joints)
- reducing stress concentrations (e.g. balanced joints, tapers, chamfers, modified lap profile)
- maintaining a continuous, uniform and (generally) thin bondline.

As mentioned in the Introduction, one of the most simple and common joint designs found in a wide variety of engineering applications is that of the overlap joint. At first sight, this joint appears quite simplistic yet the state of stress along the overlap is actually quite complex. The finite element model of a simple lap joint shown in Fig. 1 is a result of a number of parameters including differential straining of the adjacent substrates loaded in tension, highly contrasting and potentially non-linear material properties and induced peel forces due to the initial eccentricity of the load path.

Given the complexity of the simple overlap joint significant peel stresses are still generated, even in a well-designed balanced joint. It is, therefore, important to recognise the presence of peel stresses in a joint and understand the effect they will have on joint performance. In composite joints it is likely that high peel stresses are the main contributor to the failure of the adherends [14]. When the stress concentrations are significant, improvements can be made by adding fillets of adhesive and tapering the adherends, as shown in Fig. 5. This reduces the stiffness locally and provides a larger volume of the low modulus adhesive to promote a

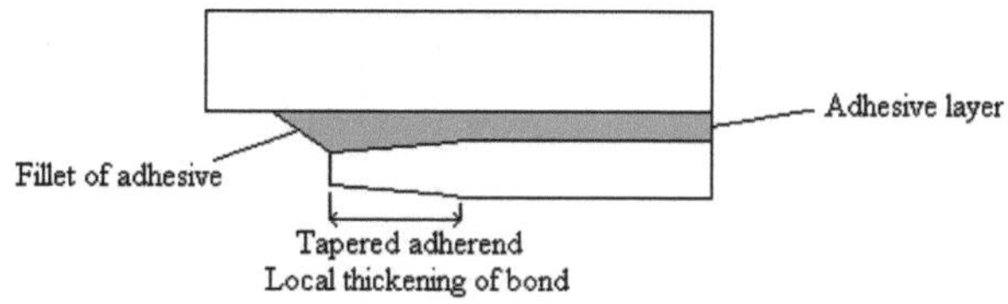

Fig. 5 Joint optimisation using a fillet of adhesive and tapered adherend

smoother transfer of the load [2]. Tapering of composite adherends is however difficult, often costly and, if done mechanically, can introduce weaknesses, such as cracks, into the composite [18].

A fact that is often forgotten in adhesive joint design is the importance of adherend properties on joint strength. It can be shown by simple theory [28] and experimental work that the strength of a joint can be increased by using stiffer adherends. Gordon and Fakley [13] have shown from physical tests that the joint strengths of various adhesives can be correlated with the elastic modulus of the adherend. Using stiff adherends lowers the bondlines stresses in two ways. Firstly the adherends deform less in tension, which reduces differential shear, and secondly there is less joint rotation, which reduces peel stresses.

However, a related factor is the surface strength, or through-thickness strength, of the adherends. Materials such as ceramics, timber, concrete and polymer composites may contain planes of weakness that will initiate failure if large transverse stresses are imposed, even indirectly, on the joint. Using analytical expressions first established by Goland and Reissner [12] it is possible to show that peel (and shear) stress concentrations in joints can also be reduced by lowering the adhesive modulus. However, the advantage lower modulus adhesives bring, through the development of a more uniform stress distribution, is counteracted by their inherent low strength. Fitton [9] tested a series of Carbon FRP single lap joints bonded using a selection of epoxy adhesives with a range of tensile moduli between 0.4 and 6 GPa. Figure 6 shows the average shear stress to failure of these joints plotted against the tensile modulus of the adhesive used to bond them. An optimum modulus of between 2 and 3 GPa appears to result in the highest joint capacity for the specific composite used. Joints made with a relatively low modulus adhesive (less than 2 GPa) failed cohesively. The peel stresses were low enough not to cause through-thickness failure of the composite adherend. The higher modulus adhesives, although exhibiting higher strength, developed high peel stresses which ultimately led to the adherend failing through-thickness.

3.2 Mixed-Adhesive Joints

A mixed-adhesive joint refers to a bonded overlap joint in which a low modulus and a high modulus adhesive are used to reduce the stress concentrations at the overlap ends. Normally, this requires the low modulus/low strength adhesive to be located at the outer regions of the overlap, acting as a stress reducer, and the high

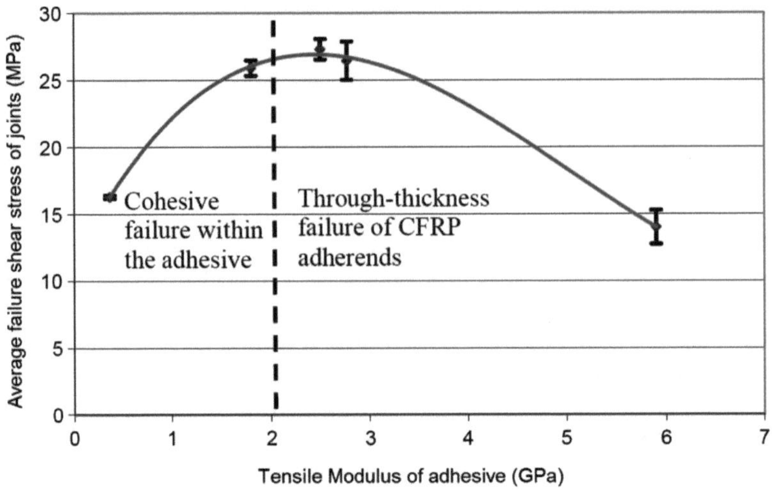

Fig. 6 Adhesive modulus versus joint strength for a single lap joint

Fig. 7 Placement of two different adhesives on one adherend separated by a thin wire spacer

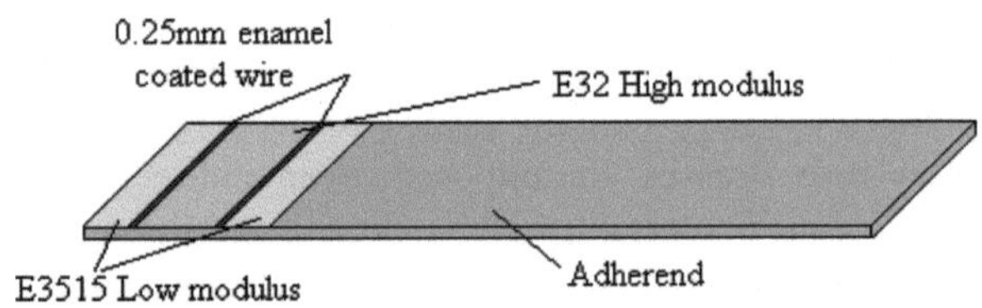

modulus/high strength adhesive positioned in the majority of the central region of the overlap, away from the stress concentrations. This is normally achieved by applying two adhesives separated by a spacer, such as a wire as in the case of Fitton [9], or silicone rubber as in the case of da Silva and Lopes [5]. Figure 7 illustrates the placement of the adhesive on one adherend. Sancaktar and Kumar [24] used only one adhesive and placed carboxylterminated butadiene acrylonitrile (CTBN) rubber particles at the end of the overlap in order to locally reduce the stiffness of the adhesive. Pires et al. [22] used two differing epoxy adhesives but neglected to control the separation of the adhesives. Only marginal gains were achieved (at best 20% increase in joint capacity) in either *Pires* or *Sancatar* and *Kumar's* work. Fitton [9] suggested this was because the joint capacity was limited by adherend yielding, which he demonstrated in tests performed on joints similar to that of Pires et al. [22]. He compared aluminium–aluminium, FRP–FRP and aluminium–FRP overlap joints. The total overlap length was 20 mm, which was split into (5–10–5) mm regions. Figure 8 compares the joint strengths of the mixed-adhesive joints with those made using each of the adhesives alone. Improvements were marginal for the metallic joints and all were limited by yielding of the aluminium, see Fig. 9. With the exception of the low modulus adhesive only, the FRP joints exhibited a marked increase in joint strength.

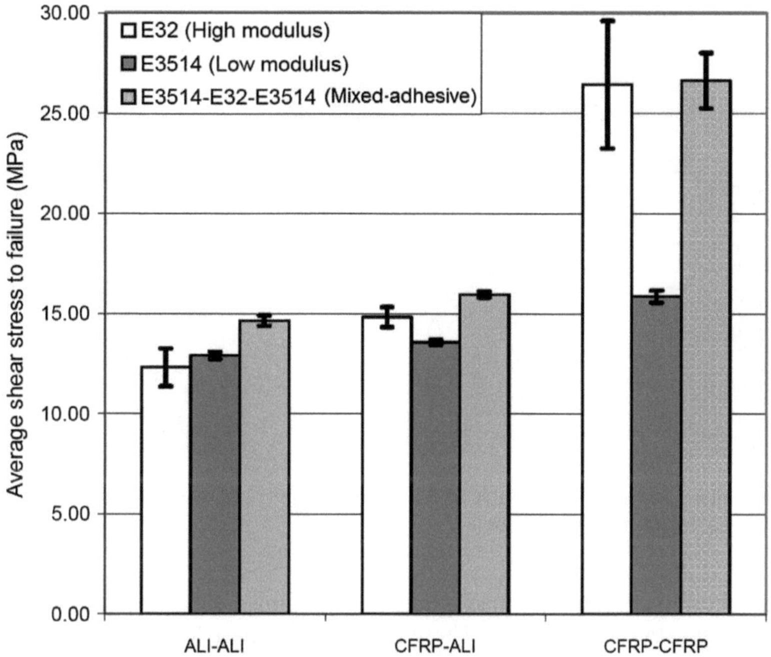

Fig. 8 Joint strengths of mixed-adhesive and single adhesive joints

Fig. 9 Failure surfaces of CFRP–aluminium lap joints (*left*: LM adhesive only; *middle*: HM adhesive only; *right*: mixed-adhesive)

In the case of the single high stiffness adhesive, the capacity of the FRP–FRP joints was limited by the adherend's through-thickness strength. Figure 10 clearly illustrates delamination of the FRP. In contrast, the mixed-adhesive joints attained the same strength but failed cohesively within the bondline. It appears that the

Fig. 10 Failure surfaces of CFRP–CFRP lap joints (*left*: LM adhesive only; *middle*: HM adhesive only; *right*: mixed-adhesive)

mixed-adhesive combination had attained the ultimate shear capacity of the high modulus/high strength adhesive. These two sets of data reflect a situation similar to the top of the curve in Fig. 6, exhibiting a transition region from cohesive failure in the adhesive through to through-thickness failure of the adherend, resulting from a given stiffness of the adhesive and strength of the adherend.

Following on from these tests, Fitton and Broughton [10] modified the surface strength of the FRP to demonstrate the capability of an epoxy mixed-adhesive bondline to reduce the peak stress concentrations at the overlap ends. By grit-blasting the surfaces of two different FRP composites with different grit media and pressures the surface was effectively damaged (in a consistent manner), reducing the through-thickness strength of the FRP. The overlap was increased from 20 to 50 mm, mainly to ease assembly of the joints. The single adhesive bondline, containing a high modulus adhesive, failed at much lower loads than previously due to the low surface strength of the composite. In the case of the mixed-adhesive bondline the strength more than doubled exhibiting a mostly cohesive failure within the adhesive, see Fig. 11. Moreover, the average shear stress of the bondline attained the maximum shear strength of the high modulus adhesive (28 MPa), which suggests that the joints had reached full shear capacity. Indeed, Fitton [9] postulated that the upper shear capacity of a mixed-adhesive (τ_{ult}) joint can be predicted by simply substituting in the values for adhesive shear strength ($\tau_{\max}$) and area (A) of the individual [high modulus (HM) and low modulus (LM)] adhesives into the equation below:

$$\tau_{\text{ult}} = \frac{(\tau_{\max}A)_{\text{HM}} + (\tau_{\max}A)_{\text{LM}}}{A_{\text{Total}}} \tag{1}$$

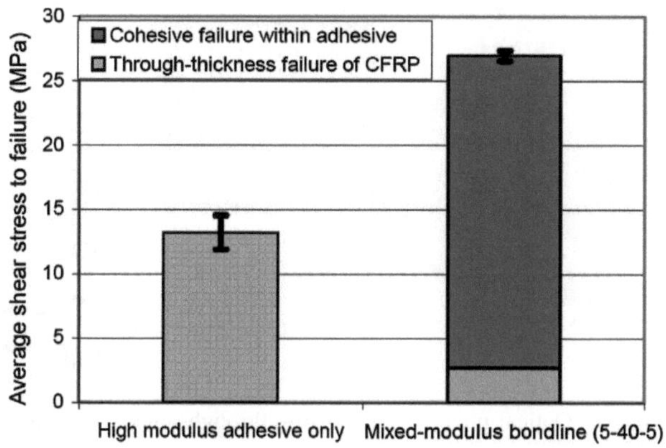

Fig. 11 Comparison of single versus mixed-adhesive bondline shear strength (also indicating percentage failure mode)

Table 1 Experimental versus predicted joint capacity

Overlap (mm)	Configuration	Predicted shear capacity (MPa)	Average shear strength (MPa)
50	High modulus only (HM)	27.6	26.5 ± (1.42)
50	5–40–5	25.3	26.2 ± (1.33)
50	10–30–10	23.0	22.9 ± (1.65)
50	15–20–15	20.7	20.4 ± (1.03)
50	Low modulus only (LM)	16.1	16.3 ± (0.14)

For the above expression to be true the stress concentrations at the ends of the overlap must be eliminated. Fitton and Broughton [10] attempted to demonstrate this by varying the ratio of the two adhesives along the bondline. The results, shown in Table 1, closely matched the estimations obtained from the above expression. Perhaps more importantly, these results demonstrated the need to optimise the ratio of the two adhesives in order to maintain high joint strength.

da Silva and Lopes [5] evaluated the performance of mixed-adhesive single lap joints experimentally using two structural epoxy adhesives. High strength steel adherends were carefully selected to prevent any yielding during loading. An increase of 52% was achieved over the better of the single-adhesive joints. Analytical predictions of strength were in close agreement with the experimental results using a method similar to Fitton and Broughton [10], based on global yielding, but also included a Volkersen type stress distribution for the central brittle adhesive, assuming the brittle adhesive would not display any plastic behaviour.

4 Failure Modes

The fractured joint specimens shown in Figs. 9 and 10 reveal the extent to which the failure mode can be altered using mixed-adhesive lap joints, subject to the adherend materials employed. Representative specimens of CFRP–aluminium and

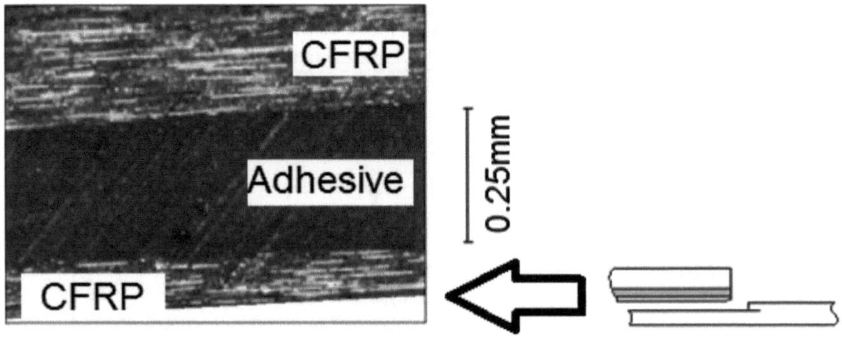

Fig. 12 Optical microscopy showing through-thickness failure of CFRP–CFRP joint

CFRP–CFRP are shown and correspond to the strengths illustrated in Fig. 8. In the case of a single low modulus adhesive, E3514, the ultimate capacity of either adherend combination was limited by cohesive rupture of the adhesive. The low modulus adhesive, although performed poorly, had actually attained its ultimate shear strength (16 MPa). The load capacity of joints made with a single high modulus adhesive, E32, was clearly limited by adherend failure, either initiated by aluminium yielding or fracture of the composite. The CFRP–CFRP joints provided the highest strength but showed the smallest improvements over a single adhesive. The mixed-adhesive bondline mainly failed in a mixed interfacial/cohesive failure in the adhesive. There was almost no visible through-thickness failure. Notably, in the case of the single high modulus adhesive joints the shear capacity was well below that of the adhesive when at least one of the adherends was aluminium. Nevertheless, the CFRP–CFRP joints produced a similar failure load to that of the mixed-adhesive combination and corresponded to a joint strength similar to that of the adhesive shear strength (28 MPa), albeit exhibiting failure within the composite. Optical microscopy of the failure, shown in Fig. 12, demonstrated through-thickness failure of the composite located at a significant distance from the adhesive layer.

Clearly, the through-thickness (or surface interface) strength of the adherend is an important factor in determining the strength of a joint. Fitton [9] used various abrasive treatments to affect the surface interfacial strength of two different composites (CFRP-1 and CFRP-2), as shown in Table 2, which indicates the pull-off strength recorded as a direct result of the various pretreatments.

Figure 13 shows the average shear stress to failure of the joints against the pull-off strengths of the composite. This is shown for joints made with the high modulus adhesive (E32) alone, the low modulus adhesive (E3514) alone, and a mixed-adhesive combination. Also included in this figure is the ultimate shear strength of each of the adhesives, which was determined from the thick-adherend shear test (TAST). The numbers above this figure correspond to the surface preparations given in Table 2. Joints made with the mixed-adhesive bondline reached a stress to failure almost equal to the TAST joint strength of the high

Table 2 Average pull-off strengths of adherends

	Composite and surface preparation	Pull-off test-average surface strength (MPa)	Coefficient of variation
1	CFRP(1)—P60 gritblast	2.03	0.09
2	CFRP(1)—P180/220 gritblast	3.93	0.18
3	CFRP(2)—P60 gritblast	5.05	0.23
4	CFRP(2)—P180/220 gritblast	6.24	0.19

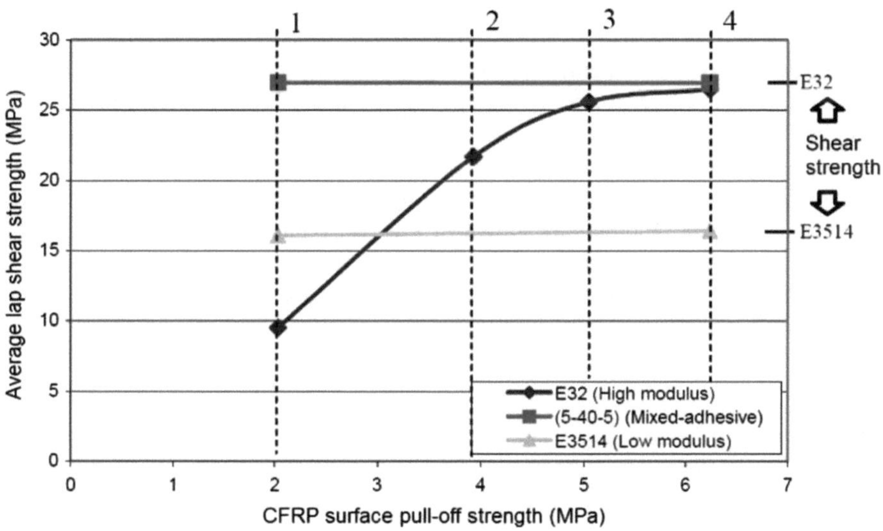

Fig. 13 Lap shear strength versus adherend pull-off strength

modulus adhesive, E32. This was achieved for composites exhibiting both the highest and weakest surface interfacial strengths. Again, this suggests that the mixed-adhesive combination reduced the stress concentrations in the adherends sufficiently to prevent premature failure in the through-thickness direction, allowing the adhesive's ultimate shear capacity to be obtained. In the case of the joints made with the composite exhibiting the weakest surface interfacial strength, the average shear stress to failure was increased by over 100%, compared with the single high modulus adhesive, or 75% increase compared to the low modulus adhesive. Clearly, it is sometimes possible to obtain higher joint strengths using a lower modulus adhesive when the adherend is weak in the through-thickness direction, despite the adhesive having lower strength. However, as discussed earlier, the strength of a joint made with a low modulus adhesive is usually limited by a lower adhesive shear strength.

5 Modelling Approaches

Modelling approaches presented in the literature have either considered relatively simple theoretical approaches, in which the shear stress distribution along the overlap is defined by distinct modulus changes at the relevant position along the overlap, or, numerical methods have been employed with varying degrees of complexity. The former approach has, in many cases, concentrated on shear stresses alone, while the latter has generally reported the effects of both shear and peel stresses.

5.1 Analytical

The earliest theoretical attempt to investigate varying the modulus of an adhesive along the overlap was reported by Raphael [23]. The work was undertaken as part of a programme to design and test bonded rocket motor cases. The aim was to obtain the highest possible joint strength for a simple overlap. Raphael approached the problem initially with a qualitative analysis by splitting the bondline up into a finite number of discrete elements as shown in Fig. 14. Each of the adhesive elements was described by their length $L_i - L_{i-1}$ and the adhesive shear modulus G_i. Considering small slices through the joint the shear stress distribution is given by:

$$\tau_i = a_i + b_i \cosh\left(\frac{8G_i}{ETt}\right)^{0.5} x + c_i \sinh\left(\frac{8G_i}{ETt}\right)^{0.5}_{-} x \tag{2}$$

The three unknown coefficients a_i, b_i and c_i were solved using equilibrium conditions and the continuity of strain theory, E is Young's modulus of the adherends, T and t were the adherend and adhesive thickness respectively. Continuity of strain dictates that the strain at the interface between the adhesives must be the same in each adhesive. The boundary conditions employed to determine the unknown coefficients give rise to similar limitations to that found for Volkersen [28]. Figure 15 shows the shape of the resultant elastic stress distribution for a three-step mixed-adhesive elastic analysis using the method given by Raphael. The theory suggests placing a more flexible adhesive in the areas of high stress should

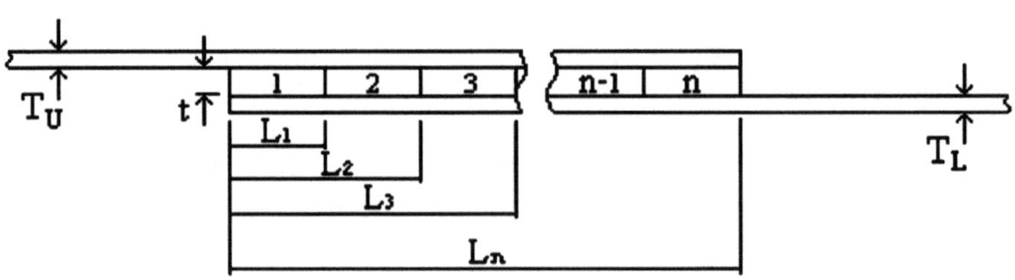

Fig. 14 Graded modulus bondline [23]

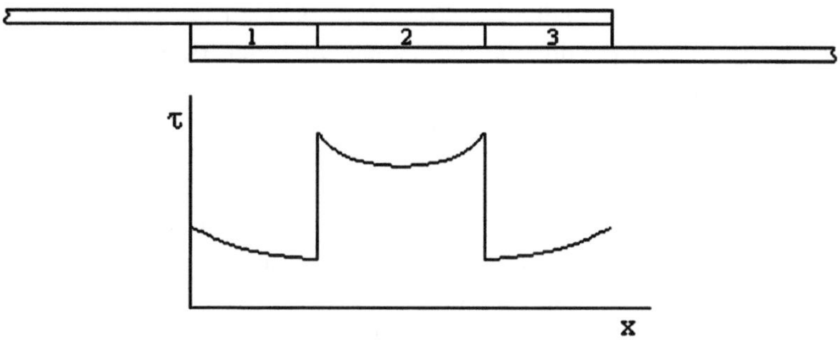

Fig. 15 Three step mixed-modulus bondline [23]

increase joint strength. Raphael also pointed out that in reality the analysis was more complex than this as the stress analysis and adhesive selection are mutually dependent (e.g. it is not possible to select the required adhesives without knowing the stress distribution, but the stress distribution depends on the selected adhesives' properties). Raphael also considered how to optimise a mixed-adhesive joint for maximum strength. He suggested that the ideal joint would vary the properties of strength and stiffness throughout the joint as required, so that the shear stress in the bondline attained its maximum throughout the joint at the same time. Raphael's work showed the possible benefits of using a mixed-modulus bondline, but did not consider the effect on peel stresses.

Semerdjiev [26] reported the strength of simple lap joints could be increased by approximately 20% by varying the adhesive rigidity along the overlap length. It is assumed that this value is purely speculation or from some non-reported research and refers to work undertaken by Hauch [16], which in turn refers to Raphael [23]. Other references to the mixed-adhesive bondline concept have been found in composite materials textbooks such as Schwartz [25], though again this makes reference to an earlier work cited in DARCOM-P [6]. The concept of using more than one adhesive to create a 'graded seal' was also mentioned by Patrick [21]. It was suggested as a solution for joining of materials with dissimilar stiffness and coefficients of thermal expansion. Possible benefits were discussed and the concept was recommended for further research. Semerdjiev, Hauch, Patrick and Raphael did not report any experimental work nor, indeed, quantify the possible benefits of a variable modulus bondline.

More recently, theoretical studies have been conducted by Fitton and Broughton [10] for single lap joints and tubular adhesive joints with a functionally graded bondline by Kumar [20]. The latter is based on a variational principal that minimized the complementary energy of the bonded system. Shear and peel stresses were determined along the bondline providing a means to optimise the joint strength accordingly. das Neves et al. [7] provided a detailed analytical model for both single and double lap joints with one or more adhesives along the overlap. The main focus of the work was to investigate the improvement of joint strength at high and low temperatures, and notably accounted for the variation of peel stresses through the thickness of the adhesive. To include peel stresses the adherends were modelled as linear elastic thin beams subject to Euler–Bernoulli assumptions. Shear deformation of the adherends was also taken into account based upon Timoshenko's beam theory, and non-linear geometric effects were included by considering transverse displacements in the bending moment expression outside the overlap region. Stress distributions throughout the overlap, which including different adhesives, were found to be in good agreement with a FE analysis. Parametric studies on the effect of differing ratios of the adhesive stiffness on peel stress were undertaken for both single and double lap joints in a following paper, das Neves et al. [8]. These are reported in more detail in the following chapter. Notably, the peel stresses generated in the single lap joint configuration were determined solely by the lower modulus adhesive, conferring with earlier work by Fitton and Broughton [10]. A further observation to be made from the numerical results suggests that the ratio of low to high adhesive modulus (E_{LM}/E_{HM}) should be approximately 0.2. In the double lap joint configuration 0.2 appeared to be the limit below which the peel stresses increased. Plots of shear stress against adhesive modulus ratio also demonstrated it was possible to obtain equal shear stresses in both adhesives at the transition point between the two adhesives (giving a ratio of 0.32 for the materials and geometry combinations employed). The experimental work undertaken by Fitton and Broughton [10] included a mixed-adhesive combination with a E_{LM}/E_{HM} ratio approximately equal to 0.14.

da Silva and Lopes [5] reported an expression for the failure load (P) of a mixed-adhesive lap joint based on global (shear) yielding of the ductile adhesive (τ_y) and maximum shear stress of the brittle adhesive (τ_r), where:

$$P = (\tau_y bl)_{\mathrm{LM}} + \left[\tau_r \frac{2bl\sinh(\lambda l)}{\lambda l(1+\cosh(\lambda l)}\right]_{\mathrm{HM}} \quad \text{and} \qquad (3)$$

$$\lambda = \frac{G}{t_a}\left(\frac{2}{Et_s}\right)$$

The notation refers to overlap length (l), joint width (b), adhesive shear modulus (G), adhesive thickness (t_a), adherend Young's modulus (E), and adherend thickness (t_s). Failure loads were reported to be in close agreement with experimental mixed-adhesive joints. Predictions for the single-adhesive lap joints were less accurate which, to a large extent, reflects the limitations of not including peel stresses in the calculations when brittle adhesives are employed at the ends of the overlap.

5.2 Numerical Analyses

Srinivas [27] investigated the application of combined flexible and stiff adhesives in the bondline using a numerical technique in which the material properties were assumed to be linear elastic. It was reported that the maximum peel and shear

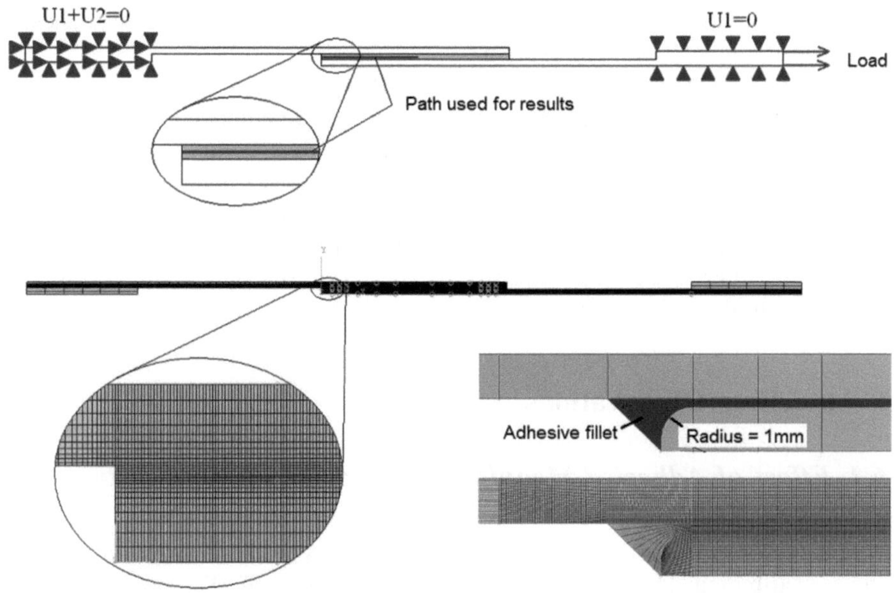

Fig. 16 Finite element model of mixed-adhesive single lap joint (square or modified fillet) geometry

stresses in a bondline consisting of both flexible and stiff adhesives were less than the stresses in a joint made from the single adhesives alone, confirming earlier analytical approximations. Srinivas did not consider the strength of the adhesives as previously discussed by Raphael [23]. Instead Srinivas concluded that optimum lengths of stiff and flexible bonds could be chosen to assure the lowest possible stresses in the bondline. Despite this, the matching of stress distribution to adhesive strengths recommended by Raphael seems to be a more reasonable approach as the strengths of stiff and flexible adhesives can vary greatly. As in a number of other papers the analysis was not validated by experiment.

Fitton [9] carried out a numerical analysis of mixed-adhesive bondlines using the ABAQUS 6.4 finite element (FE) code. The joints were modelled using 2D plane strain quadratic elements (CPE8), linear and non-linear material properties for the adhesive and linear orthotropic material properties for the CFRP adherends. Importantly, large displacement theory (NLGEOM) was employed in order to properly accommodate the adherend rotation under load that generates the peel stresses. Figure 16 shows the boundary conditions and meshes (including a fillet) at one end of the overlap region. Apart from demonstrating the variation in shear and peel stresses according to the adhesive stiffness the study also focused upon understanding the optimal distance (amount) the lower modulus adhesive should extend into the overlap. This was found to be critical in determining the effectiveness of the mixed-adhesive placement in the joint along with the ratio of adhesive stiffness (E_{LM}/E_{HM}). Significantly, the peak peel stress was solely determined by the lower modulus adhesive, given that the low modulus adhesive extended far enough into the overlap.

da Silva and Adams [4] also used the ABAQUS FE method to predict the strength of mixed-adhesive titanium–titanium and titanium–CFRP double lap joints. The criterion for joint failure was initial yielding of any part of the joint. This included shear yielding of the adhesive (mid-plane), interlaminar (tensile) failure of the composite and tensile yielding of the adherends at the overlap. For large temperature variations and high temperatures the mixed-adhesive joints performed better when mixed adherends were used, but only some benefit was found for similar-adherend joints at low temperatures.

6 Modelling Observations

6.1 Effect of Adhesive Modulus on Shear Stress Distribution

Figure 17 shows the effects of increasing the modulus of the adhesive on the shear stress distribution, located in the end regions of the joint. The modulus of the adhesive in the centre is fixed. As the modulus is increased at the ends of the joint the more shear stress is carried by that adhesive and less is carried by the high modulus adhesive in the centre of the joint. As shown in Fig. 17b the lowest shear stresses in a joint are obtained when the peak shear stress in each adhesive is equal.

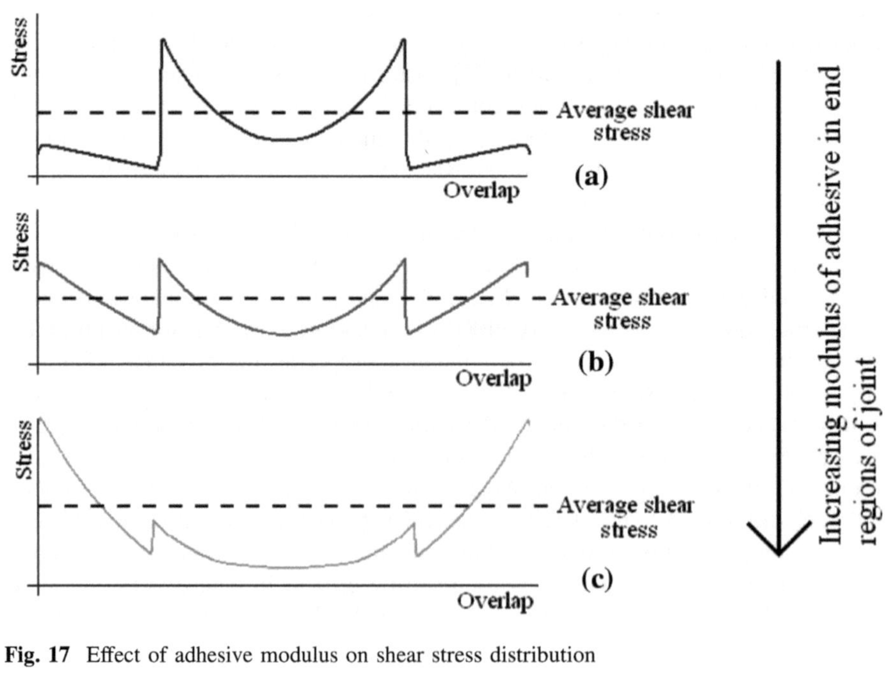

Fig. 17 Effect of adhesive modulus on shear stress distribution

This was suggested by Raphael [23] and numerically presented by das Neves et al. [8]. Whilst this might be the optimum situation to obtain the lowest shear stress distribution it is not the optimum for joint strength, given the earlier observations that strength generally reduces as the adhesive modulus reduces. Therefore, a mixed-adhesive joint designed to have equal stresses in each adhesive is likely to fail *prematurely* in the low modulus adhesive. The optimum stress distribution for joint strength would, therefore, look more like the example in Fig. 17a, where the shear stresses are shown to be lower in the low modulus adhesive.

Fitton [9] showed that even this assumption has to be treated with caution when both non-linear material and geometric properties are taken into account along with the proportion of the two adhesives employed. Combined, these factors have a significant effect on the peak shear stress in each of the adhesives, as can be seen in Fig. 18, which is taken from the FE analysis carried out at low (5 kN) and high (25 kN) loads respectively. Once the adhesive yields, the stress in that area is effectively limited. At low load, where there is only a small amount of material non-linearity, the peak shear stress in the high modulus adhesive is effectively lower if more low-modulus adhesive is employed. At a higher load the trend is reversed and the joint with the least amount of low modulus adhesive (5–40–5) has the lowest peak stress. Taking into account the ratio of the actual adhesives' ultimate shear strengths ($\tau_{\mathrm{ULT\text{-}LM}}:\tau_{\mathrm{ULT\text{-}HM}}$), which was 1:1.7 in this case, the ratios of peak shear stresses from the FE analysis were significantly higher than the strength ratio of the adhesives; between 1:2 and 1:4 at low load and 1:16 at high load. In effect, the high modulus adhesive is carrying most of the shear stress and would fail first in shear, indicating the shear stress distribution was not optimal. In fact, the peak shear stress in the 5–40–5 mixed-adhesive joint at 25 kN was actually higher than that of the high modulus adhesive alone. However, the shear stress distribution was not the primary concern for optimisation of the joints used in this study as the mode of failure in the un-optimised joints was predominantly through-thickness failure of the composite. If this joint was to be optimised for shear it is clear that the modulus of the adhesive used in the end regions of the joint must be increased. As mentioned, balancing shear stresses with the strengths of the adhesive is a difficult task. The importance of the non-linear behaviour of the adhesive and joint has shown that it is not sufficient to compare joints at a load less than the expected failure load of the joints, due to the significant change in stress distribution as the adhesive yields.

6.2 Effect of Adhesive Modulus on Peel Stress Distribution

Linear elastic FE analyses have shown that the peak peel stress at the end of the overlap is determined by the modulus of the adhesive used in the end region of the joint. In addition to this Fitton [9] showed that when an insufficient amount of low modulus adhesive was used in a mixed-adhesive joint, the peak peel stress re-located to the point of modulus transition. To avoid this, he suggested that

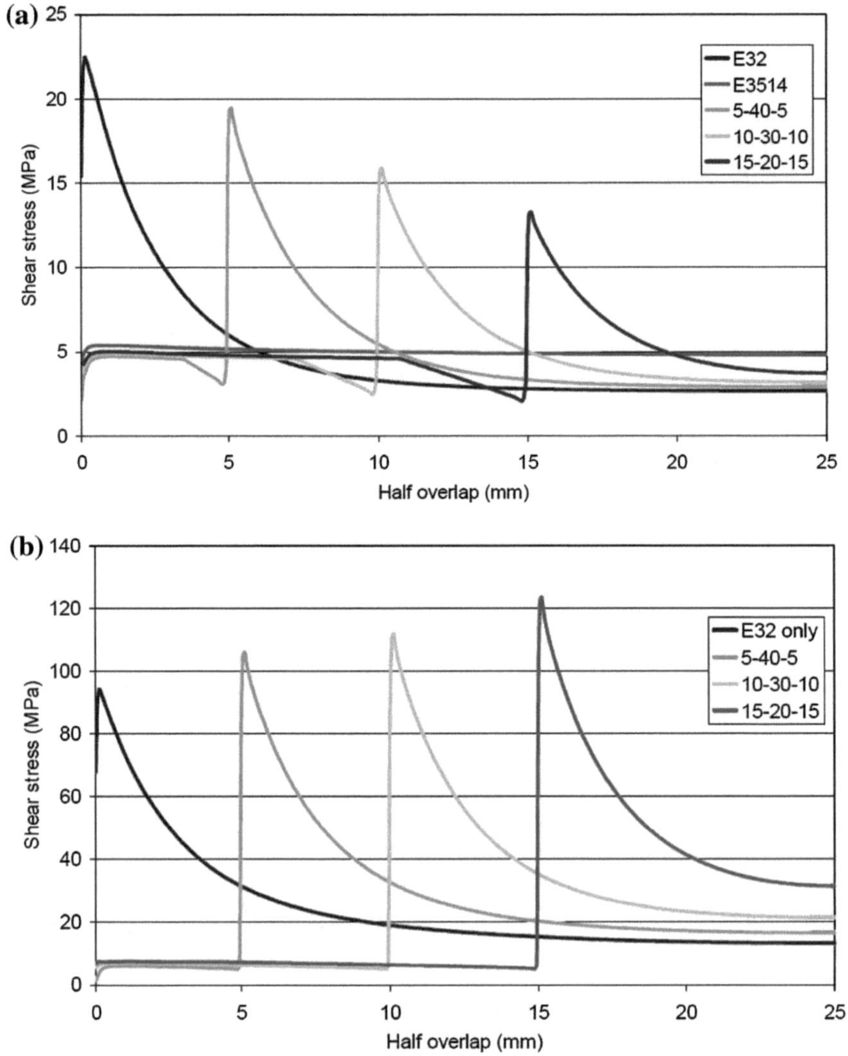

Fig. 18 Shear stress distributions at low (**a**) 5 kN and high (**b**) 25 kN loads

the transition between the low and high modulus adhesives should be positioned in the region where compressive stresses would normally be present for a single high modulus adhesive, as shown in Fig. 19.

When the analysis was extended to include non-linear behaviour of the low modulus adhesive, as shown in Fig. 20, the modulus transition point also has some effect on the peak peel stress in the joint, according to the magnitude of the load. Notably, this is not as dramatic as it was in the case for the shear stress distribution. For the mixed-adhesive joints with a low–high–low modulus distribution of 15–20–15, the peak peel stress was relatively unchanged irrespective of

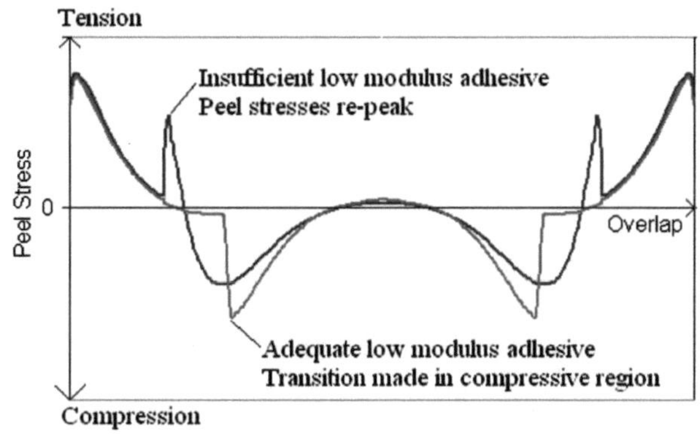

Fig. 19 Effect on modulus transition location on peak peel stress

the magnitude of the load (5 or 25 kN). In contrast, the mixed-adhesive joints with a 5–40–5 distribution exhibited a doubling of the peel stress at the higher load. This shows that the reduction in peak peel stress from a linear-elastic analysis will be under estimated if the low modulus adhesive used exhibits significant non-linear behaviour. Fitton [9] attempted to demonstrate the effect of reducing the peel stress using adhesives–adhesive CFRP joints with varying amounts of the two adhesives. The joint capacity decreased as the shear capacity of the lower modulus adhesive became more prominent (see Fig. 21), matching Fitton's simple analytical strength predictions discussed earlier. As the content of the low modulus adhesive increased, much less through-thickness composite failure was observed, confirming a reduction in peel stress.

6.3 Comparison of FE Analysis with Modified and Unmodified Model Geometry

Real joints tend to have small adhesive fillets and the FE technique generates stress singularities when a squared overlap is modelled, making it impossible to ascertain the true stresses at the ends of the overlap. Indeed, all of the FE analyses mentioned above were not predictive but comparative at best with respect to analytical solutions, as the FE method is inherently mesh dependent. Fitton [9] undertook a method to get around this that included a small fillet at the end of the adhesive overlap and a rounded end to the unloaded adherend, essentially reducing the effect of the stress singularity. Figure 22 shows the resulting peel stress distribution of a single high modulus adhesive joint and a mixed-adhesive joint when this modification is included. The high stress concentration at the interface of the loaded adherend and adhesive fillet is clearly apparent in the single adhesive joint. This is what contributes to through-thickness failure in weak adherend materials. In contrast the peel stresses in the mixed-adhesive joint are located within the overlap region and spread over a larger area. More importantly, the magnitude of the peel stress is also less than half that found in the single adhesive joint, and lower than the transverse tensile strength of the composite.

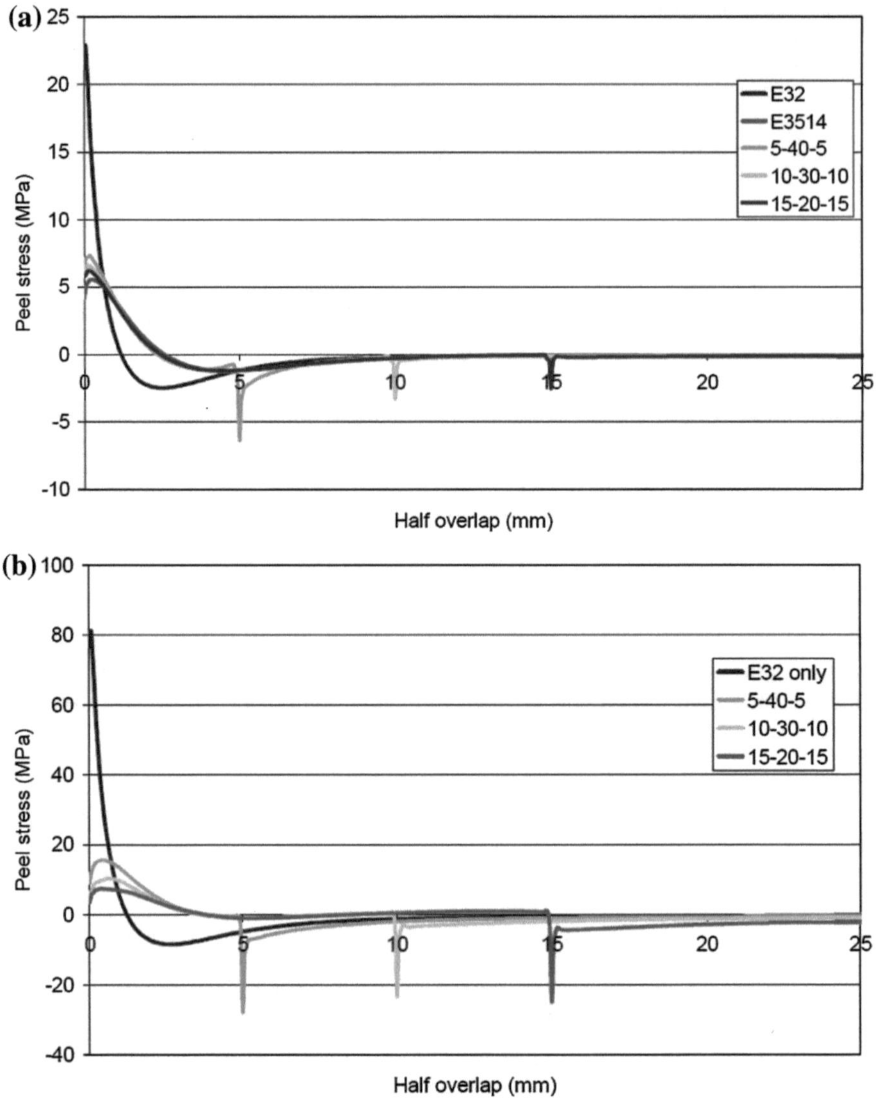

Fig. 20 Peel stress distributions at low (**a**) 5 kN and high (**b**) 25 kN loads

7 Conclusions

Various mixed-adhesive overlap joints have been investigated analytically and verified numerically in the literature. Only latterly have significant improvements in joint capacity been demonstrated experimentally; particularly for cases where the adherends were relatively weak in the through-thickness direction, as is the

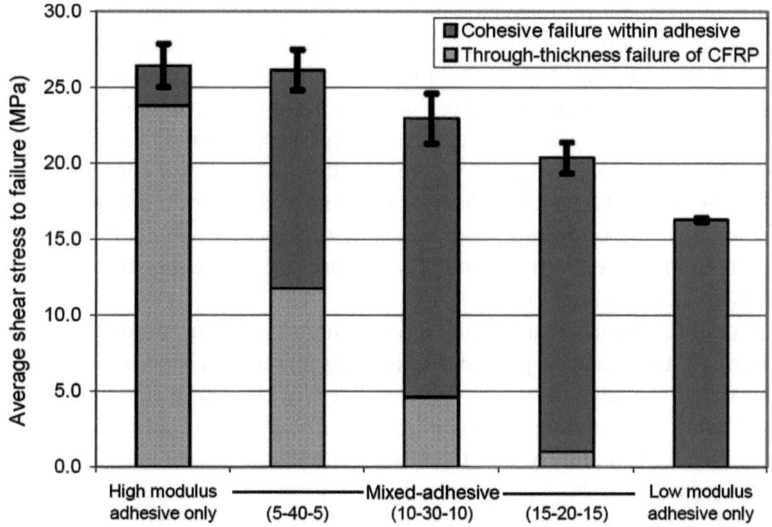

Fig. 21 Failure modes and joint strengths of CFRP single lap joints with varying mixed-adhesive content (also indicating percentage failure mode)

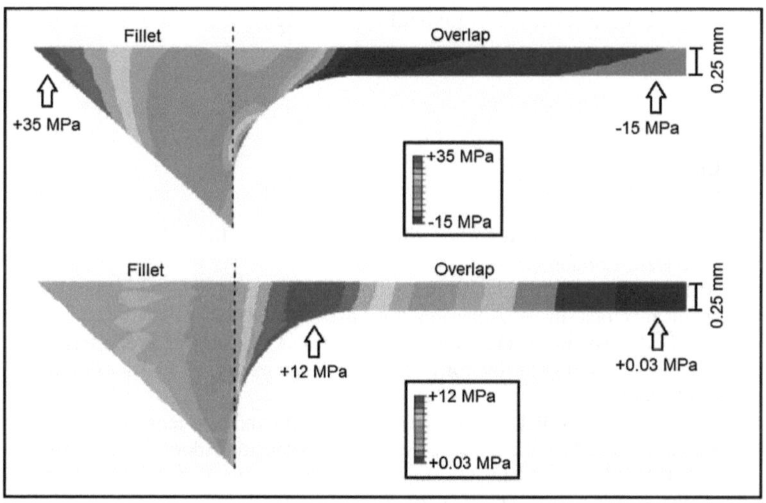

Fig. 22 Peel stress at end of modified overlap for single (*top*) and mixed-adhesive (*bottom*) joints

case with FRP composites. In contrast, metallic mixed-adhesive joints have demonstrated limited improvements due to adherend yielding. In the case of FRP adherends, the correct combination of low modulus to high modulus adhesive has prevented premature through-thickness failure of the composite, allowing the combined shear strengths of both adhesives to be realised.

On the other hand, performance has been shown to be somewhat compromised should the mixed-adhesive joint not be optimally designed, leading to a reduction in the overall shear capacity of the joint. It is important, therefore, to select the appropriate combination of adhesives and adherend properties for the given joint. For example, by employing an excessively large proportion of the lower modulus adhesive along the outer regions of the overlap, the inherent low shear strength delivered by these types of adhesive will invariably limit the joint capacity. Because of this, the optimum shear stress distribution is not likely to be the one that results in the lowest distribution of stress but one where the strength of the two adhesives is met simultaneously.

Analytical and numerical studies have demonstrated the important parameters that contribute to improved joint strength using mixed-adhesive joints. Indeed, the ultimate shear capacity of adhesives employed in the same bondline have been achieved using this technique, involving joints made from adherends that otherwise would fail prematurely due to through-thickness failure.

Whilst significant short-term improvements have been demonstrated and key material and geometric parameters identified, future development and exploitation of the mixed-adhesive joint must focus on novel methods that generate variable stiffness along the adhesive bondline using only one generic adhesive formula. Until then this technique will rely upon the selective placement of at least two different adhesives, which ultimately leads to processing compromises during the cure process. These, and other, industrial requirements are discussed in more detail in Chap. 11.

References

1. Adams, R.D., Comyn, J., Wake, W.C.: Structural Adhesive Joints in Engineering, 2nd edn. Chapman & Hall, London (1997)
2. Adams, R.D., Harris, J.A.: Strength predictions of bonded single lap joints by non-linear finite element methods. Int. J. Adhesion Adhesives **4**(2), 65–78 (1984)
3. da Silva, L.F.M., Adams, R.D.: Measurement of the mechanical properties of structural adhesives in tension and shear over a wide range of temperatures. J. Adhesion Sci. Technol. **19**, 109–142 (2005)
4. da Silva, L.F.M., Adams, R.D.: Adhesive joints at high and low temperatures using similar and dissimilar adherends and dual adhesives. Int. J. Adhesion Adhesives **27**, 216–226 (2007)
5. da Silva, L.F.M., Lopes, M.J.C.Q.: Joint strength optimization by the mixed adhesive technique. Int. J. Adhesion Adhesives **29**, 509–514 (2009)
6. DARCOM-P: Engineering Design Handbook, Joining of Advanced Composites. DARCOM-P 706 316, U.S. Army Material Development and Readiness Command (1979)
7. das Neves, P.J.C., da Silva, L.F.M., Adams, R.D.: Analysis of mixed adhesive bonded joints—part I: theoretical formulation. J. Adhesion Sci. Technol. **23**, 1–34 (2009)
8. das Neves, P.J.C., da Silva, L.F.M., Adams, R.D.: Analysis of mixed adhesive bonded joints—part II: parametric study. J. Adhesion Sci. Technol. **23**, 35–61 (2009)
9. Fitton, M.D.: Multi-modulus adhesive bonding of advanced composite materials. Ph.D. Thesis, Oxford Brookes University, UK (2004)

10. Fitton, M.D., Broughton, J.G.: Variable modulus adhesives: an approach to optimised joint performance. Int. J. Adhesion Adhesives **25**, 329–336 (2005)
11. Fessel, G., Broughton, J.G., Fellows, N.A., Durodola, J.F., Hutchinson, A.R.: Evaluation of different lap shear joint geometries for automotive applications. Int. J. Adhesion Adhesives **27**, 574–583 (2007)
12. Goland, M., Reissner, E.: J. Appl. Mech. Trans. ASME **66**, A17–A27 (1944)
13. Gordon, T.L., Fakley, M.E.: The influence of elastic modulus on adhesion to thermoplastics and thermoset materials. Int. J. Adhesion Adhesives **23**, 95–100 (2003)
14. Hart-Smith, L.J.: An engineer's viewpoint on design and analysis of aircraft structural joints. J. Aerospace Eng. G IMechE **209**, 105–129 (1995)
15. Hart-Smith, L.J.: Adhesive bonded double lap joints. NASA CR-112235 (1973)
16. Hauch, J.E. (ed.): Answers to Five Common Adhesive Problems, Materials in Design Engineering (1966)
17. Hildebrand, M.: The Strength of Adhesive-Bonded Joints Between Fibre-Reinforced Plastics and Metals, Analysis, Shape Optimisation and Experiments, vol. 192. Technical Research Centre of Finland, VTT Publication (1994)
18. Hutchinson, A.R.: Joining of fibre-reinforced polymer composite materials. Project Report 46, Construction Industry Research and Information Association (CIRIA), London (1997)
19. Kinloch, A.J.: Adhesion and Adhesives, Science and Technology. Chapman & Hall, London (1987)
20. Kumar, S.: Analysis of tubular adhesive joints with a functionally modulus graded bondline subjected to axial loads. Int. J. Adhesion Adhesives **29**(8), 785–795 (2008) (December 2009)
21. Patrick, R.L. (ed.): Treatise on Adhesion and Adhesives—Structural Adhesives with Emphasis on Aerospace Applications, vol. 4. Marcel Dekker, Inc., New York (1976)
22. Pires, I., Quintino, L., Durodola, J.F., Beevers, A.: Performance of bi-adhesive bonded aluminium lap joints. Int. J. Adhesion Adhesives **23**, 215–223 (2003)
23. Raphael, C.: Variable-adhesive bonded joints. Appl. Polym. Symp. **3**, 99–108 (1966)
24. Sancaktar, E., Kumar, S.: Selective use of rubber toughening to optimize lap-joint strength. J. Adhesion Sci. Technol. **14**, 1265–1296 (2000)
25. Schwartz, M.M.: Composite Materials Handbook. McGraw-Hill Book Company, New York (1984) (Chapter 3)
26. Semerdjiev, S.: Metal to Metal Adhesive Bonding. Business Book Limited, London (1970)
27. Srinivas, S.: Analysis of bonded joints. NASA TN D-7855 (1975)
28. Volkersen, O.: Die nietkraftoerteilung in zubeanspruchten nietverbindungen mit konstanten loschonquerschnitten. Luftfahrtforschung **15**, 41–47 (1938)

Technology of Mixed Adhesive Joints

Lucas F. M. da Silva

Abstract The technique of mixed adhesive joints has been originally proposed more than 40 years ago. However, its practical implementation is still very limited. Aspects associated with the automotive, aeronautical and aerospace industries that could benefit from this technique are described. There are basically two types of problems that the mixed-adhesive technique can solve. The first is related to improvements of joint strength in relation to joints with stiff and brittle adhesives. The second is to extend the temperature range of adhesive joints. Results that illustrate the potentiality of this technique in terms of these two aspects are presented. Simple and elaborated predictive analytical models are described and practical manufacture solutions are proposed to ease the implementation of this technique in practice.

1 Introduction

Mixed-adhesive or mixed modulus joints have been proposed [20, 22, 24, 25] to improve the stress distribution and increase the joint strength of high modulus adhesives. This technique consists in using a stiff and strong adhesive in the middle of the overlap and a flexible and ductile adhesive at the ends of the overlap to relieve the high-stress concentrations at the ends of the overlap. As shown schematically in Fig. 1, this allows to have a more uniform stress distribution which leads to joint strength increases in relation to a stiff adhesive alone [2, 10, 20, 21, 25, 26]. Although this approach has been discussed theoretically, there

L. F. M. da Silva (✉)
Department of Mechanical Engineering, Faculty of Engineering of the University of Porto, Rua Dr. Roberto Frias, 4200-465 Porto, Portugal
e-mail: lucas@fe.up.pt

Adv Struct Mater (2011) 6: 283–309

DOI: 10.1007/8611_2010_28

Published Online: 2 July 2010

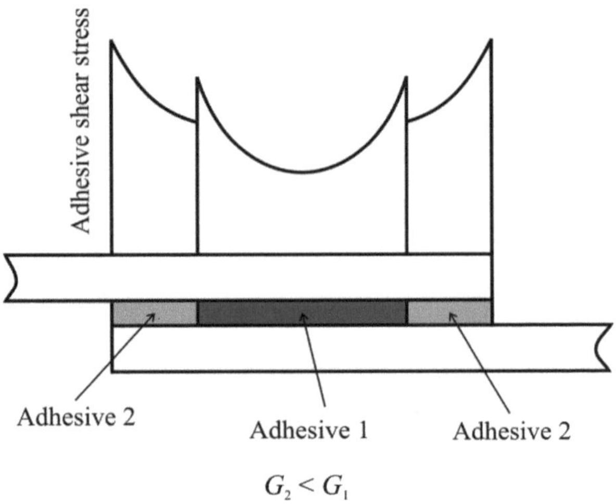

Fig. 1 Mixed adhesive joint concept for improved joint strength (*G* is the shear modulus)

have been relatively few published experimental demonstrations of a practical method that yields significant improvements in the joint performance.

The technique of using multi-modulus adhesives has been used for increasing the joint strength of single adhesive joints, although the published work is scarce and needs more development. In effect, that concept has not been extended to solve the problem of adhesive joints that need to withstand low and high temperatures. The first to recognise the use of the mixed adhesive joint for low and high temperature applications was Hart-Smith [15]. At high temperatures, a high temperature adhesive (HTA) in the middle of the joint retains the strength and transfers the entire load while a low temperature adhesive (LTA) is the load bearing component at low temperatures, making the high temperature adhesive relatively lightly stressed. Such a mixed adhesive joint would have a good strength from low to high temperatures as shown schematically in Fig. 2.

This chapter describes specific requirements in industry (aeronautical, aerospace and automotive) for the application of the mixed-adhesive technique. The adhesive to apply along the overlap may vary from stiff to ductile. Also, when temperature is involved, the variation of the adhesive properties with temperature and the thermal stresses should be taken into account. The manufacture is an important issue because it is necessary to avoid mixing of the adhesive. Results of joint strength and durability are described to illustrate the potentiality of this technique. The failure mechanisms associated with the mixed-adhesive technique are described. Finally, some examples of applications are discussed.

2 Specific Requirements in Industry

The mixed-adhesive technique may be useful in a variety of industrial applications. Examples in the automotive, aerospace and aeronautical industries are

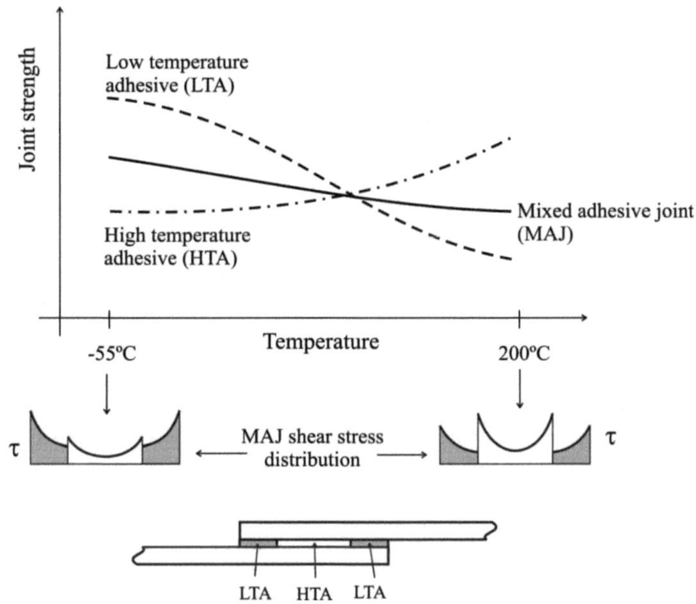

Fig. 2 Mixed adhesive joint (MAJ) concept for low and high temperatures

given next, with the specific requirements associated with each case. There are basically two types of problems that the mixed-adhesive technique can solve. The first is related to improvements of joint strength in relation to joints with stiff and brittle adhesives. The second is to extend the temperature range of adhesive joints.

The transport industry is concerned with fuel consumption. If the weight can be reduced, then there is less fuel consumption. However, this requires light multi-material structures that are difficult to assemble using traditional methods of joining such as welding. Adhesive bonding can successfully bond different materials and is therefore increasingly being used. Basically, the vehicle needs strength, stiffness, low cost and security for passengers. Impact tests are not so important for airplanes because there are few events. However, for cars, the probability of impact loads is high. Previous results show that adhesively bonded joints with metals can allow for large energy dissipation in impact conditions [1, 14]. The adhesive has to retain its strength so that the metal can dissipate the energy through plastic deformation. For that, adhesives must be ductile and not brittle. It is known that flexible and ductile adhesives are more suitable for impact behavior. However, rigid and strong adhesives are necessary for the structure stiffness and static strength. One possibility is to bond with two adhesives, one that gives stiffness and strength and another one that gives flexibility and impact strength.

Damage to aircraft is more common than usually thought and appears in various forms of varying severity. 'Ramp rash' is a general term used to describe impacts against the surface of the aircraft occurring while on the ground. Hail damage occurs when the aircraft flies through hail storms. The aeronautical industry also faces problems of ageing such as fatigue and corrosion that require a particular attention. The metallic components of the aircraft develop cracks under fatigue loading or stress corrosion. These defects initiate in places of high stress concentration such as rivet holes or non-uniform geometry. Repair of the damage is usually possible by means of a patch applied over the crack. This type of repair is very common and is economically very important, as it allows the safe operation of damaged or older aircraft for an increased amount of time. The current repair technique for aircraft with aluminium fuselages uses riveting to install the repair patch. In this procedure, the aluminium patches are placed over the damaged zone and completely riveted to the aircraft skin. The rivets are evenly distributed over the patch to provide better connection between the patch and fuselage while ensuring lower stress concentrations. Adhesive patches have gained popularity in the last decade as the use of composite parts increased. Suited to almost every material and able to handle complex geometries, this technique is known for its versatility and ease of application. The repair patch acts as a bridge between the two sides of the cracked component, transferring the loads over a gap. The design of the adhesive layer and the patch is optimized to smooth the transfer of load. The geometry of the joints present in the borders of the patch influences directly the behaviour and durability of the patch. One of the main functions of the joint is to avoid the peeling of the patch. This can be achieved in various ways. In many cases this is done by tapering the surfaces of the patch, but tapering the adhesive is also a possibility. The main idea is to lower the stiffness at the ends of the overlap for a smoother load transfer. Mixed modulus joints are another possibility to improve the stress distribution and increase the joint strength of high modulus adhesives.

Adhesive joints used in supersonic aircraft need to withstand low and high temperatures. Typically from −55 to as much as 200°C or more. As an example, if a supersonic aircraft goes to mach 2.7, the air friction at this speed would generate a surface temperature of about 232°C [16]. Adhesives suitable for high temperatures are generally too brittle for use at low temperatures. Owing to their brittleness and high stiffness, the strength in such a joint is often poor at low temperatures [4]. On the other hand, adhesives suitable for low temperatures are usually too weak or degrade at high temperatures. It may eventually be possible for chemists to develop an adhesive which will exhibit a good joint strength from −55 to 200°C, but this is unlikely in the near future. One solution therefore is to make a joint with a combination of two adhesives, consisting of a LTA which is tough at low temperatures and has a high modulus, but which does not degrade at the highest required operating temperature, and a HTA which has high strength at high temperatures but may be very brittle at low temperatures (see Fig. 2).

One of the applications of adhesive bonding in the aerospace industry is in the bonding of the ceramic tiles for temperature protection in the space shuttle. The recent disaster where some tiles detached from the internal aluminium fuselage of the shuttle shows that the technique of adhesive bonding still needs significant research. The adhesive used to bond the ceramic tiles is a RTV (room temperature vulcanizing) silicone rubber. The strength of that adhesive is just enough to hold the tiles in place at the extreme temperature reached in the re-entry of the space shuttle into the atmosphere due to the air friction. The RTV can give a bond that accommodates the different mechanical and physical behaviour of the tile and the metal, but has a low strength at high temperatures. An adhesive with a good strength at high temperatures would be too brittle at −50°C, which is the temperature attained when the shuttle is in orbit. One way to solve this problem is by bonding the tiles using two adhesives. A high temperature adhesive with a higher strength than the silicone adhesive would guarantee a better strength at high temperatures during the space shuttle re-entry into the atmosphere and the silicone adhesive would be the load carrying adhesive when the space shuttle is in orbit.

3 Adhesive Characteristics Required

Usually an increase in adhesive ductility is obtained at the expense of a decrease in strength and it is not clear where the optimum comprise is in terms of joint strength. Mixed adhesive joints can offer the best combination of strength and ductility by the use of two different types of adhesives. Various studies show that the mixed adhesive concept has little merit in comparison with a ductile adhesive alone, but gives significant joint strength improvements over a brittle adhesive alone [6, 25]. However, this depends on the type of ductile adhesive used. It is shown in Sect. 5 how the joint strength of mixed adhesive joints varies when different ductile adhesives of increasing ductility (and decreasing joint strength and stiffness) are used at the ends of the overlap.

The mechanical properties of an adhesive depend on the operational temperature in relation to the glass transition point. Below the glass transition temperature (T_g), as the temperature decreases, the modulus and strength increase while the ductility decreases. At T_g there is a rapid reduction in the modulus and strength as the temperature increases, and the adhesive can no longer carry a substantial load. An optimal joint for low and high temperatures can be obtained as long as the adhesive is loaded in the temperature range for which it has been designed. In a mixed adhesive joint at high temperatures, the LTA may be very ductile and creep, but it must not degrade. On the other hand, if the HTA fails at low temperatures, then it cannot carry any load at high temperatures. There is an uncertainty for each adhesive outside their range of application. Therefore, the mechanical properties need to be measured, from low to high temperatures, for both types of adhesives. To ensure that the mixed adhesive joint performs at high and low temperatures, and at temperatures in between the extremes, optimisation of the dimensions is necessary.

4 Manufacture

The first problem to solve when using two different adhesives with different cure schedules is what cure temperature to use. An intermediate temperature could be found which would be a compromise between the two adhesives. For mixed adhesive joints, da Silva and Lopes [7] recommend using the lowest recommended cure temperature of the two adhesives to avoid degrading the adhesive with the lowest cure temperature. The curing time was determined, using the adhesives technical data sheets, so that complete curing of the two adhesives was accomplished. However, stiff adhesives may not cure at low temperatures. In such a case, the cure of the stiff adhesive must be followed and tests must be done to make sure that the temperature does not degrade the flexible adhesive [5].

To guarantee that the two adhesives do not mix when the pressure is applied, it is necessary to have a physical separation. Pires et al. [21] did not use any separation and had problems to control the overlap length relative to each adhesive because these mixed under the application of pressure. A gap would be preferable to a physical barrier such as a metallic strip as this may introduce a crack-like defect. Therefore, a soft material such as silicone rubber may be used to separate the adhesives. The manufacture shown schematically in Fig. 3 can be followed. The joints can be prepared by bonding panels from which specimens can be later cut. The silicone rubber used to separate the adhesives needs to be first bonded to the outer lower and inner adherends to guarantee that it does not move upon the

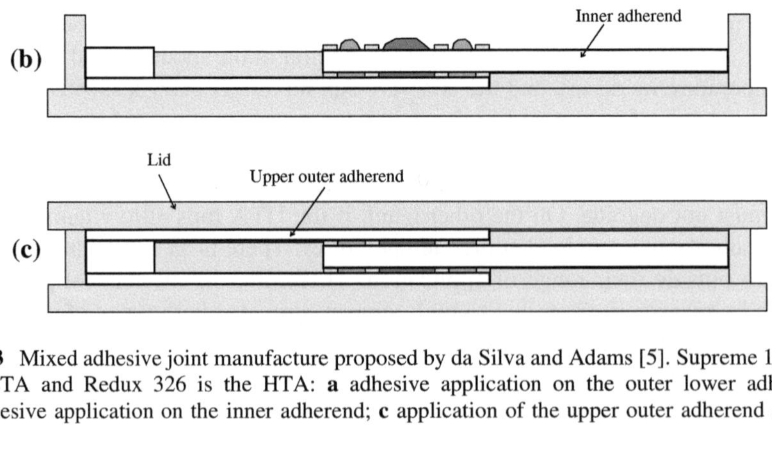

Fig. 3 Mixed adhesive joint manufacture proposed by da Silva and Adams [5]. Supreme 10HT is the LTA and Redux 326 is the HTA: **a** adhesive application on the outer lower adherend; **b** adhesive application on the inner adherend; **c** application of the upper outer adherend and lid

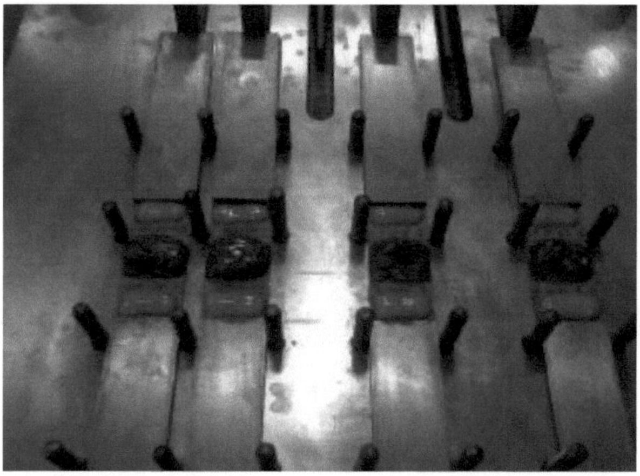

Fig. 4 Manufacture of mixed adhesive joints [7]

application of pressure. This method was also used by da Silva and Lopes [7] to manufacture single lap joints (Fig. 4).

The previous method is acceptable when long overlaps are used and where the area occupied by the silicone rubber will have a little effect on the final joint strength. However, for short overlaps and thin bondlines, silicone rubber strips might not be practical. In that case, the use of a nylon line with the same thickness of the adhesive layer between the adhesives is very efficient. The nylon line is glued using a small amount of cyanoacrylate to the substrates. This method gives a very good dimensional control and occupies a very small area when compared with other techniques tested (use of small strips of silicone or Teflon). This method was applied by Marques and da Silva [17] to manufacture double strap joints with and without tapers. A mould was used that aligned the substrates and the patches and moulds. The construction must follow a special sequence due to the number of different components that make up each specimen.

The methods described above are well adapted to paste adhesives. However, they are too laborious to be implemented in production lines. The most obvious solution is to use a film adhesive that would incorporate the two adhesives and would not need the need of any separation.

5 Strength and Durability

da Silva and Lopes [7] tested single lap adhesive joints with single and dual adhesives to failure statically at room temperature. The dual adhesive joints consisted of a rigid and brittle adhesive in the middle of the overlap and the flexible and ductile adhesive at the ends of the overlap. Four two-part paste adhesives were selected, one very stiff and brittle epoxy (Araldite AV138/HV998 from Huntsman), one very flexible and ductile acrylic (DP-8005 from 3 M) and two intermediate stiffness adhesives (epoxy Araldite 2013 and methacrylate

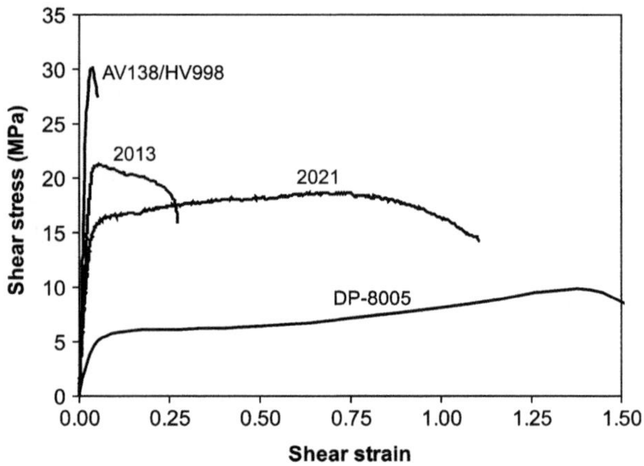

Fig. 5 Typical shear stress–shear strain curves of adhesives, obtained with the thick adherend shear test method

Araldite 2021 from Huntsman). A typical shear stress–shear strain curve for each adhesive is presented in Fig. 5. To avoid plastic deformation of the adherends, the substrates were made of DIN C75 steel, quenched and tempered to 44–46 HRC. The single lap joints had an overlap of 50 mm and a width of 25 mm (see geometry in Fig. 6). A large adhesive thickness (1 mm) was used to facilitate the manufacture, even though this thickness might not be applicable in practice. The predicted lack of adherend yielding was confirmed experimentally, i.e., there was

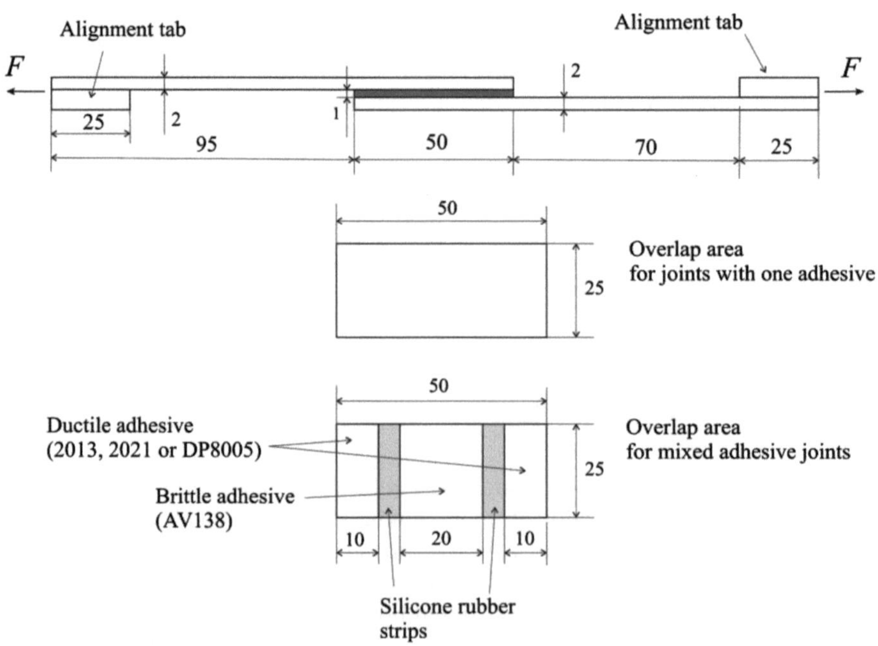

Fig. 6 Specimens geometry (dimensions in millimetre) [7]

Table 1 Experimental and predicted failure loads of the single lap joints tested by da Silva and Lopes [7]

Adhesive(s)	Failure load			
	Experimental	Predicted		
		Volkersen	Global yielding	Volkersen + global yielding[a]
AV138 (brittle)	11.3 ± 0.3	16.6	31.3	
2013 (ductile)	23.4 ± 1.3	15.0	26.3	
2021 (ductile)	28.1 ± 0.6	16.8	23.0	
DP8005 (very ductile)	7.8 ± 0.9	9.0	6.6	
AV138 + 2013	20.5 ± 0.7			22.6
AV138 + 2021	24.0 ± 0.8			21.3
AV138 + DP8005	17.2 ± 0.2			14.7

[a] Volkersen for adhesive AV138 and global yielding for the ductile adhesives

no permanent bending of the substrates following joint failure. The failure, observed with the naked eye, was cohesive within the adhesive in all cases. The experimental failure loads are presented in Table 1. The first comment is that for the joints with one adhesive, the joint strength is not proportional to the adhesive shear strength (see Fig. 5). The joint strength depends not only on the adhesive shear strength but also on its ductility. A ductile adhesive is able to redistribute the load and make use of the less stressed parts of the overlap. For the mixed adhesive joints, the first important point is that the joint strength is higher than the brittle adhesive AV138 alone in all cases, which justifies the use of mixed adhesive joints to increase the strength of joints with stiff and brittle adhesives. The mixed adhesive joints strength is lower than the strength of the joints with a single ductile adhesive for the case of ductile adhesives 2013 and 2021. There is no advantage for these adhesives in combining them with a stiffer adhesive; they have a better performance when used alone. However, the mixed adhesive joint with AV138 and DP8005 is stronger than a joint with the ductile adhesive DP8005 alone. This mixed adhesive joint gives 52% joint strength improvement in relation to a joint with AV138 alone and 121% improvement in relation to a joint with DP8005 alone. The mixed adhesive technique was known to give substantial joint strength improvements in relation to a joint with a brittle adhesive alone but here it is a case where the mixed adhesive joint is beneficial for both the brittle and the ductile adhesives. The joint strength optimization permitted by the mixed adhesive joints could be even higher if a separator between the adhesives (silicone rubber) had not been used. The results can be explained in terms of the stress distribution along the overlap.

The joint strength was predicted using a simple methodology based on the shear stress of the adhesive. The peel stresses were ignored to simplify the problem. For joints with one ductile adhesive (2013, 2021 or DP8005) along the overlap, the failure load was given by the load that causes global yielding of the overlap (see Eq. 1):

$$P_{GY} = \tau_y \cdot b \cdot l \tag{1}$$

where P_{GY} is the failure load of the adhesive due to global yielding, τ_y is the shear yield strength of the adhesive, b is the joint width and l is the overlap length. This criterion works reasonably well provided the failure shear strain of the adhesive is more than 20%, which is the case for the three ductile adhesive used in the present study. For joints with a brittle adhesive (AV138) along the overlap, the Volkersen model [27] was used and the failure occurs when the maximum shear stress at the ends of the overlap exceeds the shear strength of the adhesive. The following equation was used:

$$P = \tau_r \frac{2bl\sinh(\lambda l)}{\lambda l[1 + \cosh(\lambda l)]} \tag{2}$$

where

$$\lambda^2 = \frac{G}{t_a}\left(\frac{2}{Et_s}\right)$$

t_a is the adhesive thickness, G the adhesive shear modulus and E the adherend Young's modulus.

In mixed adhesive joints, it is assumed that both adhesives are at their ultimate strength when the joint fails. In this way, the joint has the maximum possible strength since the adhesives are at their maximum load bearing capacity. This criterion means that there is global yielding of the ductile adhesives (2013, 2021 or DP8005) and the maximum shear stress of the brittle adhesive (AV138) is equal to its shear strength. The shear stress distribution in the brittle adhesive was calculated using the Volkersen model. Therefore, the failure load of mixed adhesive joints is the sum of Eqs. 1 and 2:

$$P = \left(\tau_y \cdot b \cdot l\right)_{\text{ductile}} + \left(\tau_r \frac{2bl\sinh(\lambda l)}{\lambda l[1 + \cosh(\lambda l)]}\right)_{\text{brittle}} \tag{3}$$

A schematic of the shear stress distribution at failure is presented for the case of AV138 + 2013 or AV138 + 2021 in Fig. 7 and for the case of AV138 + DP8005 in Fig. 8. For the case depicted in Fig. 7, the average shear strength of the mixed adhesive joint is higher than that of a joint with the brittle adhesive alone but lower than that of a joint with the ductile adhesive alone. However, in Fig. 8 the average shear strength of the mixed adhesive joint is higher than both adhesives when used individually. When used alone, the brittle adhesive does not make use of the central part of the overlap and the load is concentrated at the edges of the overlap. However, when used in a mixed adhesive joint, the whole overlap contributes. Therefore, a mixed adhesive joint will always be stronger than a joint with a brittle adhesive alone. Note, however, that this is dependent on the ratio and position of the ductile and brittle adhesives over the overlap. If the majority of the overlap was of ductile adhesive then the overall joint strength could be lower than that exploiting a single brittle adhesive. For a mixed adhesive joint to be stronger than

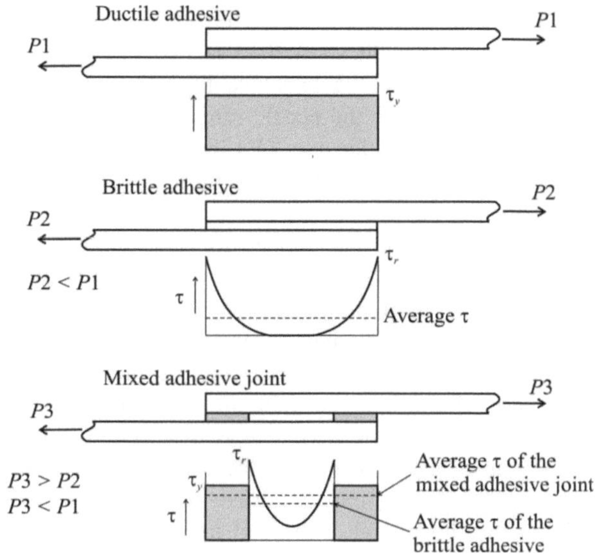

Fig. 7 Schematic shear stress distribution at failure in mixed adhesive joints with AV138 + 2013 or AV138 + 2021 [7]

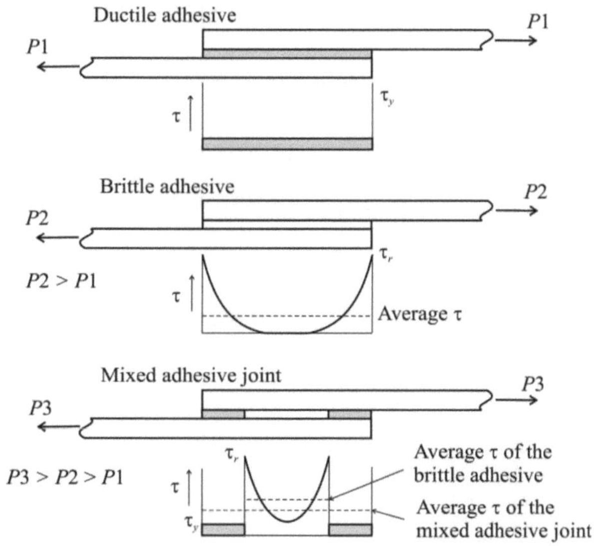

Fig. 8 Schematic shear stress distribution at failure in mixed adhesive joints with AV138 + DP8005 [7]

the brittle adhesive and the ductile adhesive used individually, the load carried by the brittle adhesive must be higher than that carried by the ductile adhesive. This situation corresponds to:

$$\tau_y < \tau_r \frac{2l_{\text{brittle}} \sinh \lambda l_{\text{brittle}}}{\lambda (l - l_{\text{brittle}}) l_{\text{brittle}} [1 + \cosh \lambda l_{\text{brittle}}]} \tag{4}$$

The joint strength predictions are presented in Table 1. For the joints with a single adhesive, the failure load was predicted using the two criteria (global yielding and Volkersen) and it is obvious that the global yielding gives better results for ductile adhesives 2013 and 2021. For adhesive DP8005, both criteria are acceptable because the adhesive high flexibility gives a nearly uniform stress distribution, even in the elastic range. The Volkersen criterion works particularly well for the brittle adhesive AV138, as expected. The simple criterion adopted for the mixed adhesive joints gives failure loads that compare quite well with the experimental results.

Adhesive double lap joints with dual adhesives to be used over a wide temperature range (−55 to 200°C) were studied theoretically [6] and experimentally [5]. The joint strength predictions have shown that for identical adherends (e.g. titanium), the mixed adhesive approach is of little benefit (see Fig. 9). However, for titanium/composite double lap joints, there is a real improvement (see Fig. 10), especially if the difference in the coefficient of thermal expansion is high. As shown in Fig. 11, the shear stress distribution due to an external tensile load is non-symmetric in dissimilar adherends. The shear stress peaks at the overlap end where the inner composite adherend has a lower stiffness. The thermal residual stresses are beneficial at the end from which the stiff adherend extends, but detrimental at the other end, creating a case where the resultant stress is biased to one end. This will result in a decrease of the load capacity of the joint. Depending on

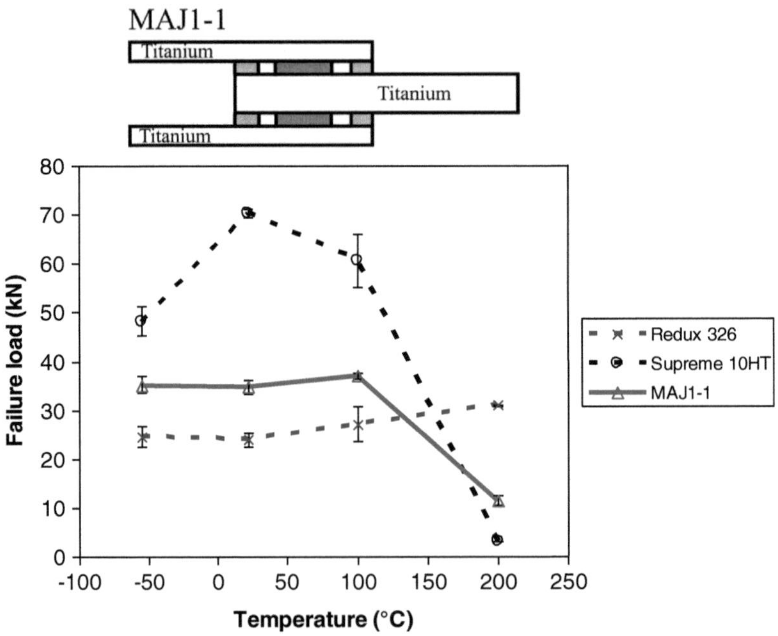

Fig. 9 Experimental failure load for titanium/titanium DLJs (total overlap of 50 mm, width of 25 mm). Redux 326 is a bismaleimide HTA (overlap of 20 mm in the middle of the joint) and Supreme 10HT is an epoxy LTA (overlap of 10 mm at each end of the joint). The *errorbars* correspond to the maximum and minimum values [5]

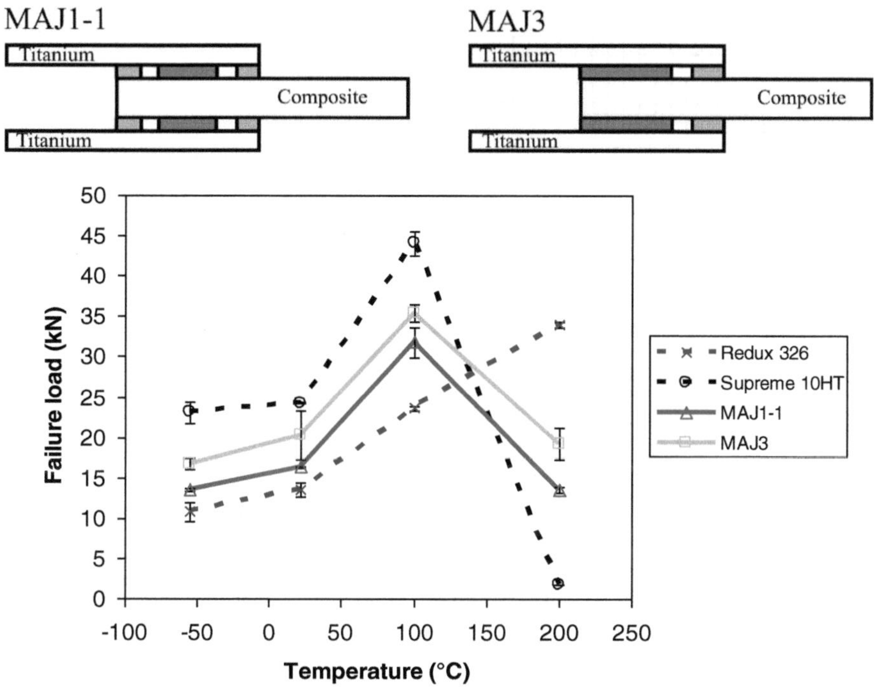

Fig. 10 Experimental failure load for titanium/composite DLJs (total overlap of 50 mm, width of 25 mm). Redux 326 is a bismaleimide HTA (overlap of 20 mm in the middle of the joint in MAJ1-1 and overlap of 30 mm at the left of the joint in MAJ3) and Supreme 10HT is an epoxy LTA (overlap of 10 mm at each end of the joint in MAJ1-1 and overlap of 15 mm at the right of the joint in MAJ3). The *errorbars* correspond to the maximum and minimum values [5]

the combination of the adherends and their relative moduli, the thermal loads can create a joint critically loaded at one end. For the joints critically loaded at one end, the classic mixed adhesive joint design (see Fig. 2) can be modified so that the ductile adhesive is located only on the critical side (design MAJ3 in Fig. 10).

Recently, an analytical analysis based on Frostig's model [11], was developed by das Neves et al. [8] to study a single lap joint (SLJ) and a double lap joint (DLJ) with one or more adhesives along the overlap. It was observed that the adhesive stress distributions of the model developed and that of a finite element analysis are in very close agreement over almost the entire overlap length. The analytical models developed for mixed adhesive single lap joints (MASJ) and mixed adhesive double lap joints (MADJ) were used to make a parametric study [9]. The effect of the ratio LTA Young's modulus/HTA Young's modulus (E_{LTA}/E_{HTA}) (which is equivalent to studying the temperature variation) on the maximum adhesive shear and peel stresses in a MASJ and in a MADJ were investigated. Only the change of E with temperature was considered, ignoring the thermal stresses. The overlap of the low and high temperature adhesives and the effect of the high temperature adhesive modulus were also studied. The main conclusion is

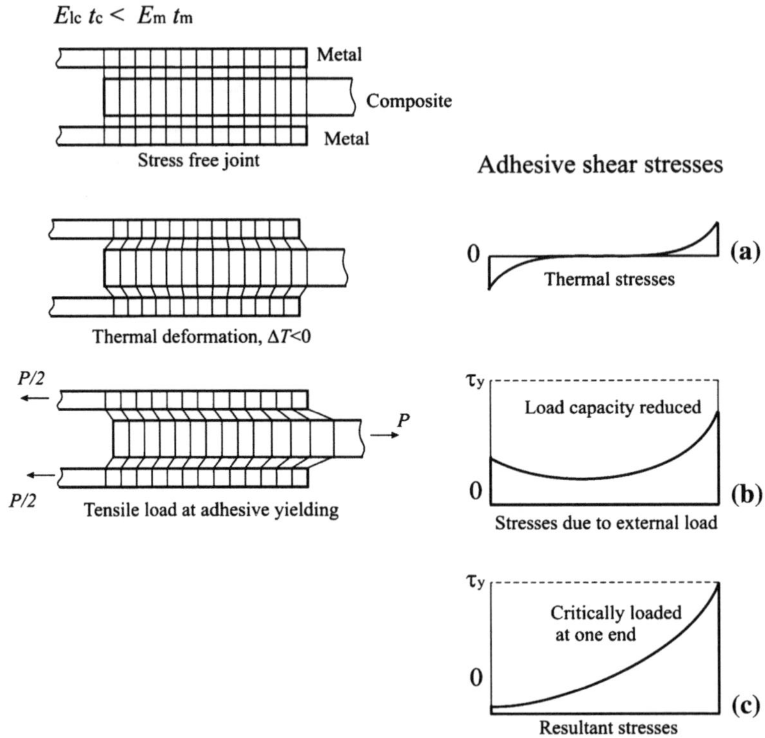

Fig. 11 Adhesive shear stresses in a metal/composite double lap joint for the case where the composite has a lower longitudinal stiffness than the metal (E_{lc} t_c < E_m t_m) under a tensile load and a thermal load, at adhesive yielding (τ_y)

that MASJs and MADJs have a similar behaviour with temperature. However, at high temperatures, the reduction of the adhesive peel stress over a joint with a HTA alone is much higher in the case of a MASJ (see Figs. 12, 13) and makes the use of mixed adhesive joints more efficient for SLJs than for DLJs. At high temperatures, the maximum stresses occur in the HTA until a certain value of E_{LTA}/E_{HTA} ratio at which they start to occur in the LTA. This transition E_{LTA}/E_{HTA} ratio, and consequently this transition temperature, corresponds to a minimum of the adhesive stresses in a mixed adhesive joint. At the transition, the maximum adhesive stresses in the LTA and in the HTA of the mixed adhesive joint are equal (see Fig. 14), making the stress distribution along the overlap length smoother.

At high temperatures, the HTA in the middle of the joint retains the strength and transfers the entire load while the LTA is the load bearing component at low temperatures, making the high temperature adhesive relatively lightly stressed. At low temperatures, the load must essentially be supported by the LTA. If its modulus is of the same order as the modulus of the HTA, most of the load will be carried by the LTA. However, if its modulus is much lower than the modulus of

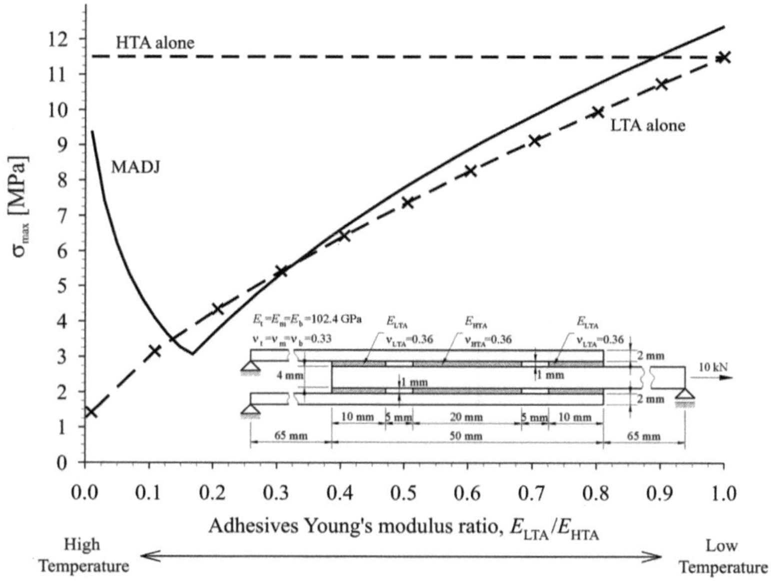

Fig. 12 Effect of the adhesives Young's modulus ratio on the maximum adhesive peel stress in a mixed adhesive double lap joint (MADJ) [9]

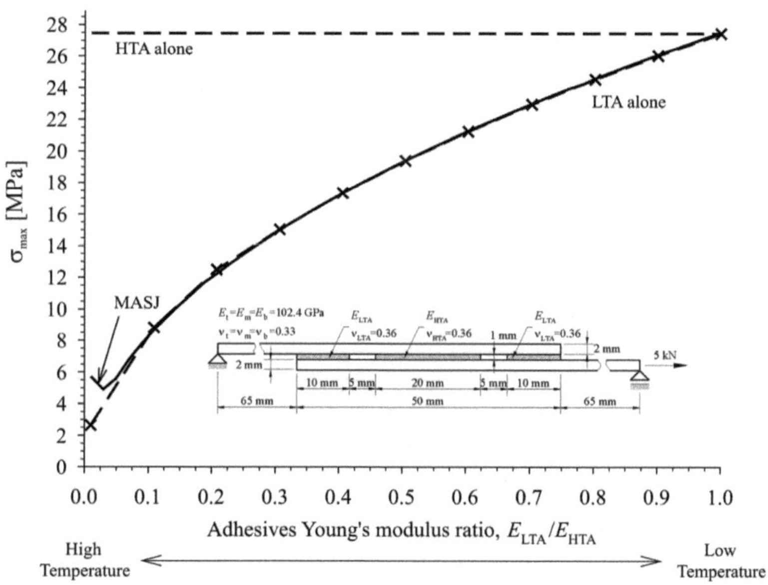

Fig. 13 Effect of the adhesives Young's modulus ratio on the maximum adhesive peel stress in a mixed adhesive single lap joint (MASJ) [9]

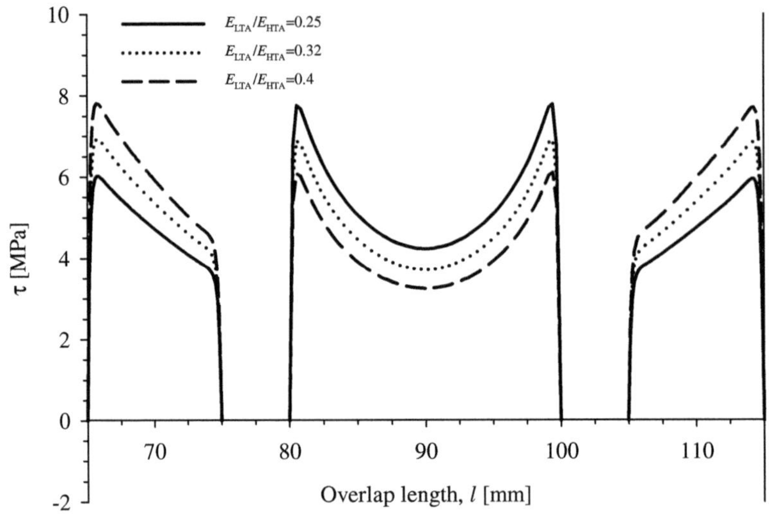

Fig. 14 Transition of the maximum overall adhesive shear stress from the HTA to the LTA in a mixed adhesive double lap joint [9]

the HTA, then the HTA might still be critically loaded. One possibility is to stiffen the ends of the overlap where the LTA is as shown in Fig. 15.

A mixed adhesive joint is expected to work at high and low temperatures. At high temperatures, the LTA must not fail so that it can sustain the load when the joint is cooled down to low temperatures. On the other hand, at low temperatures, the HTA must not fracture so that the joint can be operated when the temperature is raised. This is an important matter and it must be demonstrated experimentally that the whole concept works when temperature cycling the mixed adhesive joint. The mixed adhesive joints were cycled according to the scheme presented in Fig. 16. Cycle 1 is to check if the LTA is not damaged when loaded at high temperature. In a mixed adhesive joint at high temperatures, the load is primarily supported by the HTA and the joint deformation will be dictated by the HTA. However, it is

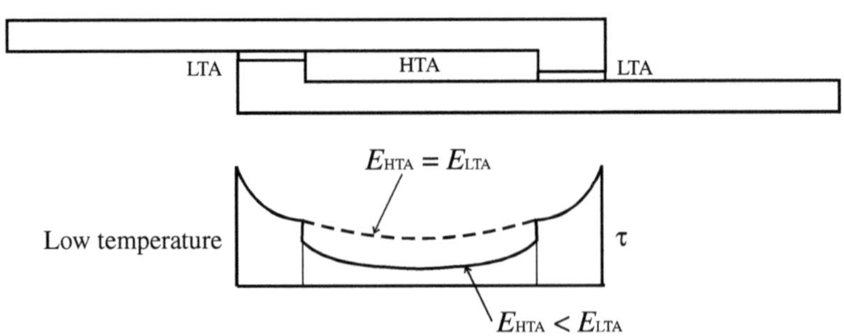

Fig. 15 Stiffening of the ends of the overlap in a mixed adhesive joint

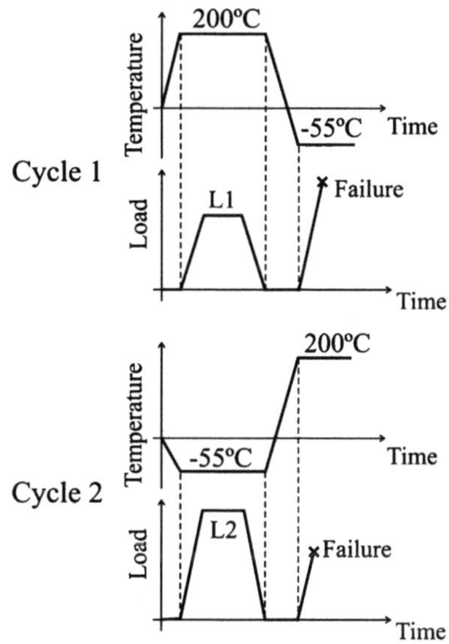

Fig. 16 Mixed adhesive joint cycling procedure (L1: load higher than the failure load of the LTA alone at 200°C, L2: load higher than the failure load of the HTA alone at −55°C) [5]

necessary to verify that the loading of the mixed adhesive joint at high temperatures to a load above the maximum capacity of a joint with only a LTA will not damage the LTA so that it can perform when the mixed adhesive joint is cooled and loaded at low temperatures. Therefore, the joint should be loaded at 200°C at a load higher than the failure load of the LTA alone at 200°C (load L1 in Fig. 16), and then loaded to failure at −55°C. If the failure load at −55°C, after the loading at 200°C, is lower than the failure load without cycling at 200°C, then the LTA has been damaged by the loading cycle at 200°C. Cycle 2 is to check if the HTA fractures when the joint is loaded at low temperatures. At low temperatures, the failure load of a mixed adhesive joint is higher than the failure load of a joint with a HTA alone. But it must be demonstrated that when the mixed adhesive joint is loaded to a load higher than the failure load of a joint with a HTA alone, the loading does not damage the HTA so that it can support the load when the mixed adhesive joint is heated to 200°C and loaded. The joint should be loaded at −55°C at a load higher than the failure load of the HTA alone at −55°C (load L2 in Fig. 16), and then loaded to failure at 200°C. If the failure load at 200°C, after the loading at −55°C, is lower than the failure load without cycling at −55°C, then the HTA has been damaged by the loading cycle at −55°C. The mixed adhesive joints (MAJ1-1 for titanium/titanium DLJs and MAJ3 for titanium/composite DLJs) were loaded to L1 at 200°C; they were then unloaded and loaded to failure at −55°C as described in cycle 1. Before loading the joints at −55°C, they were checked visually and there was no presence of cracks. The failure load at −55°C, after the cycle at 200°C, was nearly the same as without cycle 1, as shown in Fig. 17a for titanium/titanium DLJs and in Fig. 17b for titanium/composite DLJs. Therefore, the Supreme 10HT adhesive (LTA) was not damaged when designs MAJ1-1 or MAJ3 were loaded close to their failure load at 200°C. Designs MAJ1-1 (for titanium/titanium

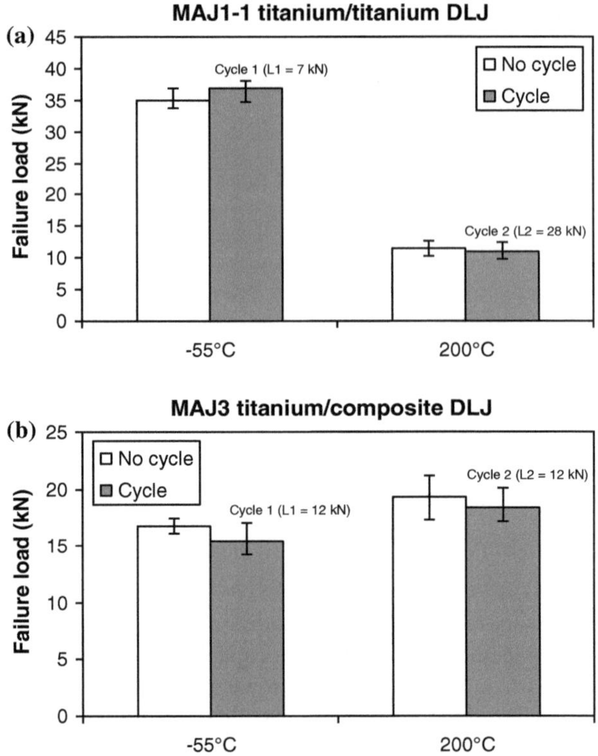

Fig. 17 Failure loads after cycling of mixed adhesive joints: **a** titanium/titanium MAJ1-1 after cycles 1 and 2; **b** titanium/composite MAJ3 after cycles 1 and 2 [5]

DLJs) and MAJ3 (for titanium/composite DLJs) were loaded to L2 at −55°C; they were then unloaded and loaded to failure at 200°C as described in cycle 2. There were no noticeable cracks in the specimens after the loading at −55°C. The failure load at 200°C, after the cycle at −55°C, was approximately the same as without cycle 2, as shown in Fig. 17a for titanium/titanium DLJs and in Fig. 17b for titanium/composite DLJs. This proves that the adhesive Redux 326 (HTA) was not damaged when designs MAJ1-1 or MAJ3 were loaded close to their failure load at −55°C. The experimental thermal cycles presented here show that the adhesives do not degrade or fracture after a stage at low (cycle 1) or high (cycle 2) temperatures, even though the loads L1 and L2 were not very demanding. However, to fully prove the mixed adhesive joint concept applicability, many cycles should be carried out at low, high and intermediate temperatures but this was beyond the scope of the work described here.

6 Types of Failures

The load displacement curves of the single lap adhesive joints with single and dual adhesives tested by da Silva and Lopes [7] presented in Figs. 18, 19 and 20 are a

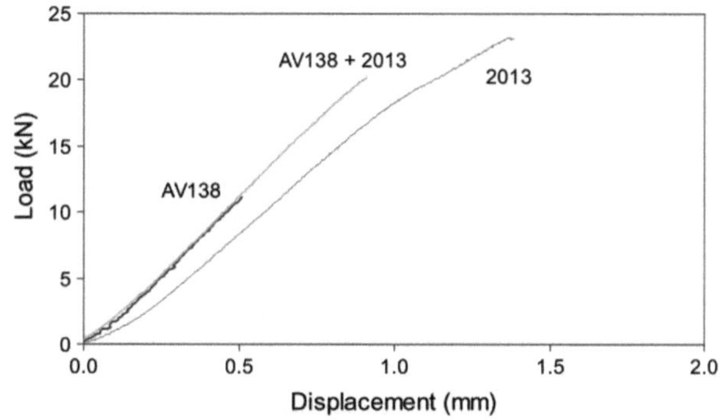

Fig. 18 Load displacement curves of joints with AV138 and 2013 [7]

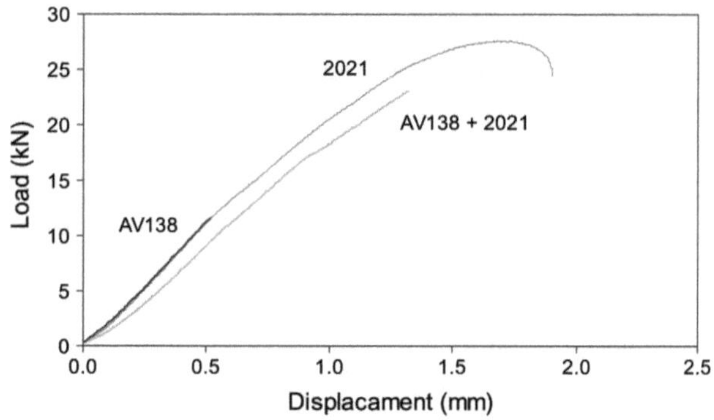

Fig. 19 Load displacement curves of joints with AV138 and 2021 [7]

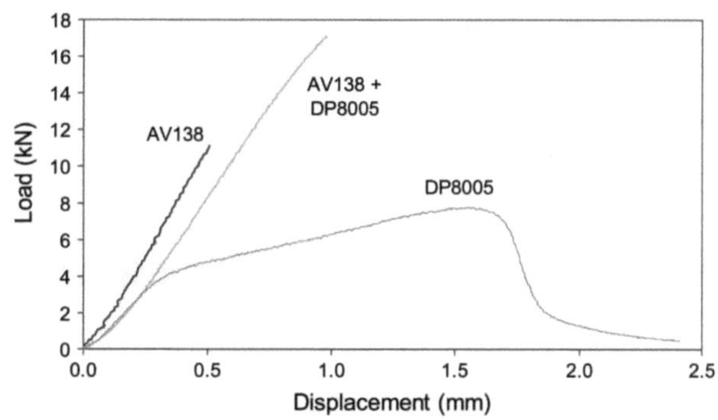

Fig. 20 Load displacement curves of joints with AV138 and DP8005 [7]

good way to visualize the difference between adhesive joints used alone and in combination. The adhesive plasticity appears clearly in the curves for 2021 alone and DP8005 alone. The AV138 curves are instead linear to fracture, as expected. In the case of mixed adhesive joints, a slight decrease in slope can be noticed before failure probably indicating plastic deformation of the ductile adhesive. The same type of load displacement curves were obtained by Marques and da Silva [17] for double strap joints and the deviation from linearity was shown to be due to

Table 2 Adhesive properties for the cohesive zone model used in ABAQUS modelling of the single lap joints tested by da Silva and Lopes [7]

Property	Adhesive AV138	Adhesive 2021
Tensile stiffness (N/mm^2)	4,590	1,500
Shear stiffness (N/mm^2)	1,559	500
Tensile strength (N/mm^2)	7	8
Shear strength (N/mm^2)	13	25
Tensile (mode I) toughness (N/mm)	0.3	2
Shear (mode II) toughness (N/mm)	3	20

adhesive plasticity. To capture the debonding failure of the single lap joints tested by da Silva and Lopes [7], the interfacial cohesive elements of ABAQUS were used. The traction versus separation approach was employed as the constitutive response of cohesive elements since the thickness of the interface is negligibly small. This constitutive relation relates cohesive tractions to the displacement discontinuities for modelling the behaviour of the material in the process zone that is located ahead of a crack tip. The elastic properties of the interface material were defined using uncoupled traction–separation behaviour. The quadratic traction–interaction failure criterion was chosen for damage initiation in the cohesive elements; and a mixed-mode, energy-based damage evolution law based on the power law (coefficient of two) criterion was used for damage propagation. The relevant material data are indicated in Table 2. The mode I values were obtained by the inverse method using the load displacement curve of a double cantilever beam specimen. The toughness in mode II was assumed to be ten times that in mode I [3]. Figure 21 shows the debonding process in the single lap joints with AV138 and 2021 alone and in a single lap joint with both adhesives. As expected, the failure process of the joint with AV138 is abrupt with the cohesive elements collapsing at the same time. On the contrary, the joint with 2021 shows a progressive debonding. A mixed-adhesive joint with AV138 and 2021 also shows a rather abrupt failure with nearly all the cohesive elements corresponding to AV138

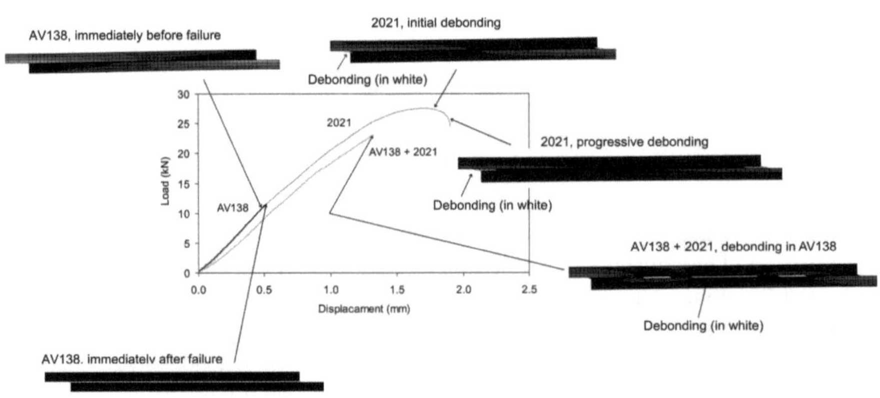

Fig. 21 Debonding in mixed adhesive single lap joints

collapsing at the same time. The failure of 2021 follows a few instants after. This is a very interesting feature since in single lap joints failure always occurs at the ends of the overlap where there are stress concentrations, however, with the mixed adhesive joint studied by da Silva and Lopes [7], it appears that the failure occurred first in the middle of the overlap where the brittle adhesive is located.

The failure surfaces of the joints tested by da Silva and Lopes [7] with one adhesive and the mixed adhesive joints are presented in Fig. 22. The failure surface of AV138 is typical of stiff and strong adhesives where the crack runs close to the interface on opposite sides and meet approximately in the middle of the overlap [13]. Adhesive 2013 is classified in the present study as a ductile adhesive, but it is relatively stiff and strong (see Fig. 5) and has a failure mode similar to that of AV138. Adhesives 2021 and DP8005 have a different failure pattern where the crack runs in the middle of the adhesive layer. This is typical of adhesives that are flexible and very ductile [12]. The failure pattern of the mixed adhesive joints is more complex to classify but generally, the failure took place in the adhesive close to an interface, even for the ductile adhesive. It is as if the brittle adhesive would have forced the ductile adhesive failure to occur close to the interface, which is in accordance with the debonding process described in Fig. 21.

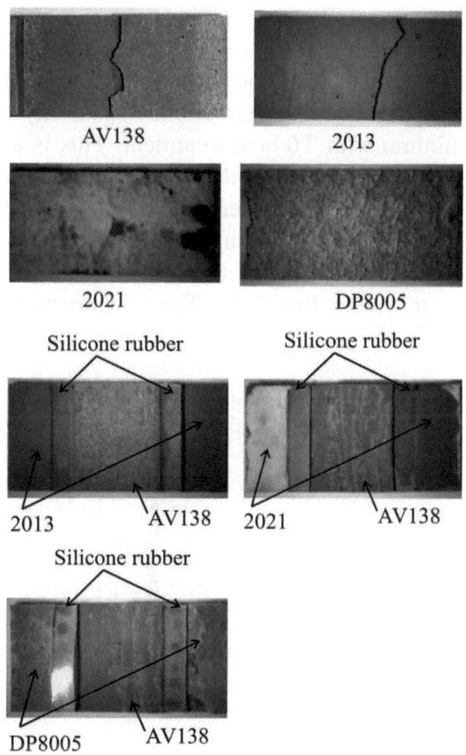

Fig. 22 Failure surfaces of the single lap joints [7]

7 Examples of Use

To the author knowledge, there is no real application of applications of the mixed adhesive joining technique in the industry. The author have shown in Sect. 5 that joints with a combination of several adhesives might be useful for the aeronautical industry in order to bear temperatures from −55 to +200°C. Another example of use in the aeronautical industry is in the application of adhesively bonded patches. It was shown in Sect. 2 that mixed modulus joints might be used to improve the stress distribution and increase the joint strength of adhesively bonded patches with a stiff adhesive. For example, Bouiadjra et al. [2] used the mixed modulus technique for the repair of an aluminium structure with a composite patch. The use of a more flexible adhesive at the edge of the patch increases the strength performance of the repair. Marques and da Silva [17] studied this technique in combination with a taper. The joint selected was a double butt strap joint, in which two aluminium plates are interconnected by means of two adhesive layers and two symmetrically placed metal patches. This standard joint is a good simplification of the case encountered in a repair situation. The mixed adhesive technique was studied with a very brittle adhesive in the middle of the joint and a ductile adhesive at the ends of the patch. Two adhesives were selected, a very stiff and brittle epoxy (Araldite AV138/HV998 from Huntsman, Salt Lake City, UT) used in aerospace applications, and a more flexible and ductile epoxy adhesive (Araldite 2015 from Huntsman). Typical stress–strain curves are shown in Fig. 23. A 6000 series alloy was used to closely represent the aluminium alloys used in the aerospace industry. Although it would be preferable to use a 7000 or 2000 series, as it is the most used in aeronautical construction, availability constraints lead to the use of 6063 aluminium with T6 heat treatment. This is a very strong aluminium alloy, as the T6 heat treatment raises the tensile strength. It was used in all of the substrates as well as in the reinforcement patches. The loads involved were not enough to deform plastically the adherends. The double strap joint (DSJ) geometry was selected so as to simulate a crack with a patch. The DSJ had an overlap of 10 mm and a width of 25 mm (see Fig. 24a). The specimens failed cohesively in the adhesive in all

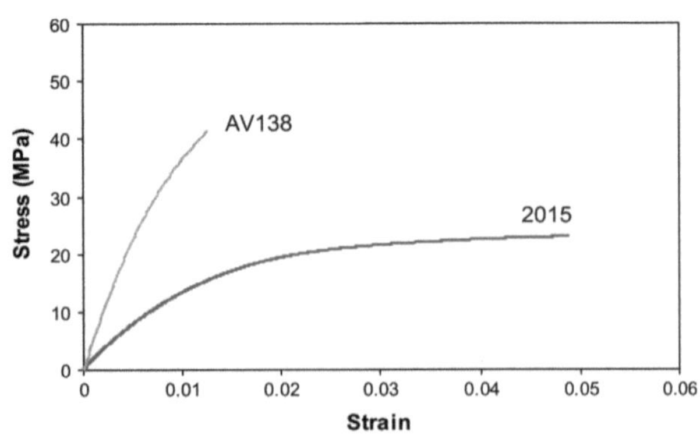

Fig. 23 Tensile stress–strain curve of adhesives AV138 and 2015

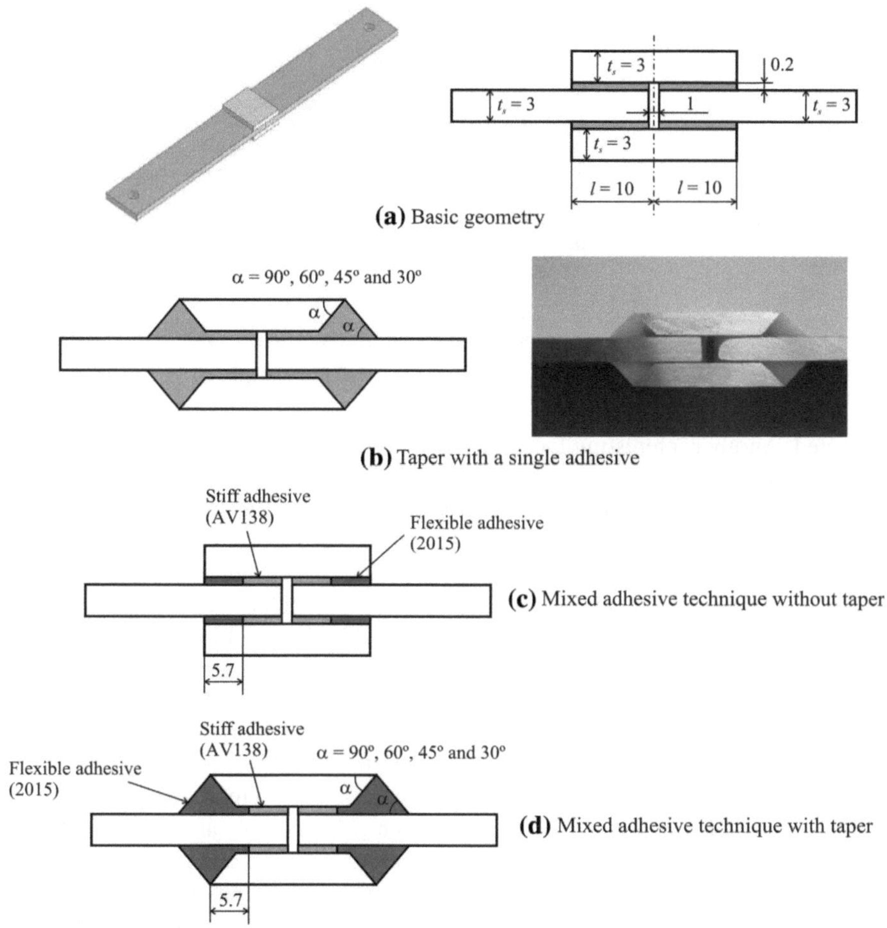

(a) Basic geometry

(b) Taper with a single adhesive

(c) Mixed adhesive technique without taper

(d) Mixed adhesive technique with taper

Fig. 24 Joint geometry (dimensions in millimetre) [17]

cases. The results of tensile testing of the single and dual adhesive specimens are presented in Fig. 25 as a function of the taper angle. Little experimental scatter (three specimens tested for each case) was found being the maximum for the dual adhesive joint without taper (standard deviation of 1.16 kN). The error bars are not represented in Fig. 25 to improve clarity. Figure 25 demonstrates that the joint strength of the stiff adhesive (AV138) can be significantly increased by use of a taper angle. This effect is not as visible when a less stiff adhesive (2015) is used alone or in the taper. It can be seen that the stiffest adhesive has a peak around 45°, falling off immediately after (Fig. 25). The gain of performance obtained for AV138 for the 45° taper is of 33% in relation to the taperless configuration. 2015 and 2015 + AV138 have almost identical distributions, owing to the fact that it is the adhesive in the taper of the joint (2015) that dictates the effectiveness of the taper. In these two cases it is the same adhesive, so similar distributions are

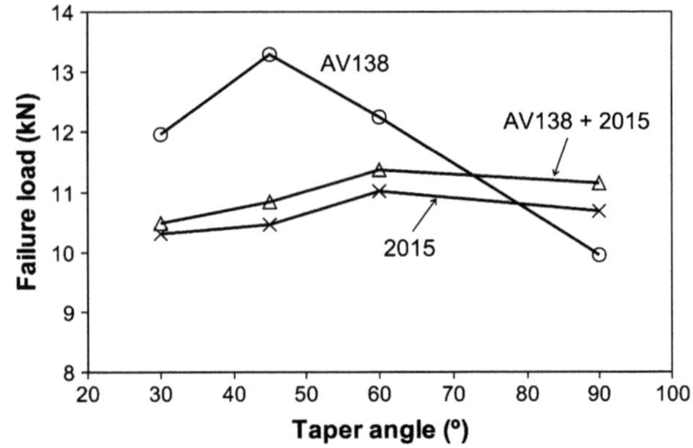

Fig. 25 Joint strength evolution with taper angle [17]

expected. Another interesting fact is the synergetic combination of adhesives in the taperless configuration (90°) (see Fig. 25). Here, it can be seen that the combination of adhesives is stronger that any of the adhesives individually. This is explained by the sensitivity to stress risers of the AV138 adhesive due to its high stiffness and brittleness. This is a very stiff and strong adhesive, but performs badly in unfavourable conditions, such as the abrupt patch termination zone. When 2015 is placed in that area it resists better than AV138 would resist itself and the result is a joint better than any of the adhesives alone could provide. A clip gauge of 50 mm was mounted on the specimen to measure the load displacement curves of the joints. The nonlinear behaviour of the joints with 2015 is clearly visible, as shown in Fig. 26. Despite the fact that joints with AV138 alone and a taper of 45° are the strongest joints, the dual adhesive system might be more appropriate when a compromise of strength and ductility is sought. The flexible and ductile adhesive at the ends of the overlap gives a higher capacity to deform and is less sensitive to defects such as cracks. Applications where dynamic loadings such as fatigue or impact loads are frequent is an example where the use of dual adhesives might be more appropriate than reducing the stress levels by geometry modifications of a joint with a stiff adhesive.

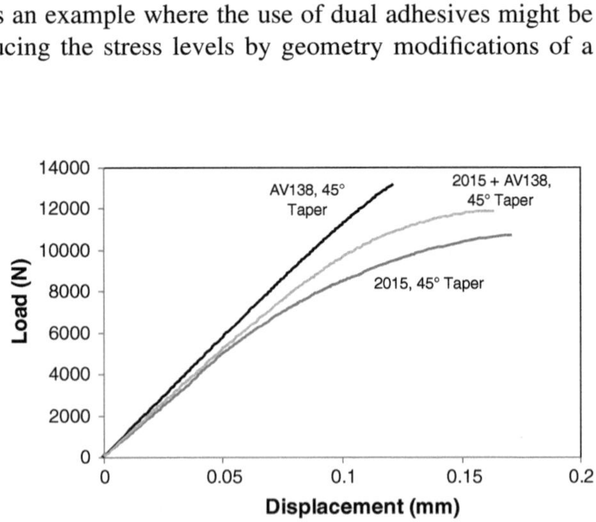

Fig. 26 Load displacement curves for joints with a 45° taper angle [17]

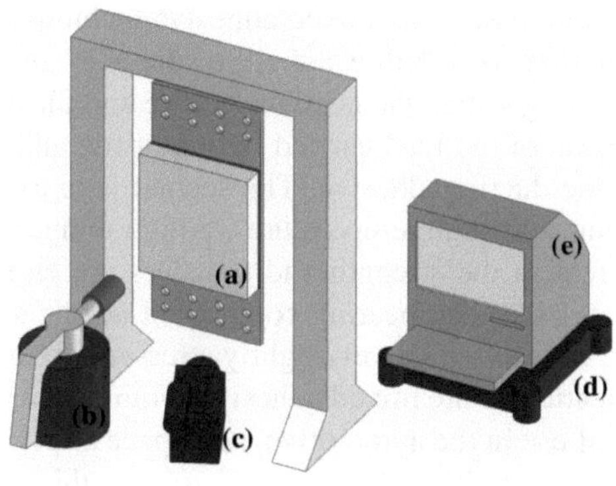

Fig. 27 Experimental set-up to heat and load ceramic–aluminium joints used in the aerospace industry. Thermal loads are applied to the specimen (**a**) by mean of a propane gas burner (**b**). An infrared sensitive camera (**c**) records the surface temperature while the digital acquisition spider (**d**) gathers the thermocouple signals. A computer (**e**) records all the data obtained [19]

It was shown in Sect. 2 that the aerospace industry is faced with problems when it comes to the bonding of the ceramic tiles to the aluminium structure of space vehicles. The use of the mixed adhesive joint might improve the strength of the ceramic–aluminium joints. Thermal and mechanical loads were applied to a model of a single tile by Marques and da Silva [18]. The thermal load was a heat flux in the top of the tile, while the mechanical load represented the pressurization stress of the vehicle structure and is applied in the metallic base. Single adhesives and combination of two adhesives were tested. Silicone, bismaleimide and ceramic adhesives were used in the simulation. It was found that ceramic adhesives must not be combined with other adhesives because large modulus transitions introduce stress concentrations. If two dissimilar adhesives are used (one for high temperature and the other for low temperature) a mixed adhesive joint with a bismaleimide and silicone does not improve the stress distribution in relation to a joint with silicone alone but gives a much higher strength at room and high temperatures. An experimental set-up has been developed by Marques et al. [19] to reproduce the heating and to load specimens containing only silicone and silicone combined with a high temperature adhesive, as shown in Fig. 27. Preliminary results confirm that mixed-adhesive joints with a silicone and a high temperature epoxy adhesive provide an improved joint strength at room and high temperature in relation to a joint with silicone alone.

8 Conclusions and Future Trends

The mixed adhesive technique may be useful in variety of industrial applications. There are basically two types of problems that the mixed adhesive technique can solve. The first is related to improvements of joint strength in the case of joints with stiff and brittle adhesives. The mixed adhesive technique gives joint strength improvements in relation to a brittle adhesive alone in all cases. For a mixed adhesive joint to be stronger than the brittle adhesive and the ductile adhesive used individually, the portion of the load carried by the brittle adhesive must be higher than that carried by the ductile adhesive. The second is to extend the temperature range of adhesive joints. At high temperatures, a high temperature adhesive in the middle of the joint retains the strength and transfers the entire load while a low temperature adhesive is the load bearing component at low temperatures, making the high temperature adhesive relatively lightly stressed. To the author knowledge, there is no real application of the mixed adhesive joining technique in the industry. However, examples of use in the automotive, aerospace and aeronautical industries are given. In order to convince the industry to use this technique, analytical methods to predict the joint strength should be available. da Silva and Lopes [7] proposed a simple predictive model and das Neves et al. [8, 9] have developed an analytical model that takes into account two adhesives along the overlap and permits to determine the best combination of adhesives and the optimum geometric factors (e.g. overlap) to have the maximum joint strength. One of the problems associated with the mixed adhesive technique is the adhesive proper separation. The best way to control the process is to use film adhesives. However, it is difficult to find compatible adhesives in the film form. This is a problem for the manufacturers to solve. Meanwhile, the adhesive separation can be done by the use of silicone strips even though a small portion of the load bearing area is reduced.

The mixed adhesive joint technique can be considered a rough version of a functionally graded material. The ideal would be to have an adhesive functionally modified with properties that vary gradually along the overlap allowing a true uniform stress distribution along the overlap. Only Sancaktar and Kumar [23] used rubber particles to modify locally the adhesive at the ends of the overlap but that is not a gradually modified adhesive. The author is presentably working on a project to develop a technological process to functionally modify the adhesive along the overlap for a uniform stress distribution, based on a differentiated cure process.

References

1. Avalle, M., Peroni, L., Peroni, M., Scattina, A.: Bi-material joining for car body structures: experimental and numerical analysis. J. Adhes. **86**, 539–560 (2010)
2. Bouiadjra, B.B., Fekirini, H., Belhouari, M., Boutabout, B., Serier, B.: Fracture energy for repaired cracks with bonded composite patch having two adhesive bands in aircraft structures. Comp. Mater. Sci. **40**, 20–26 (2007)
3. Campilho, R.D.S.G., de Moura, M.F.S.F., Ramantani, D.A., Morais, J.J.L., Barreto, A.M.J.P., Domingues, J.J.M.S.: Adhesively-bonded repair proposal for wood members damaged by horizontal shear using carbon-epoxy patches. J. Adhes. **86**, 649–670 (2010)
4. da Silva, L.F.M., Adams, R.D.: Measurement of the mechanical properties of structural adhesives in tension and shear over a wide range of temperatures. J. Adhes. Sci. Technol. **19**, 109–142 (2005)

5. da Silva, L.F.M., Adams, R.D.: Adhesive joints at high and low temperatures using similar and dissimilar adherends and dual adhesives. Int. J. Adhes. Adhes. **27**, 216–226 (2007)
6. da Silva, L.F.M., Adams, R.D.: Joint strength predictions for adhesive joints to be used over a wide temperature range. Int. J. Adhes. Adhes. **27**, 362–379 (2007)
7. da Silva, L.F.M., Lopes, M.J.C.Q.: Joint strength optimization by the mixed adhesive technique. Int. J. Adhes. Adhes. **29**, 509–514 (2009)
8. das Neves, P.J.C., da Silva, L.F.M., Adams, R.D.: Analysis of mixed adhesive bonded joints—Part I: Theoretical formulation. J. Adhes. Sci. Technol. **23**, 1–34 (2009)
9. das Neves, P.J.C., da Silva, L.F.M., Adams, R.D.: Analysis of mixed adhesive bonded joints—Part II: Parametric study. J. Adhes. Sci. Technol. **23**, 35–61 (2009)
10. Fitton, M.D., Broughton, J.G.: Variable modulus adhesives: an approach to optimised joint performance. Int. J. Adhes. Adhes. **25**, 329–336 (2005)
11. Frostig, Y., Thomsen, O.T., Mortensen, F.: Analysis of adhesive-bonded joints, square-end, and spew-fillet—high order theory approach. J. Eng. Mech. **125**, 1298–1307 (1999)
12. Giannis, S., Adams, R.D., Clark, L.J., Taylor, M.A.: The use of a modified peel specimen to assess the peel resistance of aircraft fuel tank sealants. Int. J. Adhes. Adhes. **28**, 158–175 (2008)
13. Harris, J.A., Adams, R.D.: Strength prediction of bonded single lap joints by non-linear finite element methods. Int. J. Adhes. Adhes. **4**, 65–78 (1984)
14. Harris, J.A., Adams, R.D.: An assessment of the impact performance of bonded joints for use in high energy absorbing structures. Proc. Inst. Mech. Eng. **199**, C2, 121–131 (1985)
15. Hart-Smith, L.J.: Adhesive Bonded Double Lap Joints. NASA CR-112235 (1973)
16. Hergenrother, P.M.: Development of composites, adhesives and sealants for high-speed commercial airplanes. SAMPE J. **36**, 30–41 (2000)
17. Marques, E.A.S., da Silva, L.F.M.: Joint strength optimization of adhesively bonded patches. J. Adhes. **84**, 917–936 (2008)
18. Marques, E.A.S., da Silva, L.F.M.: Stress analysis and failure properties of bonded ceramic-metal joints. International conference on advanced computational engineering and experimenting ACE-X 2008, Barcelona, Spain, 14–15 July (2008)
19. Marques, E.A.S., da Silva, L.F.M., Moreira, P.M.G.P.: Ceramic metal adhesive joints for high temperature aerospace applications. International conference on advanced computational engineering and experimenting ACE-X 2009, Rome, Italy, 22–23 July (2009)
20. Patrick, R.L. (ed.): Treatise on Adhesion and Adhesives—Structural Adhesives with Emphasis on Aerospace Applications, vol. 4. Marcel Dekker, Inc., New York (1976)
21. Pires, I., Quintino, L., Durodola, J.F., Beevers, A.: Performance of bi-adhesive bonded aluminium lap joints. Int. J. Adhes. Adhes. **23**, 215–223 (2003)
22. Raphael, C.: Variable-adhesive bonded joints. Appl. Polym. Symp. **3**, 99–108 (1966)
23. Sancaktar, E., Kumar, S.: Selective use of rubber toughening to optimize lap-joint strength. J. Adhes. Sci. Technol. **14**, 1265–1296 (2000)
24. Semerdjiev, S.: Metal to Metal Adhesive Bonding. Business Book Limited, London (1970)
25. Srinivas, S.: Analysis of Bonded Joints. NASA TN D-7855 (1975)
26. Temiz, S.: Application of bi-adhesive in double-strap joints subjected to bending moment. J. Adhes. Sci. Technol. **20**, 1547–1560 (2006)
27. Volkersen, O.: Die nietkraftoerteilung in zubeanspruchten nietverbindungen mit konstanten loschonquerschnitten. Luftfahrtforschung **15**, 41–47 (1938)

MIX
Papier aus verantwortungsvollen Quellen
Paper from responsible sources
FSC® C105338

If you have any concerns about our products,
you can contact us on
ProductSafety@springernature.com

In case Publisher is established outside the EU,
the EU authorized representative is:
Springer Nature Customer Service Center GmbH
Europaplatz 3, 69115 Heidelberg, Germany

Printed by Libri Plureos GmbH
in Hamburg, Germany